Db2 11 for LUW

Developer Training and Reference Guide

Robert Wingate

ISBN 13: 978-1-7345847-1-4

Disclaimer

The content of this book is based upon the author's understanding of and experience with the IBM DB2 product. Every attempt has been made to provide correct information. However, the author and publisher do not guarantee the accuracy of every detail, nor do they assume responsibility for information included in or omitted from it. All of the information in this book should be used at your own risk.

ISBN-13: 978-1-7345847-1-4

Contents

Introduction

Welcome

Congratulations on your purchase of **DB2 11 for LUW Developer Training and Reference Guide**. This book will help you learn the essential information you need to know about DB2 11 for Linux, UNIX and Windows so you can be productive as soon as possible. You'll receive instruction, examples and questions/answers to help you learn and to gauge your readiness for development work on a DB2 technical team.

Assumptions:

While I do not assume that you know a great deal about DB2, I do assume that you've worked in a LINUX, UNIX or Windows environment and know your way around. Also I assume that you have a working knowledge of the JAVA programming language which we will use for all the embedded SQL examples (in most cases I also created parallel c# .NET examples). All in all, I assume you have:

1. A working knowledge of Windows (or Unix/Linux) files and navigation.
2. A basic understanding of SQL.
3. Proficiency using either Java or .NET.
4. Access to a computer running Windows 7 or higher (or UNIX/LINUX).

Knowledge of DB2 11

If you are a beginner, this book should give you what you need to get started, and to develop a solid DB2 LUW foundation. Even if you have years of experience with DB2, you may find yourself challenged by some of the topics. I encourage you to fill in the gaps with the information in this text book. You may find new techniques and ways of accomplishing things in DB2.

Experience with DB2 11

Unless your shop has upgraded to DB2 11, you may not have any experience with it. My suggestion is to download the free **DB2 Express-C** product and install it on your local machine. That way you can follow along with the examples given in this text. You can also come up with your own training project and work through it at your own pace.

Note: As of this writing, IBM appears to be deemphasizing DB2 Express-C. The product links and search feature on the IBM web site redirect you to IBM Db2 Developer Edition 90 day trial. At some point the DB2 Express-C product may be unavailable. However, I was able to find the DB2 Express-C download options here:

https://www-01.ibm.com/marketing/iwm/iwm/web/pickUrxNew.do?source=swg-db2expressc

If the link above no longer works, you can try other non-IBM download sites that might still have DB2 Express-C. Your other alternative is to pursue the DB2 Developer Edition trial.

Knowledge and experience. Will that guarantee that you'll succeed as a DB2 LUW application developer? Of course, nothing is guaranteed in life. But if you put sufficient effort into a well-rounded study plan that includes both of the above, I believe you have a very good chance of excelling as an application developer in the DB2 Linux, UNIX and Windows world.

Best of luck!

Robert Wingate
IBM Certified Application Developer – DB2 11 for z/OS

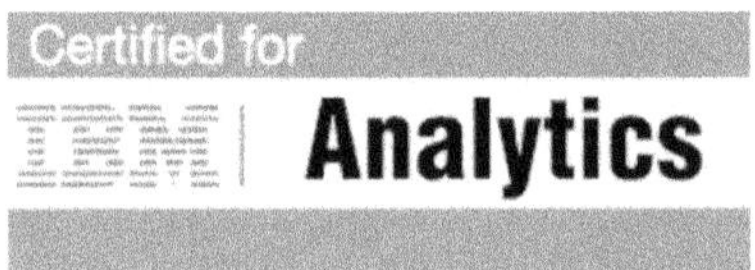

CHAPTER ONE: BASIC TOOLS FOR DB2 LUW

Welcome to **DB2 11 for LUW Developer Training and Reference Guide**! Before we get into development activities, I want to introduce you to the environment we'll be working in. If you've ever used DB2 for LUW, you've almost certainly encountered these tools. But let's make sure we're all familiar with how to install and access DB2, and how to use the basic DB2 tools available. We'll take a brief look at installing DB2 Express-C, IBM Data Studio, and two programming language IDEs which you may want to use for Java or .NET programming with DB2.

DB2 Express-C 11.1

You can download DB2 Express-C for free from the IBM web site. We'll be using that edition for this text. If you have at least 8GB of main memory on your computer (plus Windows 7 Professional or better), you can instead download the new Db2 Developer Community Edition. That product is full featured while DB2 Express-C has just a few limitations. All of the examples in this text book will work with either version of DB2.
We will install and use DB2 Express-C.

At this writing, the link to download the DB2 Express-C product is:

https://www-01.ibm.com/marketing/iwm/iwm/web/pickUrxNew.do?source=swg-db2expressc

If for some reason the link is broken, simply Google search "DB2 Express-C free download".

Installing DB2 is pretty straightforward. You'll need to decompress the download file into an installation directory of your choice. Then you can run the installer. Simply double click on the setup.exe file in the decompressed install directory.

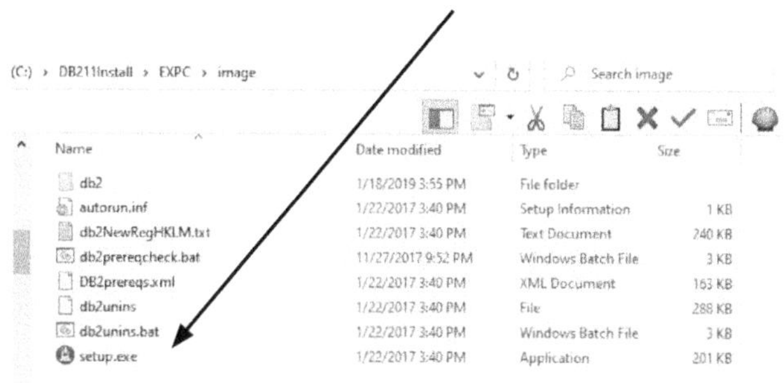

After a few moments you should see this screen. You can check out the release, planning and upgrade information if you wish. When ready to install simply click on **Install a Product**.

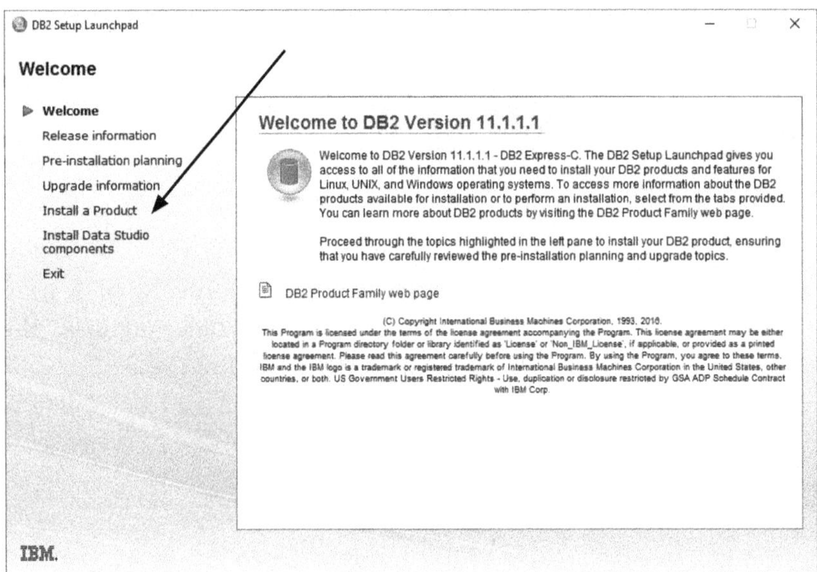

Click on **Install New.**

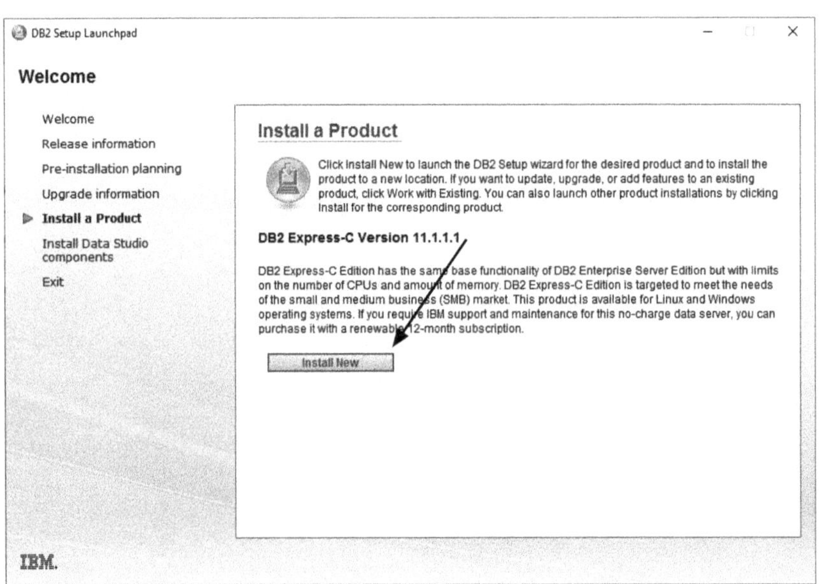

On the next screen click **Next.**

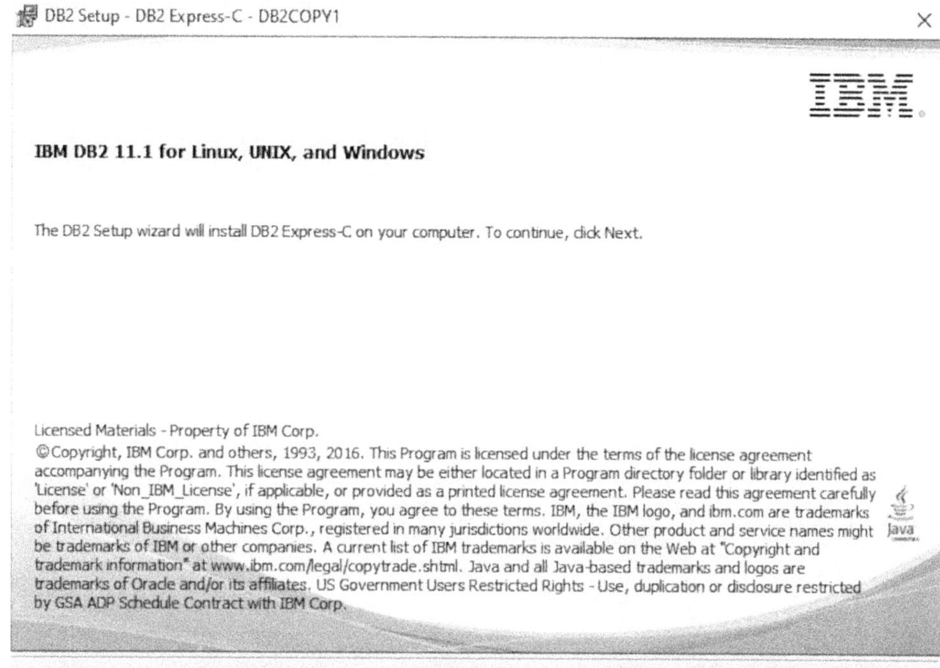

On the following panel, you'll need to accept the user agreement and then click **Next.**

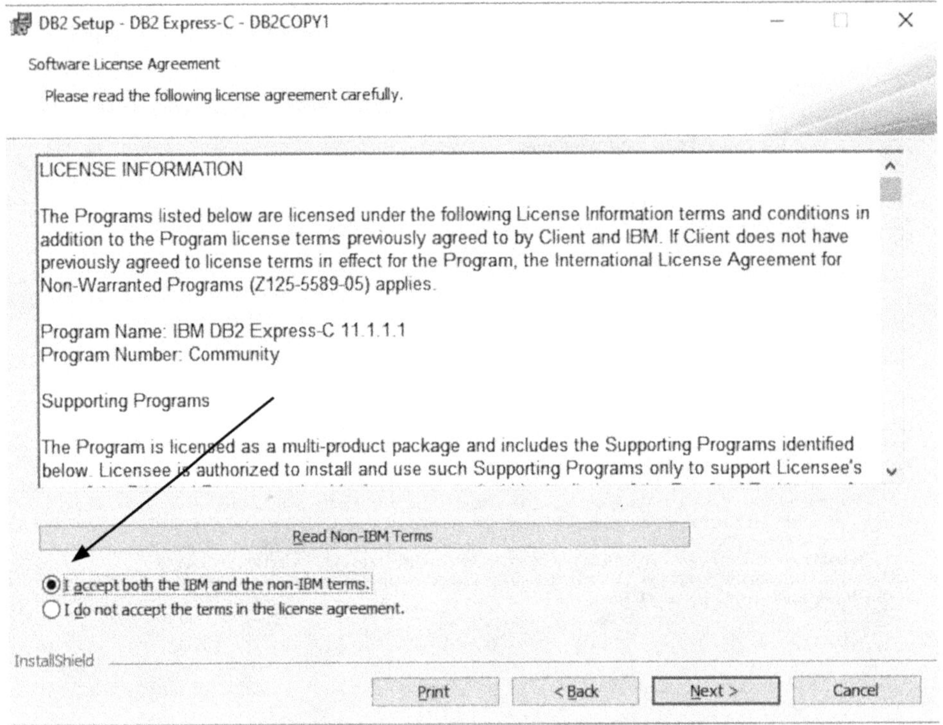

On the next panel, I recommend that you select a "typical" install, and then click **Next.**

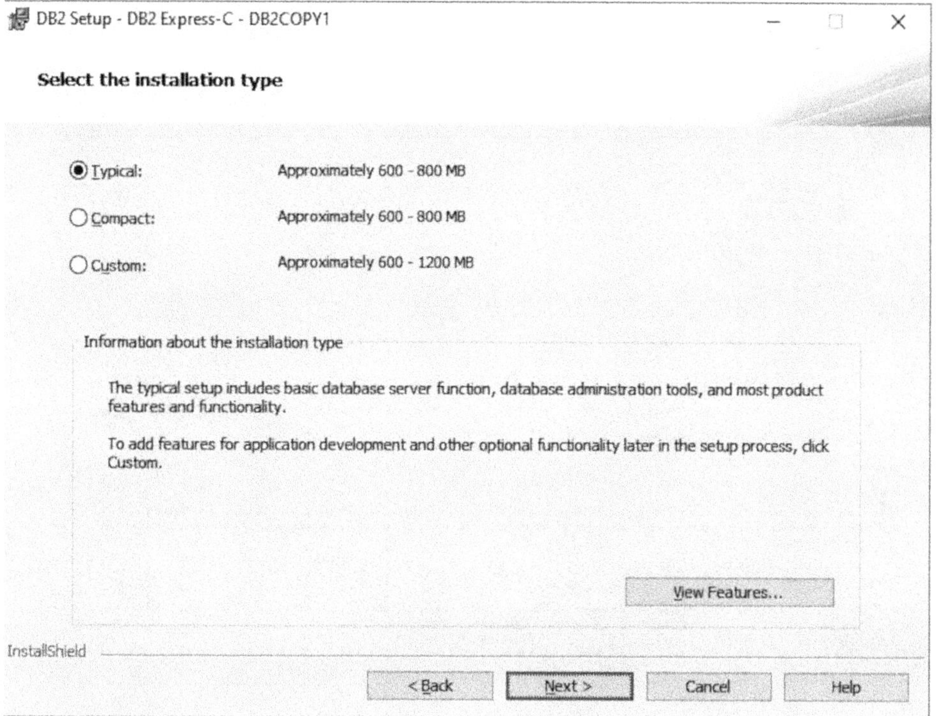

On the next panel, I recommend that you select the first option to simply install DB2. Then click on **Next.**

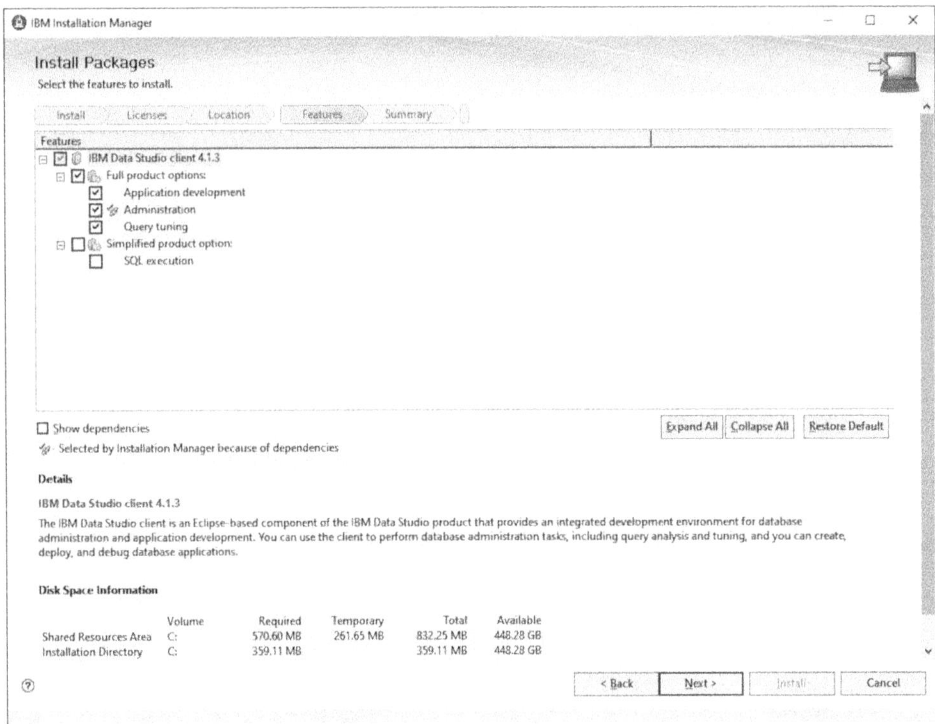

On the next panel, select an installation directory or accept the default, and then click on **Next.**

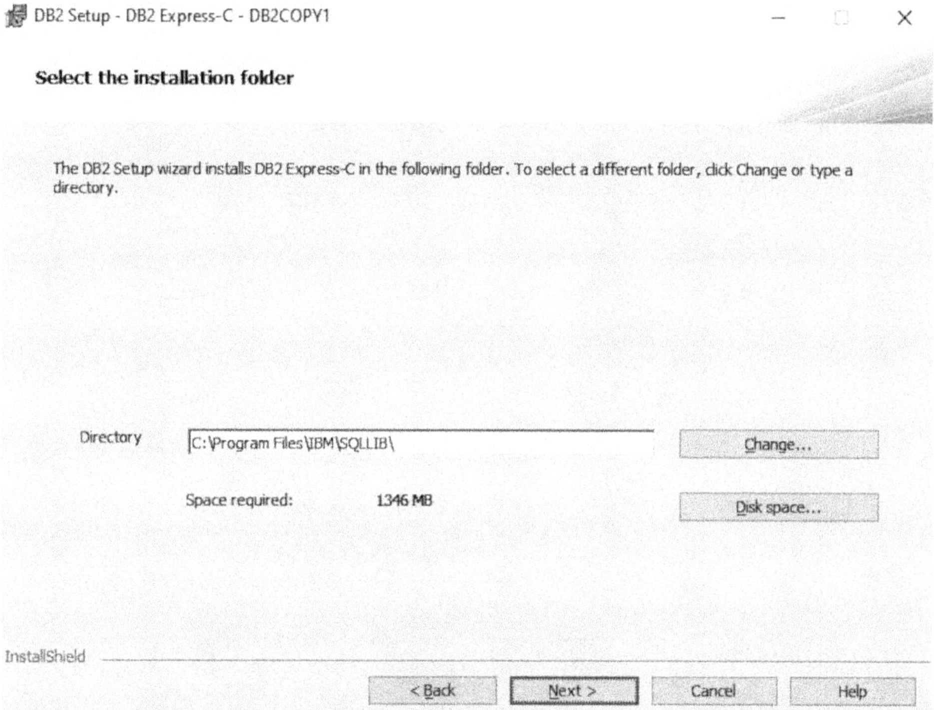

On the next panel, select the appropriate SSH server option. I recommend that you select the first option to autostart the IBM SSH server at startup. Then click **Next.**

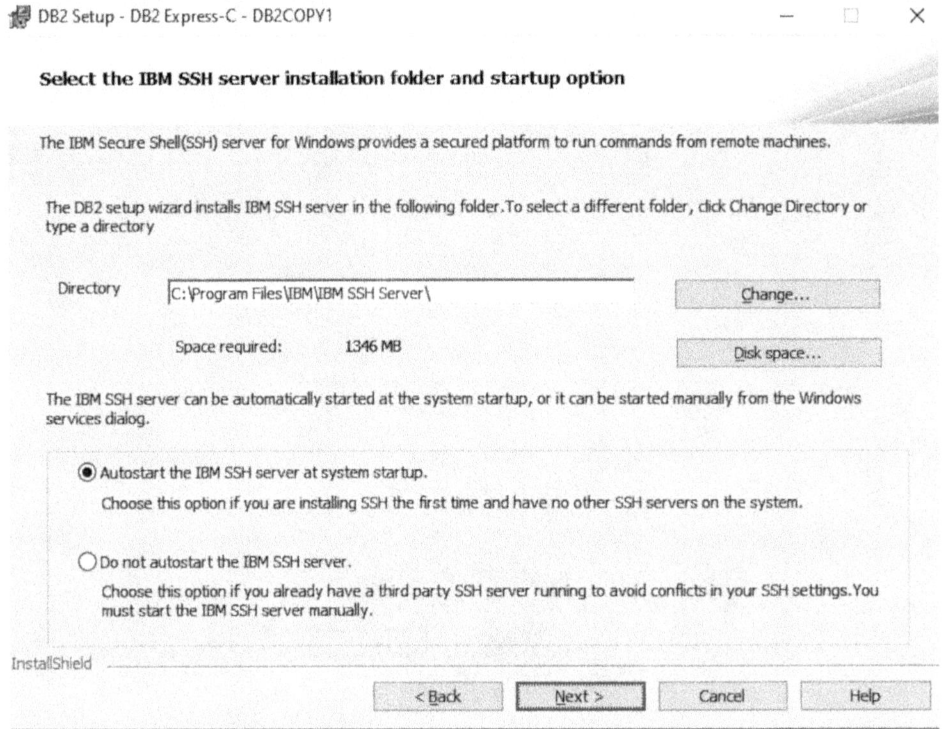

Now you must specify security for the DB2 Admin server. I strongly recommend that you use a local or domain user account and password. I recommend that you do NOT use the Local System account because it is not secure. In my case I used the local account **robert** – this is what I logon to my laptop with. Click **Next.**

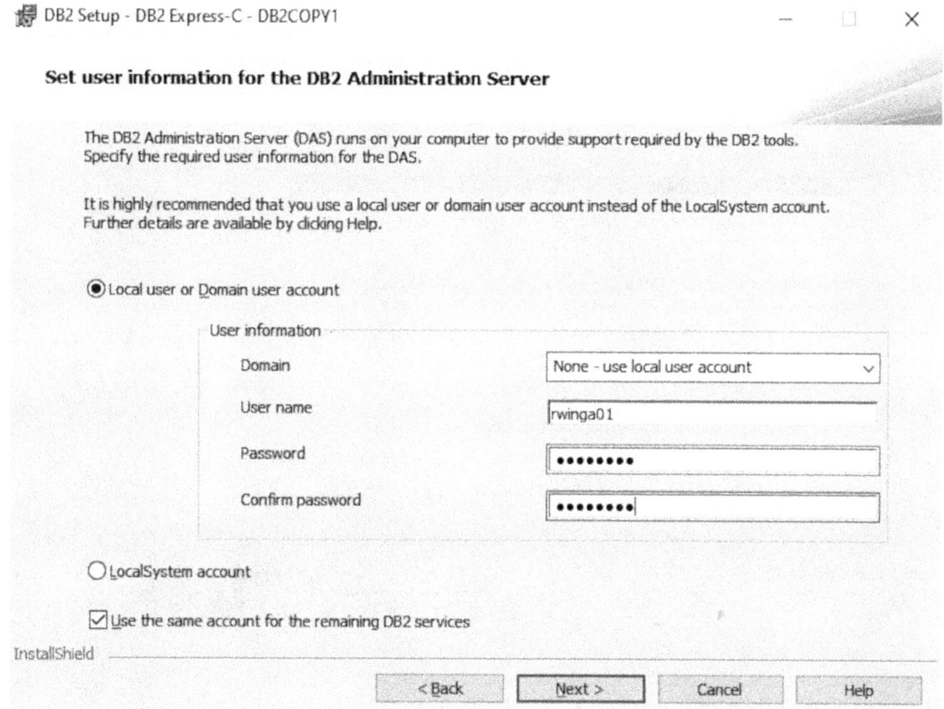

On the next panel, if you want to configure any special settings for your DB2 instance you can click on **Configure.** Otherwise simply click **Next.**

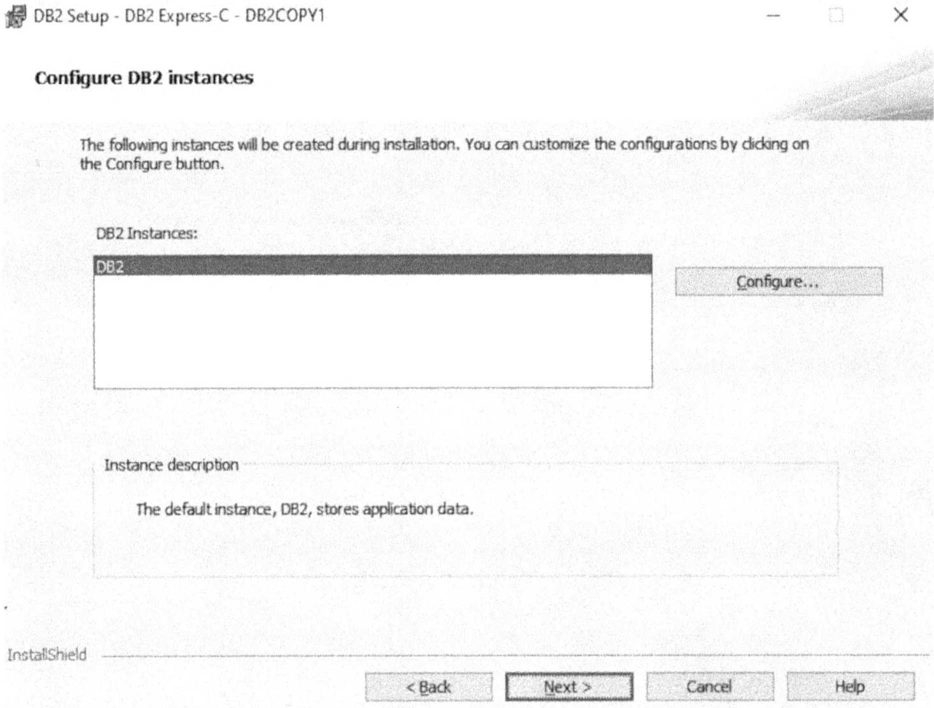

Review your choices and then click on **Install.**

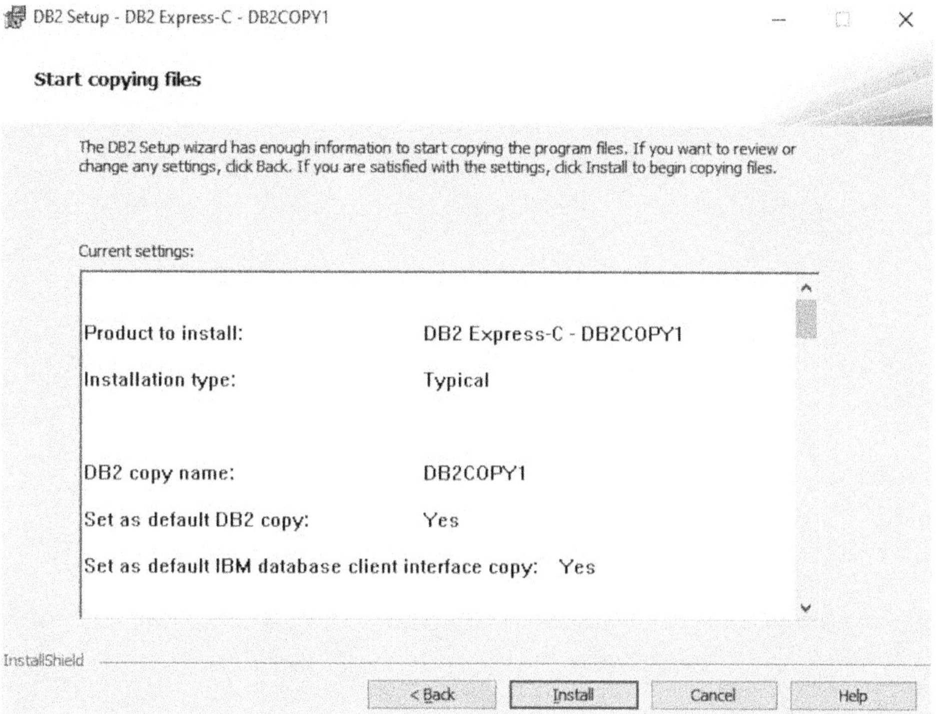

You'll see a status panel and various activities as DB2 installs.

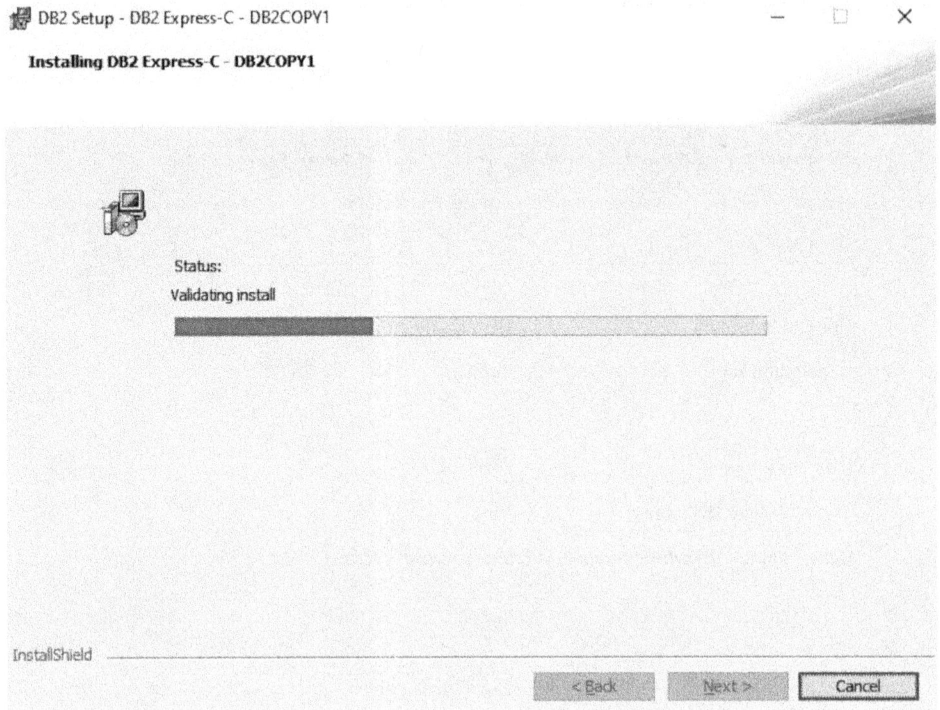

Eventually you will see a panel that says "Setup is Complete". Click **Finish.**

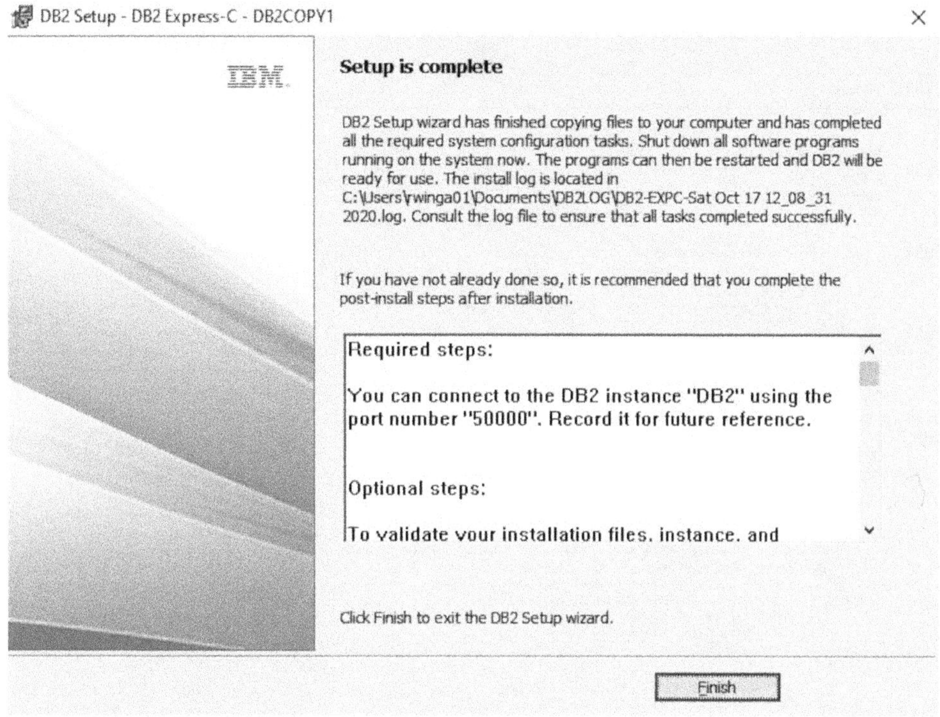

You'll see this Welcome screen. You won't need to create the SAMPLE database (that was done automatically by the install). The next section will take you through downloading and installing IBM Data Studio. You can close this window.

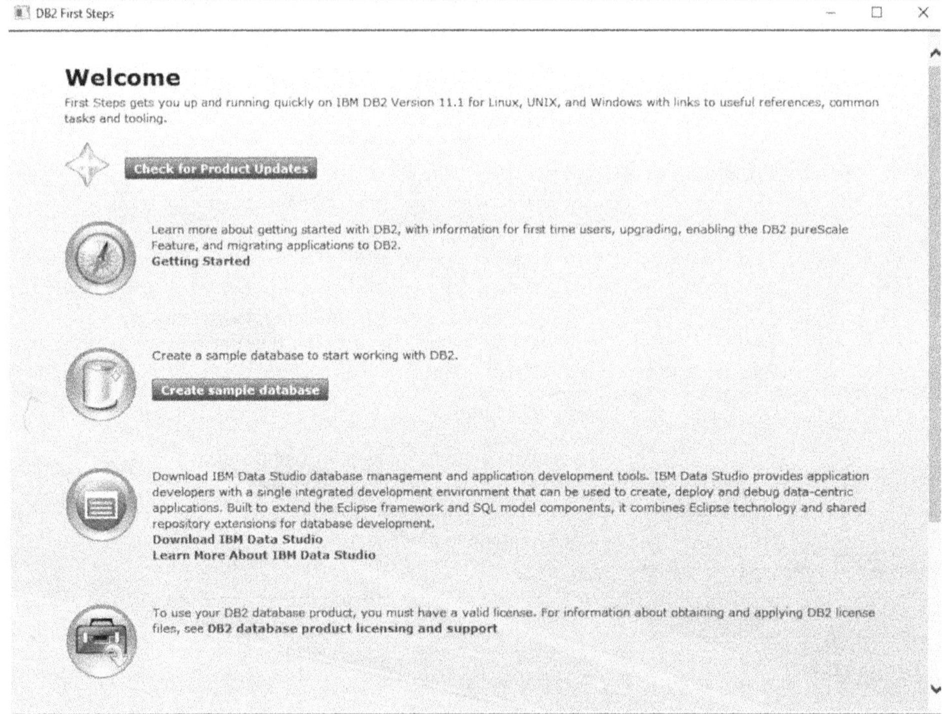

Finally, once DB2 is installed you can navigate to the basic tools that are installed with DB2 such as the command window. Select All Programs from your Windows Launcher and then click on IBM DB2 DB2COPY(Default). Right click on the DB2 Command Line Processor and select Open (or Run as Administrator if you are an admin on this computer). From here you can issue DB2 commands.

Suppose for example you want to start the DB2 Admin Server. You could issue that command as: `DB2START`

```
DB2 CLP - DB2COPY1 - C:\PROGRA~1\IBM\SQLLIB\BIN\db2setcp.bat DB2SETCP.BAT DB2.EXE
(c) Copyright IBM Corporation 1993,2007
Command Line Processor for DB2 Client 11.1.1.1

You can issue database manager commands and SQL statements from the command
prompt. For example:
    db2 => connect to sample
    db2 => bind sample.bnd

For general help, type: ?.
For command help, type: ? command, where command can be
the first few keywords of a database manager command. For example:
 ? CATALOG DATABASE for help on the CATALOG DATABASE command
 ? CATALOG          for help on all of the CATALOG commands.

To exit db2 interactive mode, type QUIT at the command prompt. Outside
interactive mode, all commands must be prefixed with 'db2'.
To list the current command option settings, type LIST COMMAND OPTIONS.

For more detailed help, refer to the Online Reference Manual.

db2 => db2start
DB20000I  The DB2START command completed successfully.
db2 =>
```

You could issue other commands, such as to create a database, but we'll do most of our work in the IBM Data Studio which is our next topic.

IBM Data Studio

I strongly recommend that you install and use IBM Data Studio. It's an excellent IDE that you really cannot afford to be without. While you could perform many operations using the Command Line Processor (CLP), I suggest you use Data Studio – it's easier.

If you did not choose the DB2 Express-C post-installation option to download IBM Data Studio (when you installed DB2 Express-C), here is a link to download it:

https://www.ibm.com/developerworks/downloads/im/data/

Once you've downloaded the zip file, you'll need to decompress it into an installation directory of your choosing. In my case, I have decompressed into a directory named DataStudioInstall.

Click on the **imLauncherWindows.bat** application file.

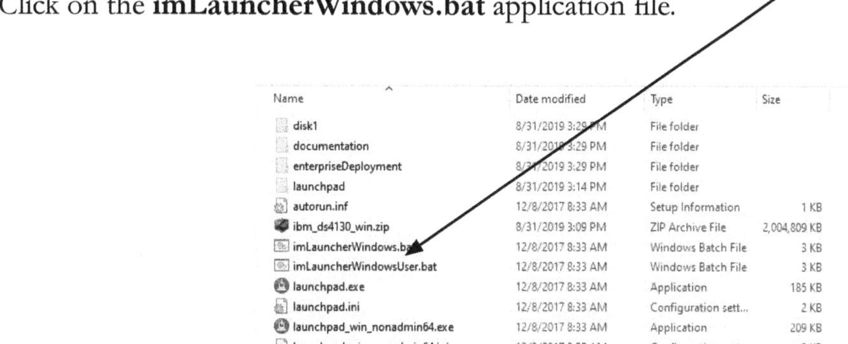

This screen will appear. Select the **Install** option.

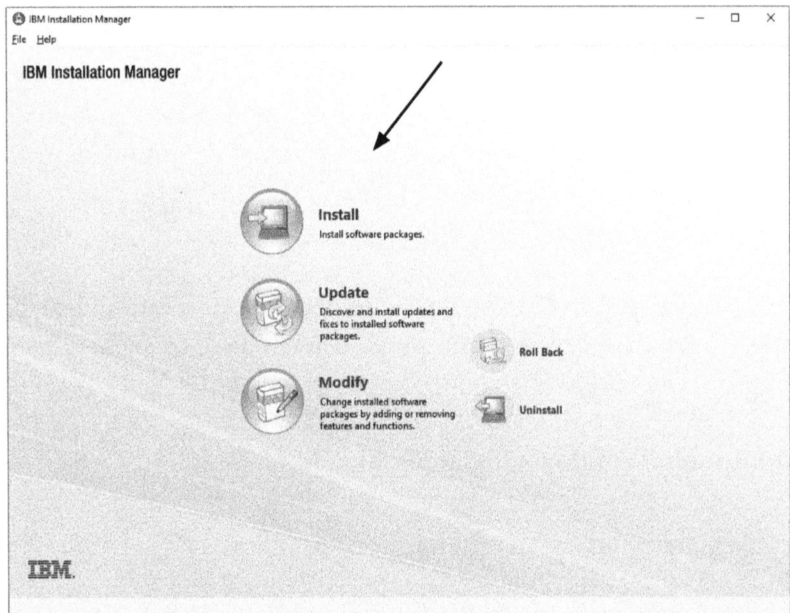

Click to check mark the IBM Data Studio client, and then click on **Next**.

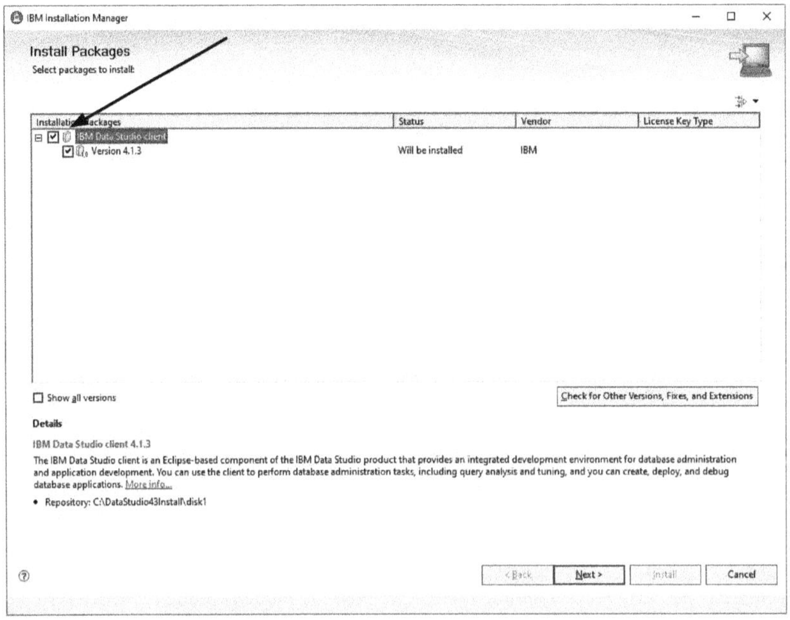

Accept the terms of the license agreement, and then click **Next.**

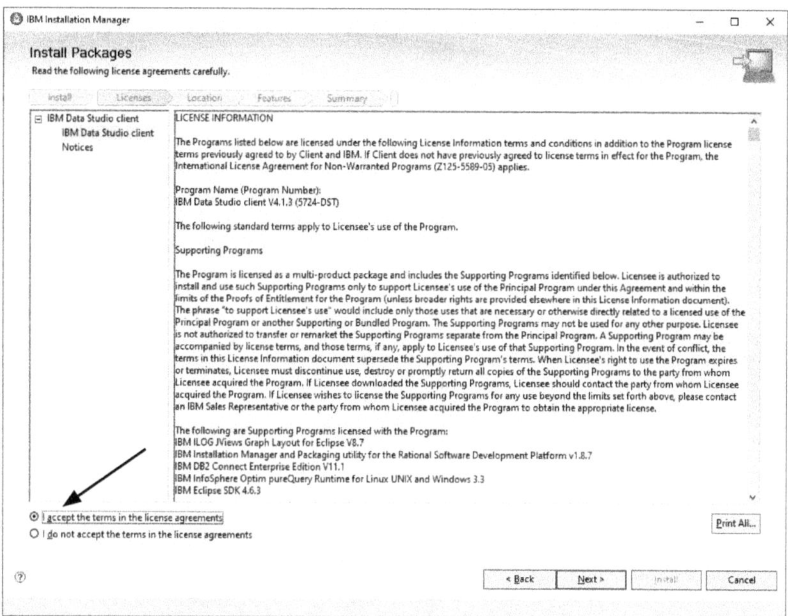

Click **Next** at this screen.

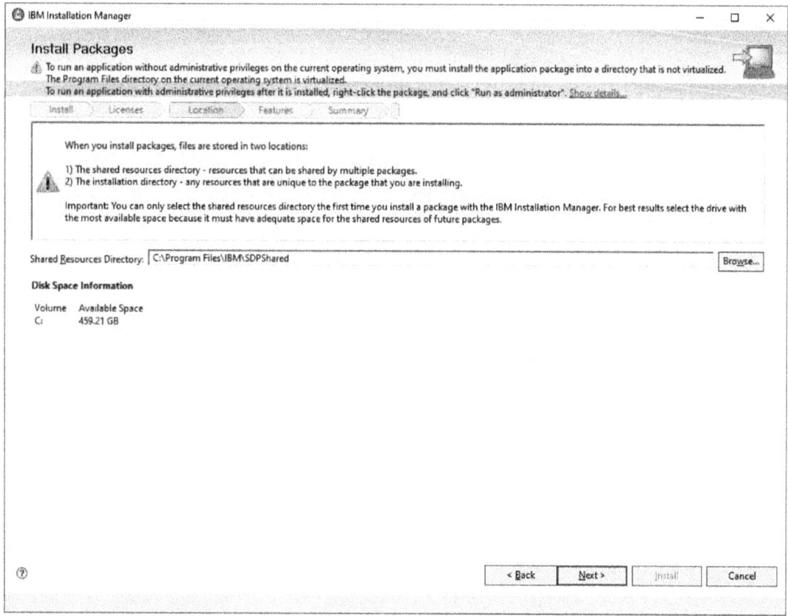

Click **Next** on this screen.

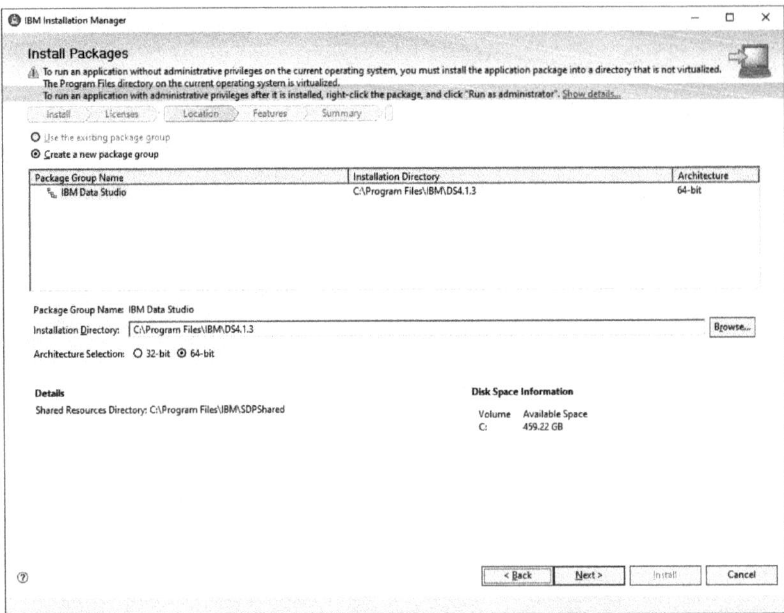

Click **Next** on this screen.

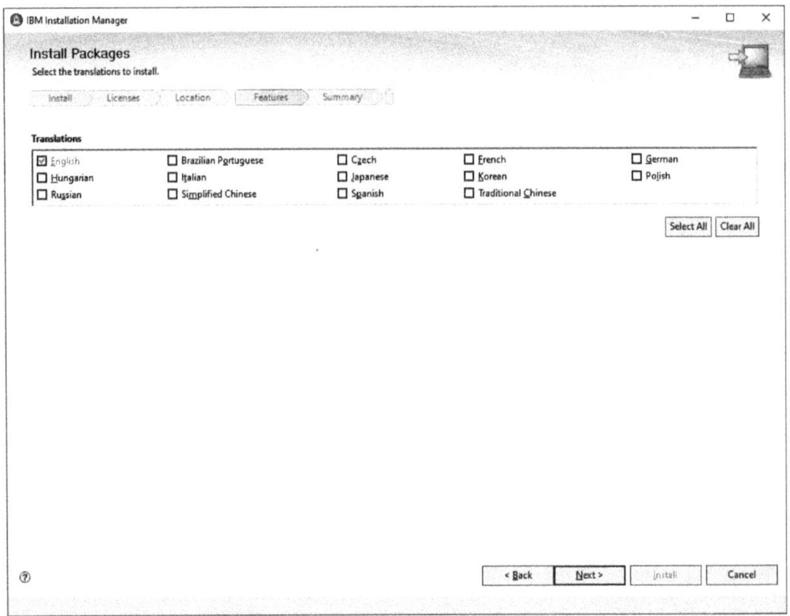

Click **Next** on this screen.

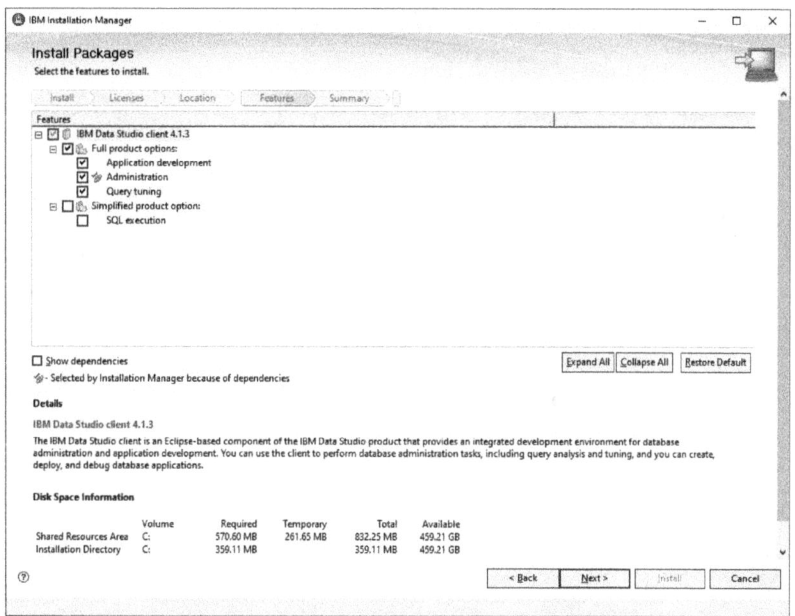

At this panel, click on **Install**.

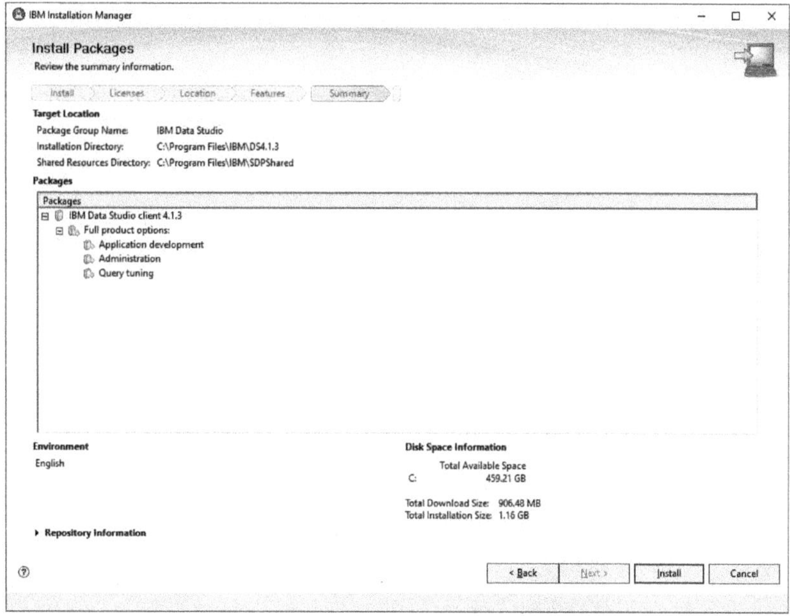

You'll see various activities while the product installs.

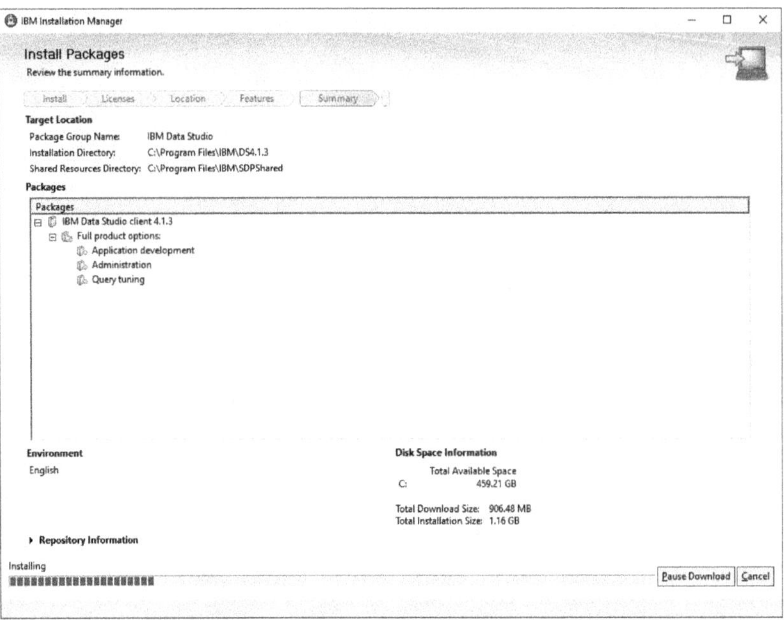

Now click **Finish.** You can also close the IBM Data Studio Client pre-installation dialog.

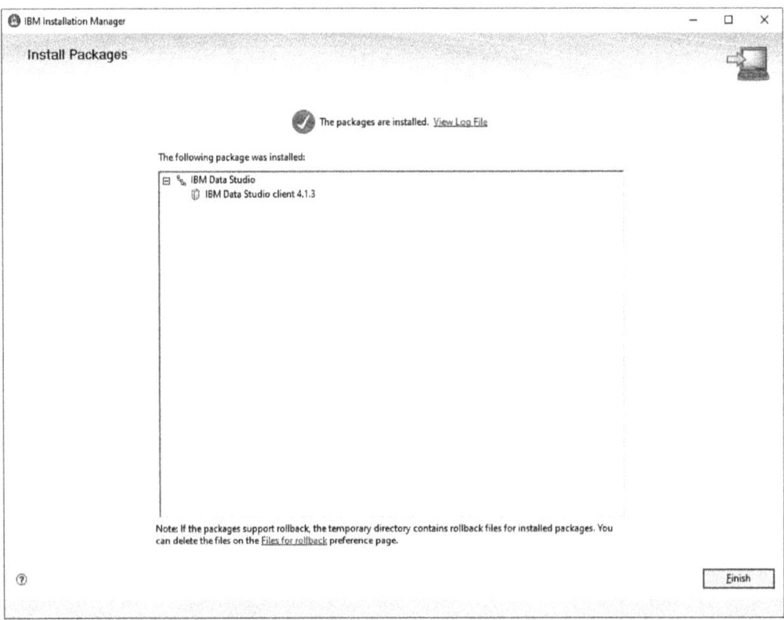

To start the Data Studio, use your program launcher and select **Data Studio 4.1.2 Client.**

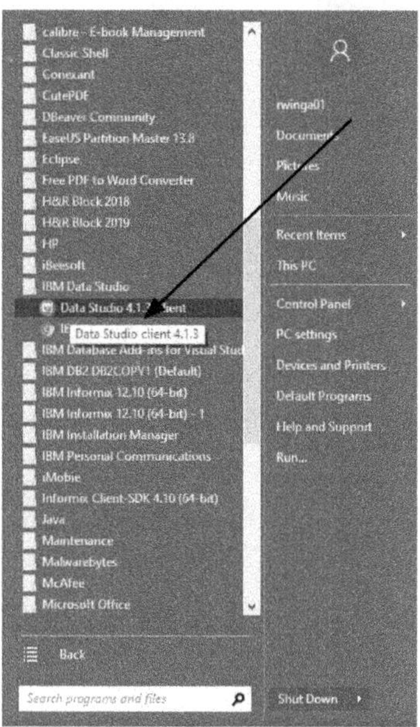

You will be asked to specify a workspace. I recommend you take the default and click **OK**.

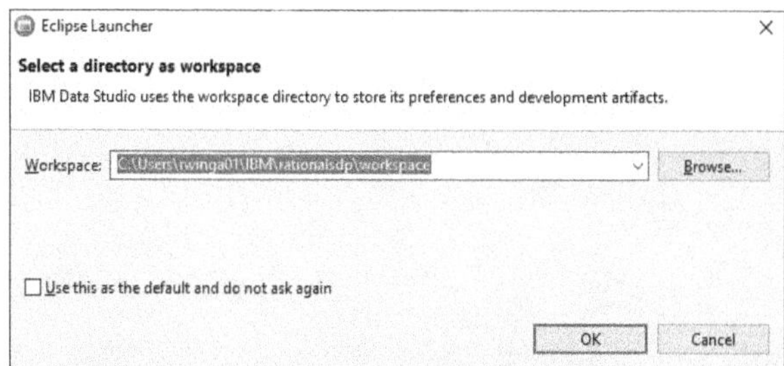

You should then see this window which means you are in the Data Studio IDE. If you are not in the **Administer Databases** perspective, select it from the dropdown show below.

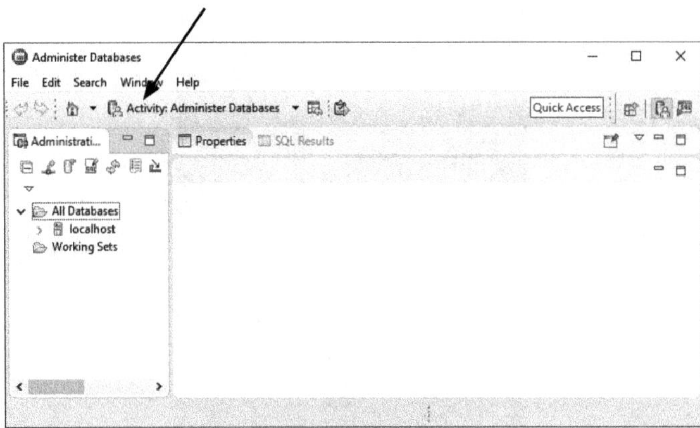

Now expand the tree by clicking on **localhost**. You should see a single node which is the DB2 instance name (which will be DB2 if you accepted the default during setup).

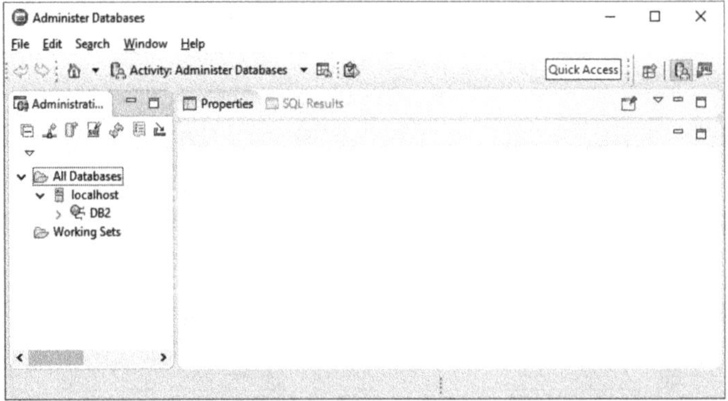

Click on the Db2 node. Assuming you created the Sample database when you installed DB2, you should see the Sample database under the Db2 node.

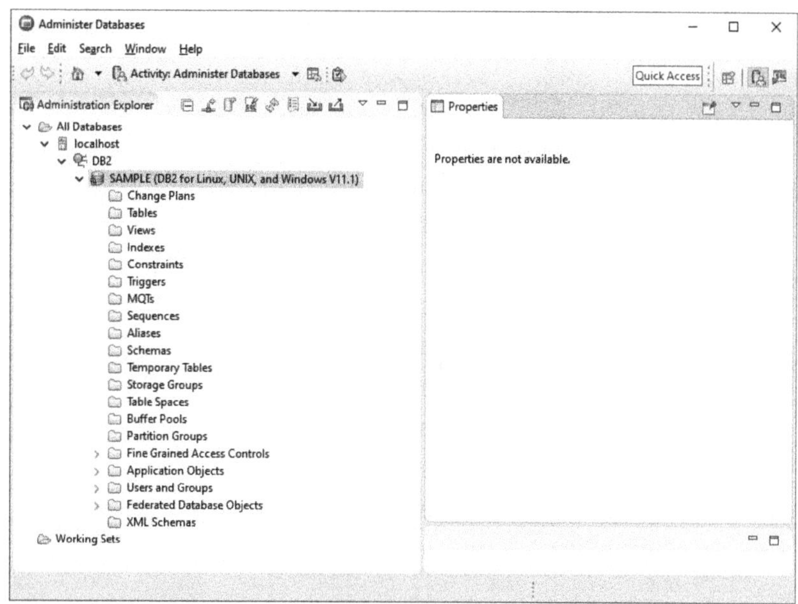

To see all the tables in the SAMPLE database, click on **Tables**.

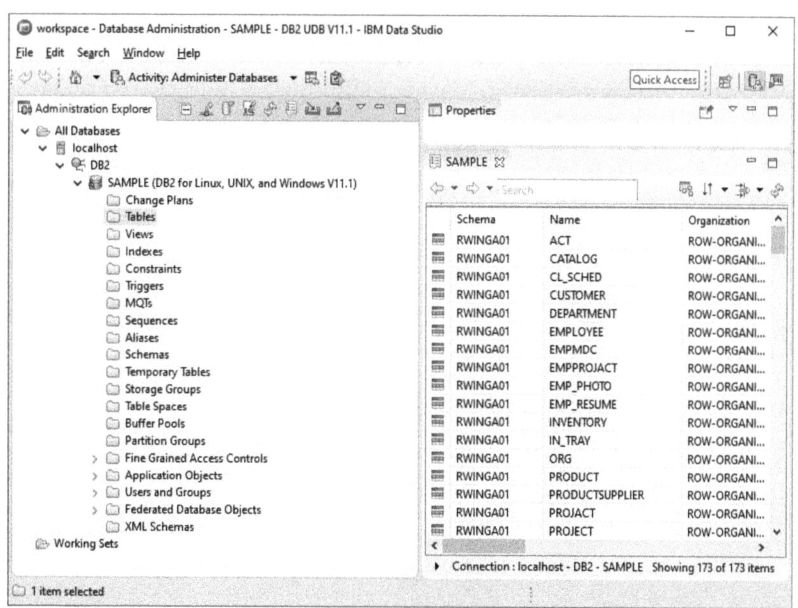

To create a new object, you can right click the folder for the object type you want to create, and then fill in the required details. For example, let's create a table named TESTTBL. Right click the **Tables** folder and then select the **Create Table** option. You'll be prompted to select a schema. For now, just click on your logonid which should be in the selection list.

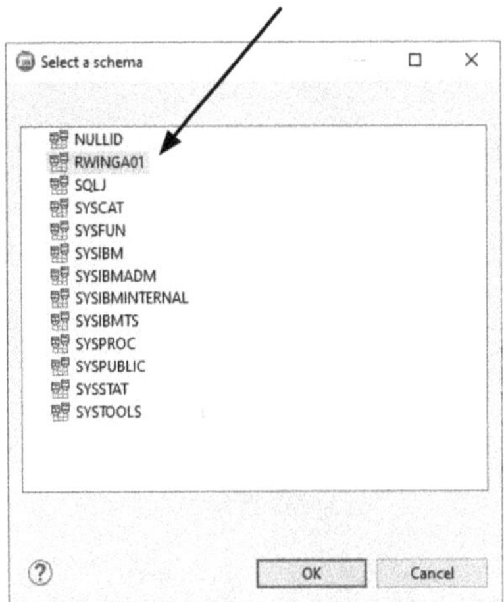

You'll be notified that you are making changes to a database object. Click on **OK.**

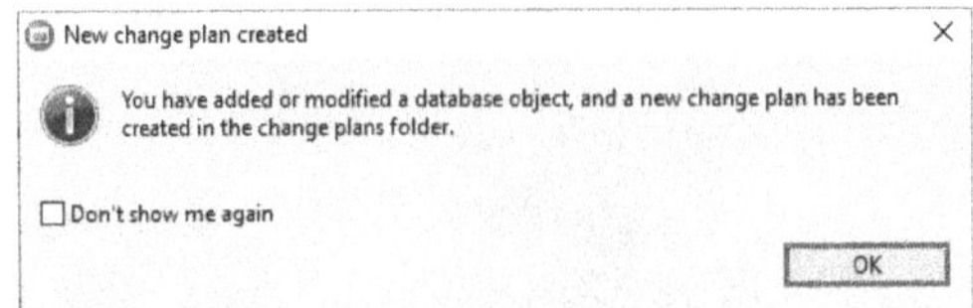

Type the name of the new table into the properties window, then click on the **Columns** button.

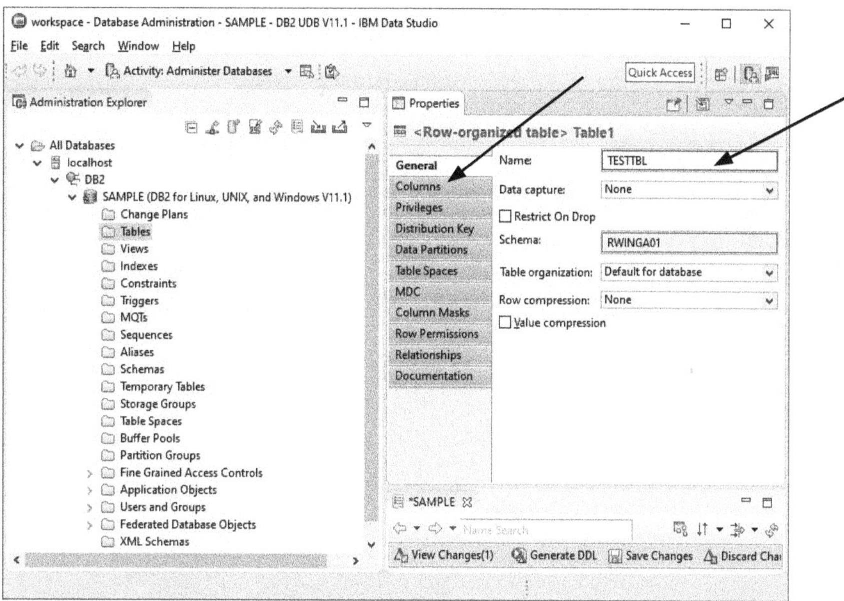

Click on the + button to create a new column, or click on the X button to delete one.

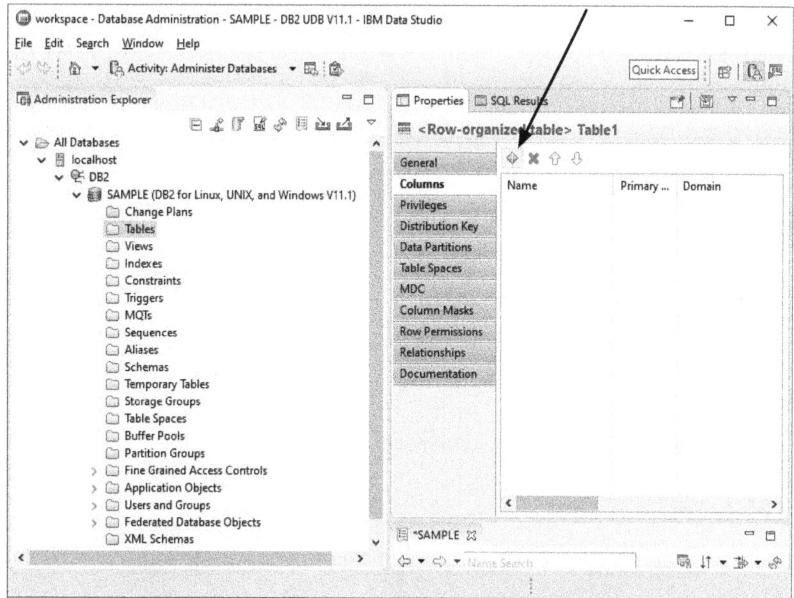

Let's add a couple of columns. One column will be named ID which will be an integer and the primary key (check the primary key check box), and the other column is named DESCRIPTION which will be a varchar (30). Both columns will be defined as NOT NULL.

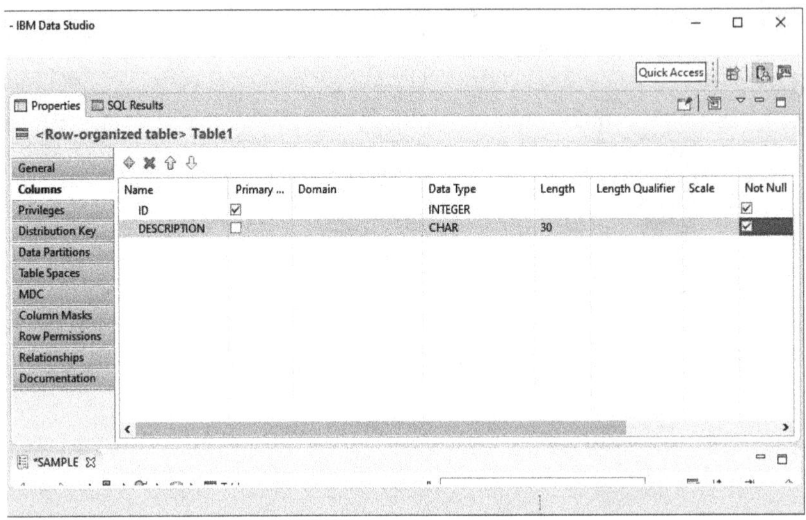

Finally, click on the **Generate DDL** button at the top of the table definitions screen. This will generate the Data Definition Language commands to create the table.

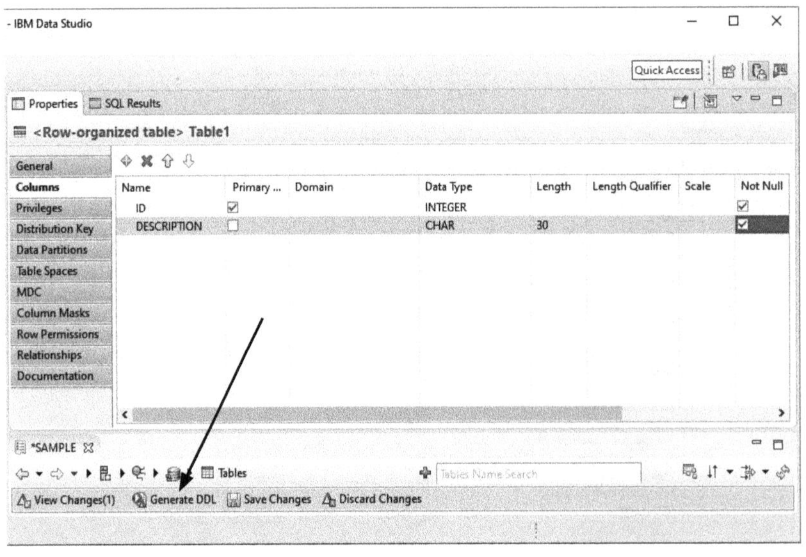

You'll see a screen that shows the DDL that was generated. If you need to change anything in the DDL, this is where you can do it. Otherwise, click on **Finish.**

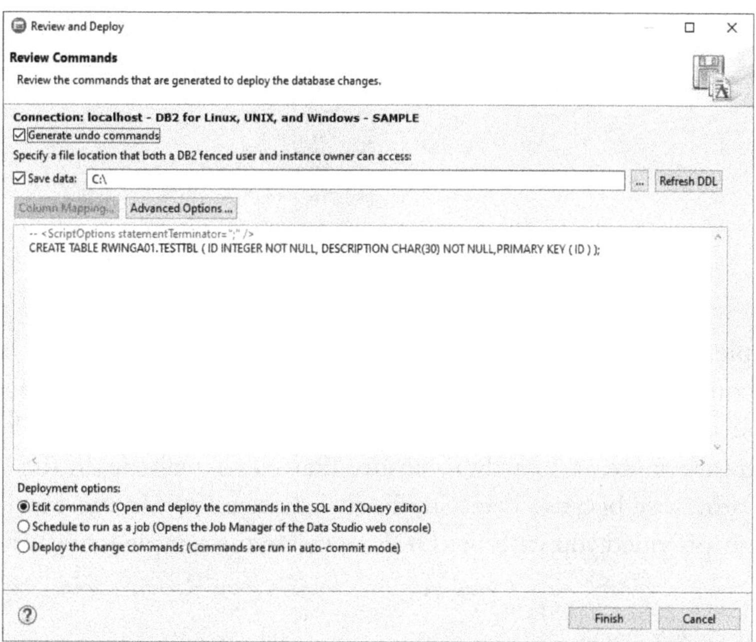

At this point we have created but not executed the DDL. You'll see UOM_REVIEW tab. To execute the DDL, click on the forward green arrow.

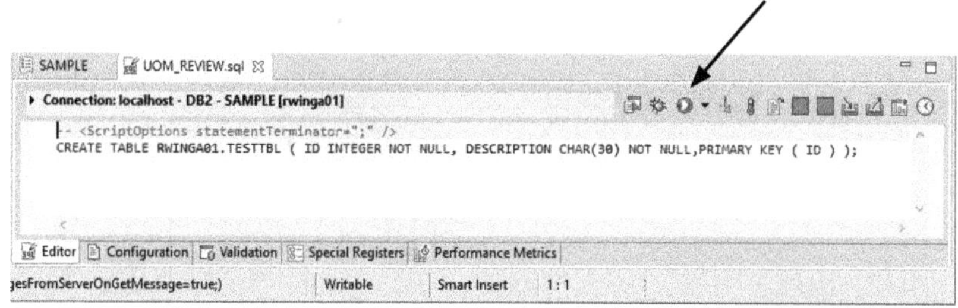

Now then you can review your results in the **SQL Status** window. As we can see, the result is successful.

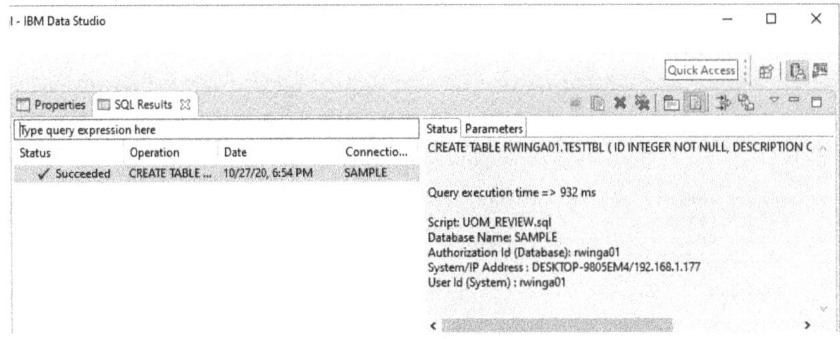

The table example shows how to create objects using the Data Studio object tree. This is very handy, and I encourage you to use the Data Studio functions. However, in this text book we'll usually create our objects by simply executing DDL through the Data Studio SQL window. This is partly to save space and partly to ensure that you actually learn the DB2 statements (you might not otherwise because Data Studio automates a lot). In the previous example we could have simply provided you with the DDL to create the sample table as follows:

```
CREATE TABLE ROBERT.TESTTBL
(ID INTEGER NOT NULL,
DESCRIPTION CHAR(30) NOT NULL,
PRIMARY KEY (ID));
```

If you'd like a good tutorial on the IBM Data Studio, there is an outstanding one on the IBM web site here:

https://www.ibm.com/support/knowledgecenter/en/SS62YD_4.1.1/com.ibm.datatools.ds.nav.doc/topics/ds_landing.html

Eclipse Java Oxygen

Eclipse is a free, open source IDE that you can use to build Java programs. I find it easy to use, and I recommend that you download and install it if you are not already using another Java IDE (such as NetBeans).

Later we'll show a basic example of creating a Java program in the Eclipse IDE, but we won't spend time showing all its features. There is plenty of user reference material on Eclipse elsewhere.

http://help.eclipse.org/oxygen/index.jsp

We'll instead focus more on DB2 concepts. To acquire and install Eclipse, go to this site:

https://www.eclipse.org/downloads/download.php?file=/oomph/epp/oxygen/R2/eclipse-inst-win64.exe

Click on the **Download** button.

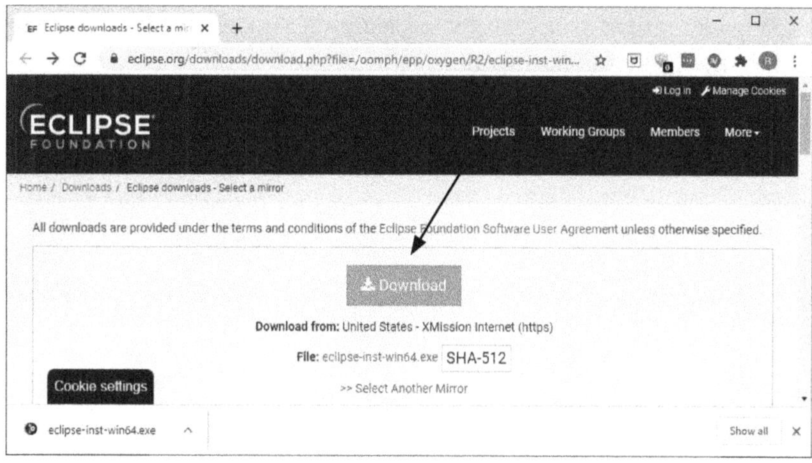

Click on **Save.**

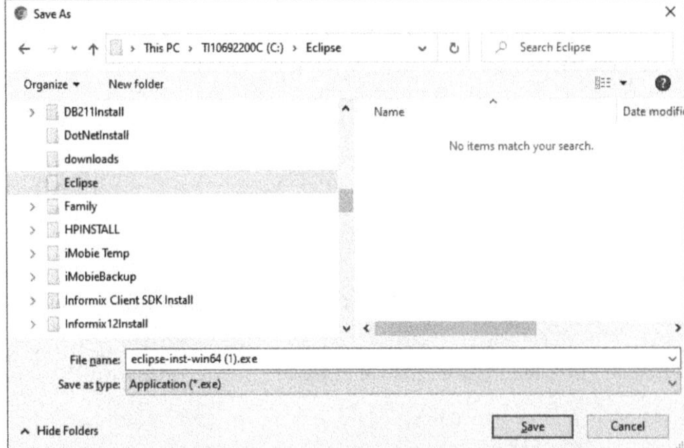

Locate file in the folder where you saved it. Assuming you chose the download for a 64-bit Windows machine, you'll download this file:

eclipse-inst-win64.exe

When you open/execute this file you will see the Eclipse installer. Select the **Eclipse IDE for Java Developers** by clicking on it.

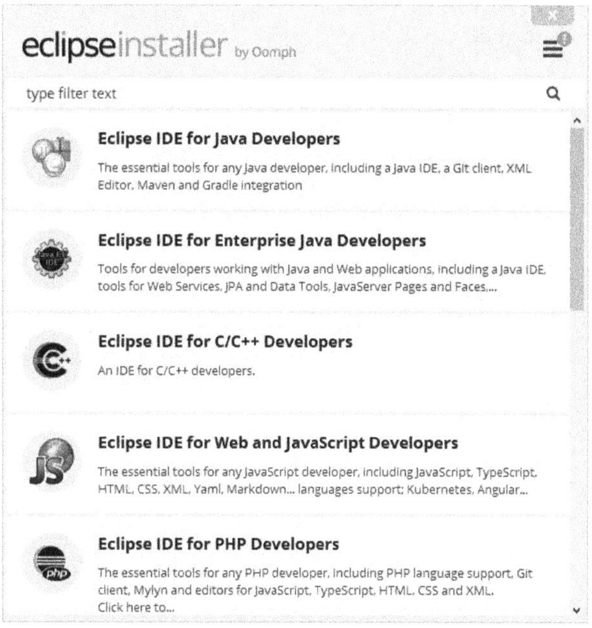

When you see the Eclipse Installer window, click on the **INSTALL** button.

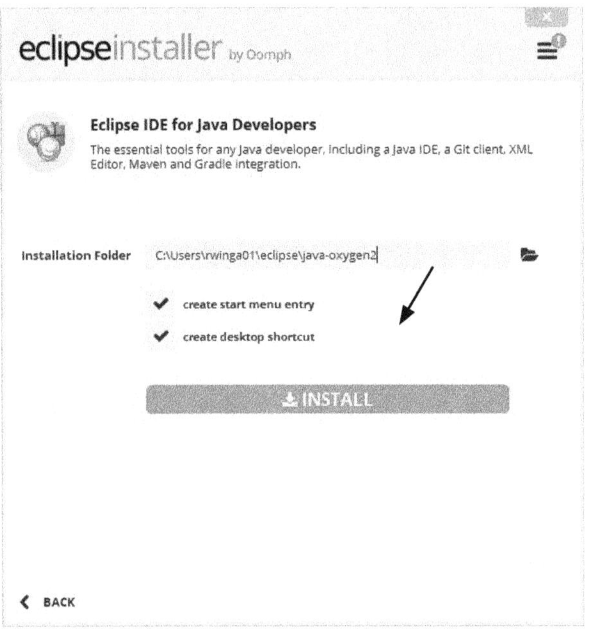

Wait a few moments while the installer runs.

Now you can either click on the **Launch** button, or come back later and use the desktop icon (or the Eclipse entry from your program launcher). Let's go ahead and launch Eclipse:

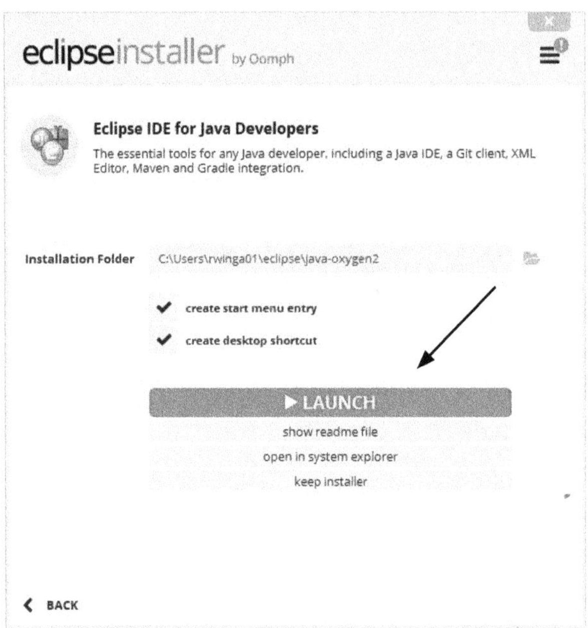

You'll need to specify a work space and then click on **Launch.**

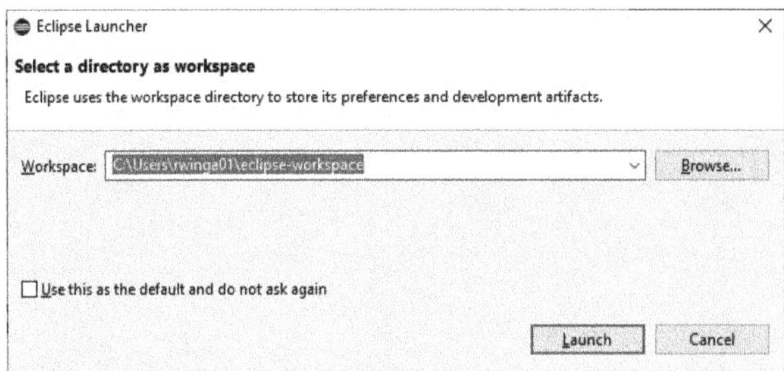

Now you'll see the Eclipse IDE. We won't go further into it now, but stay tuned. We'll get into Java in later chapters.

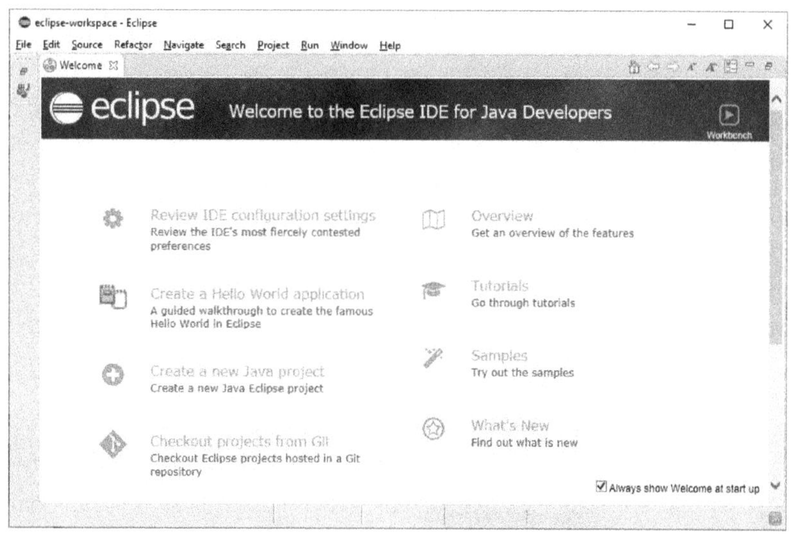

Visual Studio Community

Microsoft provides a free "community" version of its Visual Studio IDE for use with .NET. We'll show coding examples using the c# .NET language. If you are following along and doing the examples and want to use .NET, you may want to install this free IDE.

Navigate to this web site:

https://www.visualstudio.com/vs/community/

When you see this window, click on **Download Visual Studio**

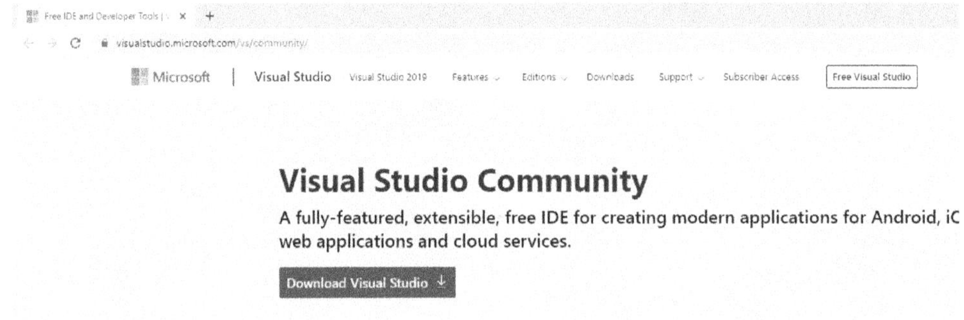

Next, go to the folder you downloaded into, and double click on the **VS Community.exe** file. In the window that opens, click on **Continue.**

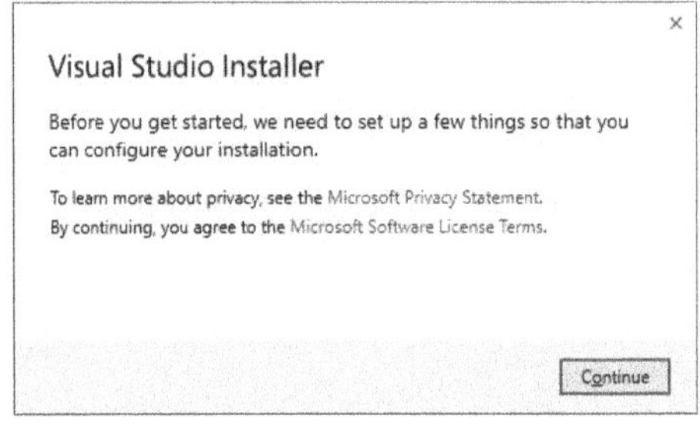

You'll arrive at this window where you can select a "workload" setting. For our purposes in this text book, we can choose **.NET desktop development.** Then click on **Install.**

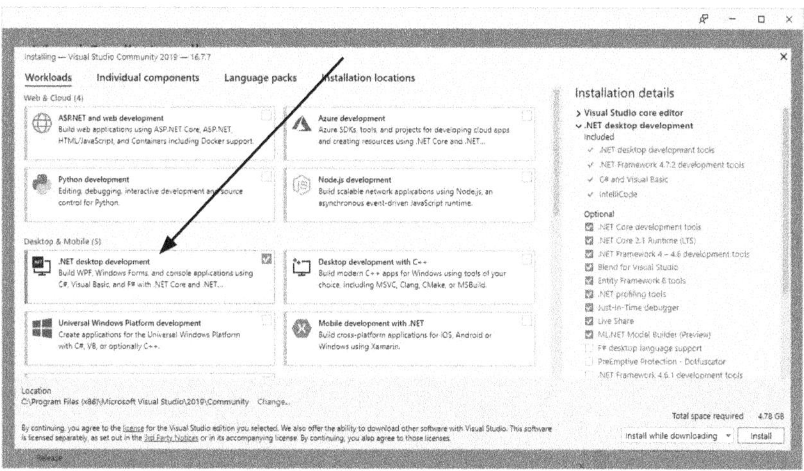

It will take a while to install the software. You might want to take a break.

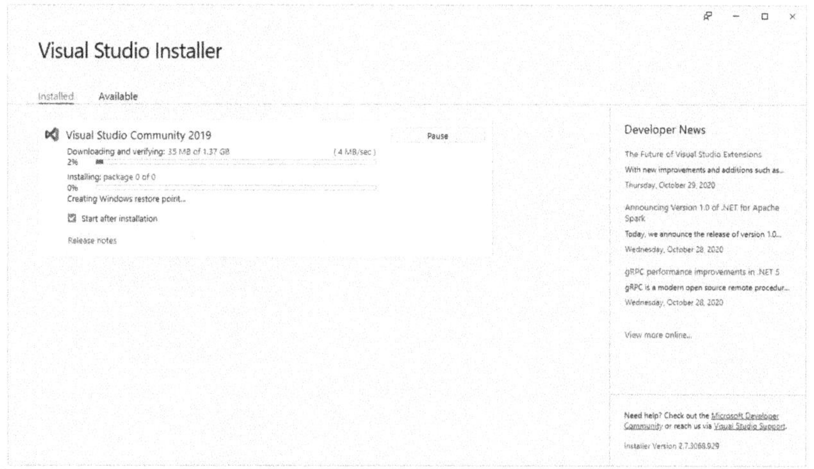

You'll see this window when the install completes. Sign into your Microsoft account (or create one if you don't have one).

Click on **Continue Without Code.**

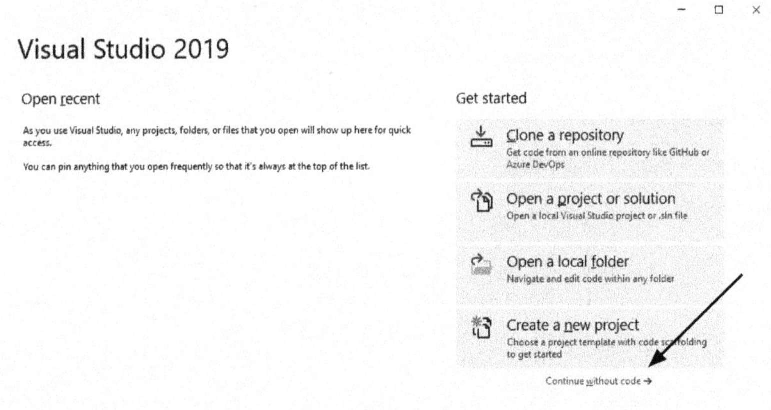

Finally you'll see the initial IDE window.

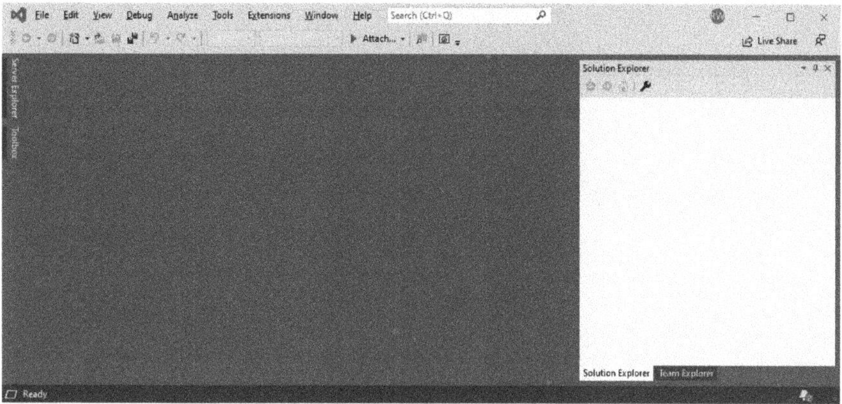

We'll do a programming example with Visual Studio in chapter five. For now, let's move onto creating some database objects with DDL.

CHAPTER TWO: DATA DEFINITION LANGUAGE

Before rushing into programming it is a good idea to understand data types, and how to create and maintain the basic DB2 objects. We'll look at the basic properties of the various objects (tables, indexes, views). Finally we'll look at the DB2 catalog and the information it provides.

Database

A DB2 database is a collection of objects including tablespaces, tables, indexes, views, triggers, stored procedures and sequences. Generally a database is concerned with a single domain such as marketing, accounting, shipping and receiving, etc. For purposes of this text book, we will be supporting a simple computerized human relations system for a fictitious company.

CREATE

You can create a database with the `CREATE DATABASE` statement. The required syntax to create the database is:

```
CREATE DATABASE <database name>
```

However you could also include other options such as storage, code page, encryption, page size and an alias. Of course there are defaults for all of these options. For example, typically DB2 databases are created with automatic storage (which is the default).

Let's name the database `DBHR` and we'll specify automatic storage. We'll keep it simple and take the defaults for the other options. The DDL to create our HR database is as follows, and you can execute it from the Command Line Processor (CLP):

```
CREATE DATABASE DBHR
AUTOMATIC STORAGE YES;
```

ALTER

If you need to add or remove storage paths for a database, you can use the ALTER statement. For example, suppose our database DBHR resides on a local drive C:, and we want to add storage using drives D: and E;. The following DDL would accomplish this:

```
ALTER DATABASE DBHR
ADD STORAGE ON 'D:', 'E:';
```

DROP

Most database objects can be removed/deleted by issuing the DROP command. The syntax to delete a database is very simple:

```
DROP DATABASE <databasename>
```

You could DROP the DBHR database by simply issuing this command:

```
DROP DATABASE DBHR;
```

We won't drop the DBHR database yet because we are going to use it throughout this chapter.

Tablespace

A tablespace is a layer between the physical containers that hold data and the logical database. In essence, a tablespace defines storage areas into which DB2 objects can be placed and maintained.

DB2 11.1 supports the following types of tables spaces:

- Large
- Regular
- System Temporary
- User Temporary

As an application developer, I doubt that you will get involved in creating or maintaining table spaces (that is normally a DBA activity). When training I suggest you specify REGULAR for any tabelspaces you create. However, here is a brief description of all four types of tablespaces available in DB2 LUW.

Tablespace Type	Description
REGULAR	Stores permanent data. Can be used with AUTOMATIC STORAGE.
LARGE	Stores permanent data and is only allowed on database managed space (DMS) table spaces. The main benefit of using LARGE tablespaces is to enable table objects to be larger than with REGULAR tablespaces. With a REGULAR tablespace a table object can be a maximum of 512 gigabytes, versus 64 terabytes in a LARGE tablespace.

SYSTEM TEMPORARY	Stores temporary tables, work areas used by the database manager to perform operations such as sorts or joins.
USER TEMPORARY	Stores created temporary tables and declared temporary tables.

For our purposes, let us create a REGULAR tablespace for our HR database, and we'll name it TSHR (tablespace HR). We'll first create storage group SGHR and bufferpool BPHR so that we can store our tablespace and relate it to a specific buffer pool.

Here is the DDL to create storage group SGHR:

```
CREATE STOGROUP SGHR ON 'C:'
OVERHEAD 6.725
DEVICE READ RATE 100.0
DATA TAG NONE;
```

Here's the DDL to create our buffer pool:

```
CREATE BUFFERPOOL BPHR
IMMEDIATE ALL DBPARTITIONNUMS SIZE 1000
AUTOMATIC PAGESIZE 4096;
```

Now we can create and use our new tablespace TSHR throughout this text book as a container for our tables and other objects.

```
CREATE REGULAR TABLESPACE TSHR
IN DATABASE PARTITION GROUP IBMDEFAULTGROUP
PAGESIZE 4096
MANAGED BY AUTOMATIC STORAGE USING STOGROUP SGHR
AUTORESIZE YES
BUFFERPOOL BPHR
OVERHEAD INHERIT
TRANSFERRATE INHERIT
DROPPED TABLE RECOVERY
ON DATA TAG INHERIT;
```

When we execute the DDL above, tablespace TSHR will be created in storage group SGHR and will use bufferpool BPHR. Don't worry if some of the values in the DDL do not make perfect sense. Again, as an application developer you don't really need to master the details of tablespaces.

Schemas

One other operation we will perform before moving on to tables is to create a schema. A schema is a qualifier used for logically grouping objects. In our case we will create a schema named HRSCHEMA and then use that to group our tables, indexes, views, etc. The schema must have an owner and let's assume we can have it owned by me (robert).

```
CREATE SCHEMA HRSCHEMA
AUTHORIZATION rwinga01;
```

Now you can create database objects such as tables, view, indexes and sequences and specify schema HRSCHEMA as the qualifier. For example you could create an EMPLOYEE table as HRSCHEMA.EMPLOYEE. If you do not specify a schema, DB2 assumes a default or current schema which is often your logon-id. Obviously if you are working as part of a group, a common schema name is a better alternative.

Tables

As I'm sure you are aware, a table is the basic structure and container for DB2 data. Let's summarize the different types of tables in DB2 11.1.

Base

A table structure which physically stores records.

History

A table that is used to store historical versions of rows from the associated system-period temporal table.

Materialized Query

A materialized query table basically stores the result set of a query. It is typically used to store aggregate results from one or more other tables.

Result

A non-persistent table that contains a set of rows that DB2 selects or generates, directly or indirectly, from one or more base tables or views in response to an SQL statement.

Temporal

A temporal table is one that keeps track of "versions" of data over time and allows you to query data according to the time frame.

Temporary

A table that is created and exists only for the duration of a session.

DDL for Tables

Now let's look at the basic DDL that is used to manage tables. As with other DB2 objects, we use the CREATE, ALTER and DROP statements to create, change and delete tables respectively.

CREATE

The basic syntax to create a DB2 table specifies the table name, column specifications and the tablespace into which the table is to be created.

```
CREATE TABLE <tablename>
(field specifications)
IN <tablespace>
```

For an example, let's create the first table for our HR application. Here are the columns and data types for our table which we will name EMPLOYEE.

Field Name	Type	Attributes
EMP_ID	INTEGER	NOT NULL, PRIMARY KEY
EMP_LAST_NAME	VARCHAR(30)	NOT NULL
EMP_FIRST_NAME	VARCHAR(20)	NOT NULL
EMP_SERVICE_YEARS	INTEGER	NOT NULL, DEFAULT IS ZERO
EMP_PROMOTION_DATE	DATE	

The table can be created with the following DDL:

```
CREATE TABLE HRSCHEMA.EMPLOYEE(
EMP_ID INT NOT NULL,
EMP_LAST_NAME VARCHAR(30) NOT NULL,
EMP_FIRST_NAME VARCHAR(20) NOT NULL,
EMP_SERVICE_YEARS INT NOT NULL WITH DEFAULT 0,
EMP_PROMOTION_DATE DATE,
PRIMARY KEY(EMP_ID))
IN TSHR;
```

Before we move on, let's create a couple more tables that we will use later. Let's say we need an EMP_PAY table to store the employee's annual pay, and an EMP_PAY_CHECK table that will be used to cut pay checks on the first and fifteen of the month. Here's the DDL for these:

```
CREATE TABLE HRSCHEMA.EMP_PAY(
EMP_ID INT NOT NULL,
EMP_REGULAR_PAY DECIMAL (8,2) NOT NULL,
EMP_BONUS_PAY DECIMAL   (8,2)) IN TSHR;

CREATE TABLE HRSCHEMA.EMP_PAY_CHECK(
EMP_ID INT NOT NULL,
```

```
EMP_REGULAR_PAY  DECIMAL (8,2) NOT NULL,
EMP_SEMIMTH_PAY DECIMAL (8,2) NOT NULL) IN TSHR;
```

ALTER

You can change various aspects of a table using the ALTER command. ALTER is often used to add an index or additional columns. Two examples follow, the first adding a column to a table, and the second version enabling a table for history.

```
ALTER TABLE HRSCHEMA.EMPLOYEE
ADD COLUMN EMP_PROFILE XML;

ALTER TABLE HRSCHEMA.EMPLOYEE
ADD VERSIONING
USE HISTORY TABLE HRSCHEMA.EMPLOYEE_HISTORY;
```

At this point we won't run this DDL because we want to do these operations later after some explanation. For now, just be aware that the way to change a DB2 table is to use the ALTER statement.

DROP

You can remove a table by issuing the DROP command.

```
DROP TABLE <table name>
```

Base Tables

The most common type of table in DB2 is a base table. This is your typical table created with the CREATE TABLE statement. A sample is our original DDL that we used for the EMPLOYEE table:

```
CREATE TABLE HRSCHEMA.EMPLOYEE(
EMP_ID INT NOT NULL,
EMP_LAST_NAME VARCHAR(30) NOT NULL,
EMP_FIRST_NAME VARCHAR(20) NOT NULL,
EMP_SERVICE_YEARS INT NOT NULL WITH DEFAULT 0,
EMP_PROMOTION_DATE DATE,
PRIMARY KEY(EMP_ID))
IN TSHR;
```

Result Tables

A result table is called that because it is simply the result set of a query. It is not persistent. To show an example, let's first add a record to the employee table. We haven't reviewed DML yet, but let's go ahead and do an INSERT with the following:

```
INSERT INTO HRSCHEMA.EMPLOYEE
(EMP_ID,
```

```
 EMP_LAST_NAME,
 EMP_FIRST_NAME,
 EMP_SERVICE_YEARS,
 EMP_PROMOTION_DATE)
VALUES (3217,
'JOHNSON',
'EDWARD',
4,
'01/01/2017');
```

Now run the following query:

```
SELECT EMP_ID, EMP_LAST_NAME,
EMP_FIRST_NAME
FROM HRSCHEMA.EMPLOYEE
WHERE EMP_ID = 3217;
```

The displayed data is a result table.

```
EMP_ID  EMP_LAST_NAME  EMP_FIRST_NAME
------  -------------  --------------
3217    JOHNSON        EDWARD
```

In most cases in this book when we run a query or DDL we'll simply show the text and the results instead of showing a screen shot each time. That will save space and enable us to include more useful content in the book.

Materialized Query Tables

A materialized query table stores the aggregate results from querying one or more other tables or views. MQTs are often used to improve performance for certain aggregation queries by providing pre-computed results. Consequently MQTs are often used in analytical or data warehousing environments.

MQTs are either system-maintained or user maintained. For a system maintained table, the data must be updated using the REFRESH TABLE statement. A user-maintained MQT can be updated using the LOAD utility, and also the UPDATE, INSERT, and DELETE statements.

Let's do an example of an MQT that summarizes monthly payroll. Assume we have a source table named EMP_PAY_HIST which will be a history of each employee's salary for each paycheck. The table is defined as follows:

Column Name	Definition
EMP_ID	Numeric
EMP_PAY_DATE	Date
EMP_PAY_AMT	Decimal(8,2)

The DDL for the table is as follows:

```
CREATE TABLE HRSCHEMA.EMP_PAY_HIST(
EMP_ID                  INT NOT NULL,
EMP_PAY_DATE            DATE NOT NULL,
EMP_PAY_AMT             DECIMAL (8,2) NOT NULL) IN TSHR;
```

Before we can use the table for an MQT we need to add some records to it. Let's run the following DDL to do that:

```
INSERT INTO HRSCHEMA.EMP_PAY_HIST
VALUES (3217,'01/15/2017',2291.66);

INSERT INTO HRSCHEMA.EMP_PAY_HIST
VALUES (3217,'01/31/2017',2291.66);

INSERT INTO HRSCHEMA.EMP_PAY_HIST
VALUES (3217,'02/15/2017',2291.66);

INSERT INTO HRSCHEMA.EMP_PAY_HIST
VALUES (3217,'02/28/2017',2291.66);

INSERT INTO HRSCHEMA.EMP_PAY_HIST
VALUES (7459,'01/15/2017',3333.33);

INSERT INTO HRSCHEMA.EMP_PAY_HIST
VALUES (7459,'01/31/2017',3333.33);

INSERT INTO HRSCHEMA.EMP_PAY_HIST
VALUES (7459,'02/15/2017',3333.33);

INSERT INTO HRSCHEMA.EMP_PAY_HIST
VALUES (7459,'02/28/2017',3333.33);
```

Now let's select the data in the EMP_PAY_HIST table is as follows:

```
SELECT * FROM HRSCHEMA.EMP_PAY_HIST;

EMP_ID        EMP_PAY_DATE   EMP_PAY_AMT
------        ------------   -----------
  3217        2017-01-15         2291.66
  3217        2017-01-31         2291.66
```

```
3217        2017-02-15        2291.66
3217        2017-02-28        2291.66
7459        2017-01-15        3333.33
7459        2017-01-31        3333.33
7459        2017-02-15        3333.33
7459        2017-02-28        3333.33
```

Finally, let's assume we regularly need an aggregated total of each employee's year to date pay. We could do this with a materialized query table. Let's first build the query that will summarize the employee pay from the beginning of the year to current date:

```
SELECT EMP_ID, SUM(EMP_PAY_AMT) AS EMP_PAY_YTD
FROM HRSCHEMA.EMP_PAY_HIST
GROUP BY EMP_ID
ORDER BY EMP_ID;

EMP_ID EMP_PAY_YTD
------ -----------
3217       9166.64
7459      13333.32
```

Now let's create the MQT with this query, using the following DDL, and we'll make it a system managed table:

```
CREATE TABLE HRSCHEMA.EMP_PAY_TOT (EMP_ID, EMP_PAY_YTD) AS
(SELECT EMP_ID, SUM(EMP_PAY_AMT) AS EMP_PAY_YTD
FROM HRSCHEMA.EMP_PAY_HIST GROUP BY EMP_ID)
DATA INITIALLY DEFERRED
REFRESH DEFERRED
MAINTAINED BY SYSTEM
ENABLE QUERY OPTIMIZATION;
```

We can now populate the table by issuing the `REFRESH TABLE` statement as follows:

```
REFRESH TABLE HRSCHEMA.EMP_PAY_TOT;
```

Finally we can query the MQT as follows:

```
SELECT * FROM HRSCHEMA.EMP_PAY_TOT;

EMP_ID EMP_PAY_YTD
------ -----------
  3217     9166.64
  7459    13333.32
```

Temporal Tables

Temporal tables were introduced to DB2 in version 10. Briefly, a temporal table is one that keeps track of "versions" of data over time and allows you to query data according to the time frame. We won't spend too much time here going into the details because we'll cover it later in the book in advanced topics. For now we'll just put forth the basic concepts.

Business Time

Business time concerns data that is valid in a business sense for some period of time. Let's go back to our employee application. An employee's pay typically changes over time. Besides wanting to know current salary, there may be many scenarios under which an HR department or supervisor might need to know what pay rate was in effect for an employee at some time in the past. We might also need to allow for cases where the employee terminated for some period of time and then returned. Or they took a non-paid leave of absence.

This is the concept of business time and it can be fairly complex depending on the business rules required by the application. It basically means a period of time in which the data is accurate from a business standpoint or business rule. You could think of the event or condition as having an effective date and discontinue date.

A table can only have one business time period. When a `BUSINESS_TIME` period is defined for a table, DB2 generates a check constraint in which the end column value must be greater than the begin column value.

System Time

System time simply means the time during which a piece of data is in the database, i.e., when the data was added, changed or deleted. Sometimes it is important to know this. For example a user might enter an employee's salary change on a certain date but the effective date of the salary change might be earlier or later than the date it was entered into the system. The system time simply records when the data was entered into the system, changed in the system, or deleted from the system. An audit trail application or table often has a timestamp that can be considered system time.

Bitemporal Support

In some cases you may need to support both business and system time in the same table. DB2 supports this and it is called bi-temporal support.

History Table

A table that is used to store historical versions of rows from the associated system-period temporal table. This will also be elaborated later in this book.

Temporary Table

Sometimes you may need the use of a table for the duration of a session but no longer than that. For example you may have a programming situation where it is convenient to load a temporary table for these operations:

- To join the data in the temporary table with another table
- To store intermediate results to be queried later in the program
- To load data from a flat file into a relational format

In all cases, it is assumed that you only need a temporary table for the duration of a session or iteration of a program (since temporary tables are dropped as soon as the session ends).

Temporary tables are created using either the CREATE statement or the DECLARE statement. The main difference between using the CREATE statement versus the DECLARE statement is the CREATE results in the table definition being stored in the system catalog. If you use the DECLARE statement, your table definition is gone at the conclusion of your session.

Here's the sample DDL we could use to create a temporary table called EMP_INFO using both methods;

```
CREATE GLOBAL TEMPORARY TABLE
EMP_INFO(
EMP_ID   INT,
EMP_LNAME  VARCHAR(30),
EMP_FNAME  VARCHAR(30));
```

```
DECLARE GLOBAL TEMPORARY TABLE
EMP_INFO(
EMP_ID   INT,
EMP_LNAME  VARCHAR(30),
EMP_FNAME  VARCHAR(30));
```

Indexes

Indexes are structures that provide a means of quickly locating a record in a table. One of the main reasons for having indexes is that they improve performance when accessing data randomly. There are other reasons as well which we'll explore now.

Benefits of Indexes

Indexes are beneficial in three ways:

1. Indexes improve performance in that it is typically faster to use the index row locator to navigate to a specific row than to do a table scan (except in cases of very small tables).

2. Unique indexes ensure uniqueness of record keys (either primary or secondary).

3. Clustered indexes enable data to be organized in the base table to optimize sequential access and processing.

Types of Indexes

The types of indexes available in DB2 11.1 are as follows.

Unique index

A unique index enforces the rule that every row must contain a unique value in the indexed column. For example, if a unique index is defined for a column containing an employee's social security number, then the value must be unique for each row in the table.

Primary index

A primary index is the **primary key** column of the table. A table can have only one primary index. If you define a primary key on a table, you must also create a unique index on the column that was chosen for the primary key (another way of saying this is that the primary key must be unique).

Clustering index

A clustering index physically groups data according to a sequence to make certain kinds of processing more efficient. For example, if you have customers who have orders that must be processed, you might prefer that orders for a particular customer be located close together in the table. You therefore might create a clustering index for the orders based on customer id.

Expression-based index

An index that is defined based on a general expression.

Secondary index

A secondary index is any index that is not a primary index. For example you might have an employee table for which the primary key is an `EMPLOYEE_ID`. However you might also have a social security number column for which you have created a unique index. The index on social security number is a secondary index.

XML index

An XML index uses an XML pattern expression to locate values in XML documents (specifically those XML document that are stored in a single column which s common).

Examples of Indexes

Unique primary

We created a unique primary index a while ago for the `EMPLOYEE` table. Recall that we defined the EMPLOYEE table with a primary key. When you create a table with a primary key, DB2 LUW automatically creates a unique index on the column(s) involved to support the primary key. The index is created under the SYSIBM schema and the index name is generated by DB2.

This is the table:

```
CREATE TABLE HRSCHEMA.EMPLOYEE(
EMP_ID INT NOT NULL,
EMP_LAST_NAME VARCHAR(30) NOT NULL,
EMP_FIRST_NAME VARCHAR(20) NOT NULL,
EMP_SERVICE_YEARS INT NOT NULL WITH DEFAULT 0,
EMP_PROMOTION_DATE DATE, PRIMARY KEY(EMP_ID)) ;
```

We can select Indexes in the Data Studio object tree to find the index created for our EMPLOYEE table. In my case the index name is SQL171102201231020. You don't need to worry about knowing the name. DB2 completely maintains and enforces the index.

Unique

While a table can only have one primary key, it can have more than one unique index. Suppose that we added a social security column to the employee table. For data security reasons (and simply to reduce the possibility of keying errors), we might want to create a unique index on the social security number column.

The DDL to add the social security number column to the table and create a unique index on it is as follows:

```
ALTER TABLE HRSCHEMA.EMPLOYEE
ADD COLUMN EMP_SSN CHAR(09);

CREATE UNIQUE INDEX HRSCHEMA.NDX_SSN
ON HRSCHEMA.EMPLOYEE (EMP_SSN);
```

Clustered

One case in which a clustered index is helpful is when sequential processing is called for. Suppose we create an EMP_PAY_EFT table and we use the EMP_PAY_CHECK content to generate electronic funds transfers on pay day. Rather than organizing the EFT data by EMP_ID, we might like to physically locate it according to pay date so the records for a pay cycle will be close together (otherwise our pay check program might have to jump all over the table to get the needed records for the pay cycle).

Here's the DDL we used to create the EMP_PAY_EFT table.

```
CREATE TABLE HRSCHEMA.EMP_PAY_EFT(
EMP_ID INT NOT NULL,
EMP_PAY_DATE   DATE NOT NULL,
EMP_REGULAR_PAY  DECIMAL (8,2) NOT NULL,
EMP_SEMIMTH_PAY DECIMAL (8,2) NOT NULL)  IN TSHR;
```

We can ensure that the data is clustered as we want it by creating this index:

```
CREATE INDEX HRSCHEMA.NDX_EMP_EFT
ON HRSCHEMA.EMP_PAY_EFT (EMP_PAY_DATE, EMP_ID) CLUSTER;
```

Now DB2 will attempt to order all records in the `EMP_PAY_EFT` table by `EMP_PAY_DATE` and `EMP_ID`.

Note: DB2 11.1 allows you to prevent the entry of NULL values in an index by including the clause `EXCLUDE NULL KEYS` when you create the index. If you do not specify that you want to exclude null keys, then DB2 will allow null values in the index (assuming the key field allows nulls).

DATA TYPES

DB2 supports the following data types. You should be familiar with all of these so that you can choose the best data type for your purpose.

Numeric Data Types

DB2 supports several types of numeric data types, each of which has its own characteristics.

SMALLINT

A small integer is an integer value between the values of -32768 and +32767.

INTEGER

An integer is sometimes referred to as a "large integer". This data type can store values between -2147483648 and +2147483647.

BIGINT

A big integer can store values between -9223372036854775808 to +9223372036854775807.

DECIMAL or NUMERIC

The decimal type is stored in packed decimal format and it has an implied decimal point. You would use this for any data values that need this precision, such as money values. For example, a column defined as DECIMAL (7,2) means a total of 7 decimal position with two values to the right of the decimal point. For example this value could be stored in a DECIMAL (7,2) column:

10,377.45

DECFLOAT
A DECFLOAT type means "decimal floating-point". The position of the decimal is stored with the value itself.

REAL
The real data type is a single-precision floating-point number. Its value can be between -7.2E+75 and 7.2E+75.

DOUBLE
The double data type is a double-precision floating-point number. Its value can be between -7.2E+75 to 7.2E+75.

String data types
DB2 supports several types of string data: character strings, graphic strings, and binary strings.

CHARACTER(n)
The character or CHAR data type stores fixed length character strings. You would use this for character data which is always the same length, such as state codes which are two bytes. The maximum length of a CHAR type is 255 bytes.

VARCHAR(n)
The VARCHAR type stores varying length character strings for which the maximum length is specified. For example VARCHAR(40) means between 1 and 40 bytes in length. A VARCHAR can be defined with a maximum of 32704 bytes. If you need more, see the CLOB data type below under LOB data types.

GRAPHIC(n)
The GRAPHIC data type stores fixed-length graphic strings of length n in double byte format. The n refers to the number of double byte characters. The maximum length is 128 double byte chars.

VARGRAPHIC(n)
The VARGRAPHIC type stores varying-length graphic strings in double byte format. The n refers to the maximum number of double byte characters. The maximum length is 16352 double byte characters.

BINARY(n)

The BINARY type stores binary strings. The n refers to the number of fixed length bytes. The maximum length is 255 bytes.

VARBINARY(n)

The VARBINARY type stores varying-length binary strings. The n refers to the maximum number of bytes, so the actual length is between 1 ad n bytes. The maximum length of a VARBINARY column is 32704 bytes.

Date, time, and timestamp data types

Date/Time Types

DATE

The DATE type represents a date in year, month and day format. It is 10 bytes long and the format depends on the date format specified when DB2 was installed. The following are two common formats:

```
MM/DD/YYYY (USA standard)

YYYY-MM-DD (ISO standard)
```

TIME

The TIME type stores a time of day in hours, minutes and second. For example, 10:34 am and 27 seconds is:

10:34:27

TIMESTAMP

The TIMESTAMP type stores a value that represents both a date and a time. It includes the year, month, date, hour, minute, seconds and microseconds. For example:

```
2017-04-17-02.17.54.142357
```

XML data type

The XML data type defines a column that will store a well formed XML document.

Large object data types

You can use large object data types to store text as well as multimedia files such audio, video, images, and other files that are larger than 32 KB and up to 2GB in size.

Character large objects (CLOBs)
The CLOB data type can be used to store text that uses a single character set. This could include reports that need to be archived or any other text that exceeds the 32K limit imposed by the VARCHAR type. The limit for a CLOB is 2 GB.

Double-byte character large objects (DBCLOBs)
The DBCLOB can be used when the data is formatted in a double byte character set.

Binary large objects (BLOBs)
The BLOB data type stored binary (non-character) data. You would use this for multimedia files such as images, audio and video files.

ROWID data type

The ROWID type uniquely identifies a row in a table, including the location of the row. Using ROWIDs enables you to navigate directly to the row without using an index.

Distinct types

A *distinct type* is a user-defined data type that is based on existing built-in DB2 data types. Here are a couple of examples

```
CREATE DISTINCT TYPE US_DOLLAR AS DECIMAL (9,2);

CREATE DISTINCT TYPE CANADIAN_DOLLAR AS DECIMAL (9,2);
```

Constraints

Types of constraints
There are basically four types of constraints, as follows:

1. Unique
2. Referential
3. Check
4. NULL

A UNIQUE constraint requires that the value in a particular field in a table be unique for each record.

A REFERENTIAL constraint enforces relationships between tables. For example you can define a referential constraint between an `EMPLOYEE` table and a `DEPARTMENTS` table, such that a DEPT field in the `EMPLOYEE` table can only contain a value that matches a key value in the `DEPARTMENTS` table.

A CHECK constraint establishes some condition on a column, such as the value must be >= 10.

A NOT NULL constraint establishes that each record must have a non-null value for this column.

Unique Constraints

A unique constraint is a rule that the values of a key are valid only if they are unique in a table. In the case of the `EMPLOYEE` table we have been working with, you can create a unique index as follows:

```
CREATE UNIQUE INDEX SSN_EMPLOYEE ON EMPLOYEE (EMP_SSN);
```

Check Constraints

A check constraint is a rule that specifies the values that are allowed in one or more columns of every row of a base table. It establishes some condition on a column, such as the stored value must be >= 10. The real worth of check constraints is that business rules such as edits and validations can be stored in and performed by the database manager instead of application programs.

If you try to INSERT or UPDATE a record, the value in the column is evaluated against the check constraint rules. If the value follows the rules, then the action is permitted; otherwise the action will fail with a constraint violation.

Let's create a new table `EMP_DATA`, and then we'll define some check constraints on it. We'll create the table and then alter it to add the two constraints:

```
CREATE TABLE HRSCHEMA.EMP_DATA
(EMP_ID    INT,
EMP_LNAME  VARCHAR(30),
EMP_FNAME  VARCHAR(20),
EMP_AGE    INT) IN TSHR;
```

Let's say we require that the employee number be a value between zero and 9999. Also the employee's age must be between 18 and 99. Now let's add the constraint on employee id.

```
ALTER TABLE HRSCHEMA.EMP_DATA
ADD CONSTRAINT  X_EMPID CHECK (EMP_ID BETWEEN 0 AND 9999);
```

Next, we'll add the constraint on age.

```
ALTER TABLE HRSCHEMA.EMP_DATA
ADD CONSTRAINT  X_AGE
CHECK (EMP_AGE >= 18);
```

Now let's try this insert:

```
INSERT INTO HRSCHEMA.EMP_DATA
VALUES
(17888, 'BROWN', 'WILLIAM', 17);
```

```
The requested operation is not allowed because a row does not satisfy the check
constraint "HRSCHEMA.EMP_DATA.X_EMPID".. SQLCODE=-545, SQLSTATE=23513, DRIV-
ER=4.18.60
```

It turns out that we mis-keyed the employee id. It should be 1788 instead of 17888. The value 17888 exceeds 9999 which is the maximum limit for employee id according to the check constraint.

Let's fix the employee id and try again:

```
INSERT INTO HRSCHEMA.EMP_DATA
VALUES
(1788,
 'BROWN',
 'WILLIAM',
 17);
```

```
The requested operation is not allowed because a row does not satisfy the
check constraint "HRSCHEMA.EMP_DATA.X_AGE".. SQLCODE=-545, SQLSTATE=23513, DRIV-
ER=4.18.60
```

Obviously 17 is less than 18 which is the lower limit on the employee age. So let's fix the age and we can see the row is accepted now.

```
INSERT INTO HRSCHEMA.EMP_DATA
VALUES
(1788,
 'BROWN',
 'WILLIAM',
 18);
```

```
Updated 1 rows.
```

Check constraints can be a very powerful way of building business logic into the database itself. No application programming is required to implement a constraint (other than trapping and handling the error). There is great consistency when using check constraints because the constraints are applied regardless of which program or ad hoc process attempts the data modification.

Referential Constraints

A referential constraint is the rule that the non-null values of a foreign key are valid only if they also appear as a key value in a parent table. The table that contains the parent key is called the parent table of the referential constraint, and the table that contains the foreign key is a dependent of that table. Referential constraints ensure data integrity by using primary and foreign key relationships between tables.

In DB2 you define a referential constraint by specifying a column in the child table that references a primary key column in a parent table. For example, in a company you could have a `DEPARTMENT` table with column `DEPT_CODE`, and an `EMP_DATA` table that includes a column `DEPT` that represents the department code an employee is assigned to. The rule would be that you cannot have a value in the `DEPT` column of the `EMP_DATA` table that does not have a corresponding `DEPT_CODE` in the `DEPARTMENT` table. You can think of this as a parent and child relationship between the `DEPARTMENT` table and the `EMP_DATA` table.

Let's create the `DEPARTMENT` table and also add the `DEPT` column to the `EMP_DATA` table.

```
CREATE TABLE HRSCHEMA.DEPARTMENT
(DEPT_CODE   CHAR(04) NOT NULL,
DEPT_NAME   VARCHAR (20) NOT NULL,
PRIMARY KEY(DEPT_CODE))
IN TSHR;

CREATE UNIQUE INDEX HRSCHEMA.NDX_DEPT_CODE
      ON HRSCHEMA.DEPARTMENT (DEPT_CODE);

INSERT INTO HRSCHEMA.DEPARTMENT
VALUES ('DPTA','DEPARTMENT A');


ALTER TABLE HRSCHEMA.EMP_DATA
ADD DEPT CHAR(04);

UPDATE HRSCHEMA.EMP_DATA
SET DEPT = 'DPTA';

Updated 1 rows.
```

Adding a Foreign Key Relationship

Now you can add the foreign key relationship by performing an ALTER on the child table.

```
ALTER TABLE HRSCHEMA.EMP_DATA
   FOREIGN KEY FK_DEPT_EMP (DEPT)
      REFERENCES HRSCHEMA.DEPARTMENT(DEPT_CODE) ;
```

Now if you try to update an EMP_DATA record with a DEPT value that does not have a corresponding DEPT_CODE in table DEPARTMENT, you'll get an SQL error -530 which means a violation of a foreign key.

```
UPDATE HRSCHEMA.EMP_DATA
SET DEPT = 'DPTB';

The insert or update value of the FOREIGN KEY "HRSCHEMA.EMP_DATA.FK_DEPT_EMP"
is not equal to any value of the parent key of the parent table.. SQLCODE=-530,
SQLSTATE=23503, DRIVER=4.18.60
```

The parent table DEPARTMENT does not have DEPT_CODE "DPTB" value in it, and our business rules (as implemented with a foreign key relationship) require that DPTB be added to the DEPARTMENT table before the EMP_DATA record can be updated with that value.

Deleting a Record from the Parent Table

Now let's talk about what happens if you want to delete a record from the parent table. Assuming no EMP_DATA records are linked to that DEPARTMENT record, deleting that record may be fine. But what if you are trying to delete a DEPARTMENT record who's DEPT_CODE is referenced by one or more records in the EMP_DATA table?

Let's look at a record in the table:

```
SELECT EMP_ID, DEPT FROM HRSCHEMA.EMP_DATA
WHERE EMP_ID = 1788;

EMP_ID  DEPT
------  ----
  1788  DPTA
```

Ok, we know that the DEPT_CODE in use is DPTA. Now let's try to delete DPTA from the DEPARTMENT table.

```
DELETE FROM HRSCHEMA.DEPARTMENT
WHERE DEPT_CODE = 'DPTA';

A parent row cannot be deleted because the relationship "HRSCHEMA.EMP_DATA.FK_
DEPT_EMP" restricts the deletion.. SQLCODE=-532, SQLSTATE=23504, DRIVER=4.18.60
```

As you can see, when we try to remove the DEPT_CODE from the DEPARTMENT table, we get a -532 SQL error telling us our SQL is in violation of the referential constraint. That's probably what we want, i.e., to have an error flagged. But there are some other options for handling the situation.

You can specify the action that will take place upon deleting a parent record by including an ON DELETE clause in the foreign key definition. If no ON DELETE clause is present, or if the ON DELETE RESTRICT clause is used, then the parent record cannot be deleted unless all child records referencing that parent record are first deleted (alternately the child reference to that parent record can be changed to some other existing parent record). RESTRICT is the most commonly used ON DELETE value (or just omitting the ON DELETE clause which has the same effect). This is the case above.

Here are the two other options:

If **ON DELETE CASCADE** is specified, then any rows in the child table that correspond to the deleted parent record will also be deleted. Wow, that is probably not what we want, but there may be cases where this function is useful. Possibly if a certain product is discontinued you might want to delete all pending SHIPPING table entries for it. I can't think of many other needs for this, but be aware that this option is available.

If **ON DELETE SET NULL** is specified, then the foreign key field will be set to NULL for corresponding rows that reference the parent record that is being deleted.

Let's redefine our constraint to use this last option:

```
ALTER TABLE HRSCHEMA.EMP_DATA DROP CONSTRAINT FK_DEPT_EMP;  COMMIT WORK;

ALTER TABLE HRSCHEMA.EMP_DATA
   FOREIGN KEY FK_DEPT_EMP (DEPT)
      REFERENCES HRSCHEMA.DEPARTMENT (DEPT_CODE) ON DELETE SET NULL;
```

Now try deleting the DPTA record from the DEPARTMENT table:

```
DELETE FROM HRSCHEMA.DEPARTMENT
WHERE DEPT_CODE = 'DPTA';

Updated 1 rows.
```

We see the delete is successful. So now let's check and see if the DEPT value for the child record has been set to NULL.

```
SELECT EMP_ID, DEPT
FROM HRSCHEMA.EMP_DATA
WHERE EMP_ID = 1788;
```

```
EMP_ID      DEPT
------      ----
  1788      NULL
```

And in fact the DEPT column has been set to NULL.

This closes our discussion of referential integrity. Do make sure you understand what a referential constraint is, the syntax for creating a foreign key relationship, and the various options/ outcomes for the ON DELETE clause.

Not Null Constraints

A NOT NULL constraint on a column requires that when you add or update a record, you must specify a non null value for that column. If you do not specify a non null value, you will get an error. Let's take an example of trying update an EMPLOYEE record without specifying a value for one of the columns.

In this case, let's leave off the EMP_FIRST_NAME value which is defined in the table as NOT NULL.

```
INSERT INTO HRSCHEMA.EMPLOYEE
(EMP_ID,
 EMP_LAST_NAME,
 EMP_PROMOTION_DATE)

VALUES (7420,
'JACKSON',
'09/01/2016')
```

```
Assignment of a NULL value to a NOT NULL column "TBSPACEID=3, TABLEID=4, COLNO=2"
is not allowed.. SQLCODE=-407, SQLSTATE=23502, DRIVER=4.18.60
```

Defining a column with the NOT NULL attribute ensures that no record can be added to the table with a NULL value in this column. The NOT NULL attribute enforces that requirement regardless of which application or user is processing the data, so it's "universal" and does not depend on program logic to enforce the business rule.

SEQUENCIES and IDENTITIES

Some scenarios require an auto-generated sequence that can be used to uniquely identify a record. DB2 provides two methods of doing this: sequences and identity columns. Although both methods provide auto-generated numbers, they work differently. The main difference is that an identity column is contained in a specific table – it cannot be shared across objects. A sequence is itself a separate database object and can be used to generate numbers for multiple objects. Let's look at examples of each.

SEQUENCES

A sequence generates a sequential set of numbers. You define a sequence with a starting value and increment it with the NEXTVAL function. You can also obtain the most recent value generated by using the PREVVAL function.

The basic syntax for creating a sequence is:

```
CREATE SEQUENCE <name of sequence>
START WITH <start value>
INCREMENT BY <increment value>
NO CYCLE   <reuse old values?>
```

Now let's create a sequence to generate new employee numbers and that we want to start with employee number 1001. The following DDL would accomplish this:

```
CREATE SEQUENCE HRSCHEMA.EMPSEQ
START WITH 1001
INCREMENT BY 1
NO CYCLE ;
```

The `NO CYCLE` clause means we will not reuse numbers when the maximum for the data type is reached. You could explicitly create the sequence as SMALLINT, INTEGER, BIGINT, or DECIMAL with a scale of zero. If you don't specify a data type, the default is INTEGER.

Now we could add a record to the `EMPLOYEE` table without specifying the employee id. Instead we would use the sequence number with the NEXTVAL option to generate an employee id.

```
INSERT INTO HRSCHEMA.EMPLOYEE
VALUES (NEXT VALUE FOR HRSCHEMA.EMPSEQ,
'HENDERSON',
'JOHN',
1,
'12/01/2016',
NULL,
345973452);

Updated 1 rows.
```

Now let's select the record with key 1001.

```
SELECT EMP_ID, EMP_LAST_NAME
FROM HRSCHEMA.EMPLOYEE
WHERE EMP_ID = 1001;
```

```
EMP_ID       EMP_LAST_NAME
------       -------------
  1001       HENDERSON
```

You can also change a sequence after it is created. For example you could change the increment of our sequence from 1 to 2 as follows:

```
ALTER SEQUENCE HRSCHEMA.EMPSEQ
INCREMENT BY 2
CYCLE ;
```

You can delete a sequence by issuing the DROP command.

```
DROP SEQUENCE HRSCHEMA.EMPSEQ;
```

IDENTITY Columns

Now let's perform a similar setup with another table and this time we will use an identity column on the table. The identity column is so named because it allows you to uniquely identify a record, i.e., it provides a unique key. Let's do this example, and we'll create a different employee table for the example:

```
CREATE TABLE HRSCHEMA.EMPLOYE2(
EMP_ID SMALLINT GENERATED ALWAYS AS IDENTITY
   (START WITH 1001,
    INCREMENT BY 1,
    NOCYCLE),
EMP_LAST_NAME VARCHAR(30) NOT NULL,
EMP_FIRST_NAME VARCHAR(20) NOT NULL,
EMP_SERVICE_YEARS INT
NOT NULL WITH DEFAULT 0,
EMP_PROMOTION_DATE DATE)
IN TSHR;
```

Now let's insert a row into the `EMPLOYE2` table:

```
INSERT INTO HRSCHEMA.EMPLOYE2
  (EMP_LAST_NAME,
   EMP_FIRST_NAME,
   EMP_SERVICE_YEARS,
   EMP_PROMOTION_DATE)
VALUES
  ('JOHNSON',
  'BILL',
   1,
  '12/01/2016');

Updated 1 rows.
```

Notice that we did not specify any value for the EMP_ID. That is because it's an identity field and DB2 will generate the value. Now we can query the table and see the contents:

```
SELECT EMP_ID, EMP_LAST_NAME
FROM HRSCHEMA.EMPLOYE2;

EMP_ID        EMP_LAST_NAME
------        -------------
  1001        JOHNSON
```

As you can see, we get the same results as using a sequence as in the earlier example. The main difference is that an identity column takes care of generating the value without any prompting. Whereas with a sequence you must specify the sequence name and request the next value. And of course the other difference between sequences and identity columns is that a sequence is a separate object in the database. It can be used for generating numbers independently of any table. But an identity column is always tied to a single table.

VIEWS

A view is a virtual table that is based on a SELECT query against a base table or another view. Views can include more than one table (or other view), including the results of a join.

DDL for Views

CREATE

The basic syntax to create a view is as follows:

```
CREATE VIEW <name of view> AS
SELECT <columns> FROM <table> WHERE <condition>
```

Optionally, you can specify the WITH CHECK OPTION. The WITH CHECK OPTION clause ensures that a record inserted via a view is consistent with the view definition. For example, let's go back to our EMPLOYE2 table. Let's say we want a view named EMP_SENIOR that shows us data from EMPLOYE2 only for senior employees, meaning employees with at least 5 years of service. We will also allow records to be inserted to the EMPLOYE2 table via the view. Here is the view definition:

```
CREATE VIEW HRSCHEMA.EMP_SENIOR AS
SELECT EMP_ID,
EMP_LAST_NAME,
EMP_FIRST_NAME,
EMP_SERVICE_YEARS,
EMP_PROMOTION_DATE
FROM HRSCHEMA.EMPLOYE2
WHERE EMP_SERVICE_YEARS >= 5;
```

Now let's insert a couple of records using this view.

```
INSERT INTO HRSCHEMA.EMP_SENIOR
  (EMP_LAST_NAME,
   EMP_FIRST_NAME,
   EMP_SERVICE_YEARS,
   EMP_PROMOTION_DATE)
VALUES
  ('FORD',
  'JAMES',
   7,
  '10/01/2015');

Updated 1 rows.
```

The first record we inserted is for an employee with more than 5 years of service. We can confirm that the record was successfully inserted by querying the view. Good.

```
SELECT EMP_ID,
EMP_LAST_NAME,
EMP_FIRST_NAME
FROM HRSCHEMA.EMP_SENIOR;

EMP_ID EMP_LAST_NAME  EMP_FIRST_NAME
------ -------------  --------------
  1002 FORD           JAMES
```

Now let's insert a record that does not fit the view definition. In this case, let's add a record for which the employee has only 2 years service:

```
INSERT INTO HRSCHEMA.EMP_SENIOR
  (EMP_LAST_NAME,
   EMP_FIRST_NAME,
   EMP_SERVICE_YEARS,
   EMP_PROMOTION_DATE)
VALUES
  ('BUFORD',
  'HOLLAND',
   2,
  '07/31/2016');

Updated 1 rows.
```

Interestingly, DB2 allowed the record to be added to the table via the view even though the data did not match the view definition. We cannot know that fact from querying the view because when we do so it only shows us data which conforms to the view definition. Notice that only the record for the person with 7 years' service is returned by the view.

```
SELECT EMP_ID,
EMP_LAST_NAME,
EMP_FIRST_NAME
FROM HRSCHEMA.EMP_SENIOR;

EMP_ID EMP_LAST_NAME  EMP_FIRST_NAME
------ -------------  --------------
  1002 FORD           JAMES
```

But when we query the base table, we see the new record was in fact added:

```
EMP_ID  EMP_LAST_NAME          EMP_FIRST_NAME        EMP_SERVICE_YEARS
------  -------------          --------------        -----------------
  1001  JOHNSON                BILL                                  1
  1002  FORD                   JAMES                                 7
  1003  BUFORD                 HOLLAND                               2
```

Maybe this is ok if it fits your overall business rules. However, if you want to prevent records that do not conform to the view definition from being inserted into the table using that view, you must define the view using the `WITH CHECK OPTION` clause. Let's drop the view and recreate it that way.

```
DROP VIEW HRSCHEMA.EMP_SENIOR; COMMIT WORK;

CREATE VIEW HRSCHEMA.EMP_SENIOR AS
SELECT
EMP_ID,
EMP_LAST_NAME,
EMP_FIRST_NAME,
EMP_SERVICE_YEARS,
EMP_PROMOTION_DATE
FROM HRSCHEMA.EMPLOYE2
WHERE EMP_SERVICE_YEARS >= 5 WITH CHECK OPTION;
```

Now let's try to insert two more records, first an employee with more 5 or more years of service.

```
INSERT INTO HRSCHEMA.EMP_SENIOR
  (EMP_LAST_NAME,
   EMP_FIRST_NAME,
   EMP_SERVICE_YEARS,
   EMP_PROMOTION_DATE)
VALUES
  ('JACKSON',
  'MARLO',
   8,
  '06/30/2015');

Updated 1 rows.
```

This still works which is fine. Now let's try one with less than 5 years of service. In this case the insert fails as we can see with a -161 SQLCODE:

```
INSERT INTO HRSCHEMA.EMP_SENIOR
  (EMP_LAST_NAME,
   EMP_FIRST_NAME,
   EMP_SERVICE_YEARS,
   EMP_PROMOTION_DATE)
VALUES
  ('TARKENTON',
  'QUINCY',
   3,
  '09/30/2015');
```

```
The resulting row of the insert or update operation does not conform to the view
definition. SQLCODE=-161, SQLSTATE=44000, DRIVER=4.18.60
```

This is how to create a view that will disallow any insert or update that does not conform to the view definition. This is one way of enforcing ever-changing business rules without changing the underlying table definition.

ALTER

Views cannot be changed. They must be dropped and recreated.

DROP

Finally, you can delete a view by issuing the DROP command.

```
DROP VIEW <view name>
```

READ-ONLY VIEWS

The following are some conditions under which a view is automatically read-only, meaning you cannot insert, update or delete any rows using it.

1. The first SELECT clause includes the keyword DISTINCT.
2. The first SELECT clause contains an aggregate function.
3. The first FROM clause identifies multiple tables or views.
4. The outer fullselect includes a GROUP BY clause.

Views for Security

For purposes of security, a view is a classic way to restrict a subset of data columns to a specific set of users who are allowed to see or manipulate those columns. If you are going to use views for security you must make sure that users do not have direct access to the base tables, i.e. that they can only access the data via view(s). Otherwise they could potentially circumvent your rules to allow only access by view.

Let's look back at our EMPLOYEE table. Suppose we add a column for the employee's Social Security number. That is obviously a very private piece of information that not everyone should see. Our business rule will be that users HRUSER01, HRUSER02 and HRUSER99 are the only ones who should be able to view Social Security numbers. All other users and/or groups are not allowed to see the content of this column, but they can access all the other columns.

We might implement this as follows, first adding the new social security column to the EMPLOYEE table:

```
ALTER TABLE HRSCHEMA.EMPLOYEE
ADD COLUMN EMP_SSN CHAR(09);
```

Next, let's update this column for employee 3217:

```
UPDATE HRSCHEMA.EMPLOYEE
SET EMP_SSN = '238297536' WHERE EMP_ID = 3217;
```

Now let's create two views, one of which includes the EMP_SSN column, and the other of which does not:

```
CREATE VIEW HRSCHEMA.EMPLOYEE_ALL
AS SELECT
EMP_ID,
EMP_LAST_NAME,
EMP_FIRST_NAME,
EMP_SERVICE_YEARS,
EMP_PROMOTION_DATE,
EMP_PROFILE
FROM HRSCHEMA.EMPLOYEE;

CREATE VIEW HRSCHEMA.EMPLOYEE_HR
AS SELECT
EMP_ID,
EMP_LAST_NAME,
EMP_FIRST_NAME,
EMP_SERVICE_YEARS,
EMP_PROMOTION_DATE,
EMP_PROFILE,
EMP_SSN
FROM HRSCHEMA.EMPLOYEE;
```

Finally, issue the appropriate grants.

```
GRANT SELECT ON HRSCHEMA.EMPLOYEE_ALL TO PUBLIC;

GRANT SELECT on HRSCHEMA.EMPLOYEE_HR
TO HRUSER01,
   HRUSER02,
   HRUSER99;
```

At this point, assuming we are only accessing data through views, the three HR users are the only users able to access the EMP_SSN column. Other users cannot access the EMP_SSN column because it is not included in the EMPLOYEE_ALL view that they have access to. To prove this:

```
SELECT EMP_ID, EMP_SSN
FROM HRSCHEMA.EMPLOYEE_ALL
WHERE EMP_ID = 3217;

"EMP_SSN" is not valid in the context where it is used.. SQLCODE=-206, SQL-
STATE=42703, DRIVER=4.18.60
```

If you are one of the HR users, you will be able to access the EMP_SSN column using the other view, EMPLOYEE_HR:

```
SELECT EMP_ID, EMP_SSN
FROM HRSCHEMA.EMPLOYEE_HR
WHERE EMP_ID = 3217;

EMP_ID EMP_SSN
------ ---------
3217   238297536
```

Here are just a few more things you should remember about views. First, a view can reference another view. So we could create an EMPLOYEE_PAY view that includes a reference to the EMPLOYEE_ALL view.

Second, when a view gets deleted, then any dependent view will be marked inopertive. If you delete the EMPLOYEE_ALL view, the the EMPLOYEE_PAY view that references it will be marked inoperstive and you won't be able to use it.

Finally, if the column size increases on a base table, then any view which references that column will be automatically regenerated.

As we conclude the DDL chapter of this study guide, I strongly recommend that you **remove all the objects we have created** thus far. This will enable us to begin with a clean slate for the next chapter on Data Manipulation Language. The easiest way to remove all the objects is to simply drop the database and then recreate it.

Issue this DDL to drop the database from the Command Line Processor.

```
DROP DATABASE DBHR;
```

Then recreate the database, storage group, bufferpool, tablespace and schema that we will use for the HR application.

```
CREATE DATABASE DBHR
AUTOMATIC STORAGE YES;
CREATE TABLESPACE TSHR;
```

Here is the DDL to create storage group SGHR:

```
CREATE STOGROUP SGHR
ON 'C:'
OVERHEAD 6.725
DEVICE READ RATE 100.0
DATA TAG NONE;
```

Here's the DDL to create our buffer pool:

```
CREATE BUFFERPOOL BPHR IMMEDIATE
ALL DBPARTITIONNUMS SIZE 1000 AUTOMATIC PAGESIZE 4096;
```

Now we can create the tablespace TSHR.

```
CREATE REGULAR TABLESPACE TSHR
IN DATABASE PARTITION GROUP IBMDEFAULTGROUP
PAGESIZE 4096
MANAGED BY AUTOMATIC STORAGE USING STOGROUP SGHR
AUTORESIZE YES
BUFFERPOOL BPHR
OVERHEAD INHERIT
TRANSFERRATE INHERIT
DROPPED TABLE RECOVERY
ON DATA TAG INHERIT;
```

Finally, let's create the schema.

```
CREATE SCHEMA HRSCHEMA
AUTHORIZATION DBA001; ← This should be your DB2 id, whatever it is.
```

The DB2 Catalog

The DB2 system catalog is a wealth of information for research and problem solving. The catalog consists of a set of tables containing information about the various objects in DB2 (tablespaces, tables, indexes, views, packages, etc). The schema for the read-only system catalog views is SYSCAT.

Here are a few sample queries you can run to get information about tables, views and index objects. This first sample displays the tables under schema HRSCHEMA, including table type, column count and tablespace name.

```
SELECT TABNAME, TYPE, COLCOUNT, TBSPACE FROM SYSCAT.TABLES
WHERE TABSCHEMA = 'HRSCHEMA';

TABNAME      TYPE COLCOUNT TBSPACE
------------ ---- -------- ----------
EMPLOYEE     T           5 TSHR
EMPRECOG     T           4 TSHR
EMP_DEPT_DATA T          5 USERSPACE1
EMP_PAY_HIST  T          3 TSHR
EMP_PAY_TOT   S          2 TSHR
EMP_SENIOR    V          5 NULL
```

Tablespace types are given as follows:

A Alias
G Created temporary table
H Hierarchy table
L Detached table
N Nickname
S Materialized query table
T Table (untyped)
U Typed table
V View (untyped)
W Typed view

In the next example, we display all views owned by the HRSCHEMA schema. We include the VIEWCHECK column which tells us whether the view definition is checked when adding or updating records using the view. We also include the READONLY column which tells us whether the view is read only.

```
SELECT VIEWNAME, VIEWCHECK, READONLY
FROM SYSCAT.VIEWS
WHERE VIEWSCHEMA = 'HRSCHEMA';

VIEWNAME    VIEWCHECK READONLY
----------- --------- --------
EMP_PAY_TOT N         N
EMP_SENIOR  C         N
```

Finally, we can look at referential constraints by querying the SYSCAT.REFERENCES view. In this case we will display the parent and child table names, plus the respective columns that are involved in the referential relationship.

```
SELECT CONSTNAME,
TABNAME, REFTABNAME,
FK_COLNAMES,
PK_COLNAMES
FROM SYSCAT.REFERENCES;

CONSTNAME    TABNAME  REFTABNAME FK_COLNAMES          PK_COLNAMES
----------- -------- ---------- -------------------- --------------------

FK_DEPT_EMP EMPLOYEE DEPARTMENT  DEPT                  DEPT_CODE
```

The above is a sampling of what is available in the DB2 catalog. I recommend that you review all views in the SYSCAT schema to familiarize yourself with them.

Chapter Two Questions

1. Which of the following is NOT a valid data type for use as an identity column?

    a. INTEGER
    b. REAL
    c. DECIMAL
    d. SMALLINT

2. You need to store numeric integer values of up to 5,000,000,000. What data type is appropriate for this?

    a. INTEGER
    b. BIGINT
    c. LARGEINT
    d. DOUBLE

3. Which of the following is NOT a LOB (Large Object) data type?

    a. CLOB
    b. BLOB
    c. DBCLOB
    d. DBBLOB

4. If you want to add an XML column VAR1 to table TBL1, which of the following would accomplish that?

    a. ALTER TABLE TBL1 ADD VAR1 XML
    b. ALTER TABLE TBL1 ADD COLUMN VAR1 XML
    c. ALTER TABLE TBL1 ADD COLUMN VAR1 (XML)
    d. ALTER TABLE TBL1 ADD XML COLUMN VAR1

5. If you want rows that have similar key values to be stored physically close to each other, what keyword should you specify when you create an index?

    a. UNIQUE
    b. ASC
    c. INCLUDE
    d. CLUSTER

6. To ensure all records inserted into a view of a table are consistent with the view definition, you would need to include which of the following keywords when defining the view?

    a. UNIQUE
    b. WITH CHECK OPTION
    c. VALUES
    d. ALIAS

7. If you want to determine various characteristics of a set of tables such as type (table or view), owner and status, which system catalog view would you query?

    a. SYSCAT.TABLES
    b. SYSIBM.SYSTABLES
    c. SYSCAT.OBJECTS
    d. SYSIBM.OBJECTS

Chapter Two Exercises

1. Write a DDL statement to create a base table named EMP_DEPENDENTS in with schema HRSCHEMA and in tablespace TSHR. The columns should be named as follows and have the specified attributes. There is no primary key.

Field Name	Type	Attributes
EMP_ID	INTEGER	NOT NULL, PRIMARY KEY
EMP_DEP_LAST_NAME	VARCHAR(30)	NOT NULL
EMP_DEP_FIRST_NAME	VARCHAR(20)	NOT NULL
EMP_RELATIONSHIP	VARCHAR(15)	NOT NULL

2. Create a statement to create a referential constraint on table EMP_DEPENDENT such that only employee ids which exist on the HRSCHEMA.EMPLOYEE table can have an entry in EMP_DEPENDENT. If there is an attempt to delete an EMPLOYEE record that has EMP_DEPENDENT records associated with it, then do not allow the delete to take place.

CHAPTER THREE: DATA MANIPULATION LANGUAGE

Overview

Data Manipulation Language (DML) is used to add, change and delete data in a DB2 table. DML is one of the most basic and essential skills you must have as a DB2 professional. In this section we'll look at the five major DML statements: INSERT, UPDATE, DELETE, MERGE and SELECT.

XML data access and processing is another skill that you need to have. DB2 includes an XML data type and various functions for accessing and processing XML data. I'll assume you have a basic understanding of XML, but we'll do a quick review anyway. Then we'll look at some examples of creating an XML column, populating it, modifying it and manipulating it using XML functions such as XMLQuery. We'll cover some more advanced XML topics in another section of the book.

Special registers allow you to access detailed information about the DB2 instance settings as well as certain session information. CURRENT DATE is an example of a special register. You can access special registers in an application program and then use the information in your processing.

Built-in functions can be used in SQL statements to return a result based on an argument. Think of them as productivity tools because they can be used to replace custom coded functionality in an application program and thereby simplify development and maintenance. Whether your role is application developer, DBA or business services professional, the DB2 built-in functions can save you a great deal of time if you know what they are and how to use them.

Database, Tablespace and Schema Conventions

Throughout this book we will be using a database called DBHR which is a database for a fictitious human relations department in a company. We will use storage group SGHR and bufferpool BPHR. The main tablespace we will use is TSHR. Finally, our schema will be HRSCHEMA.

If you are following along and creating examples on your own system, you may of course use whatever database and schema is available to you on your system. At the end of the last chapter we dropped and recreated these objects. If you missed that and want the basic DDL to create the objects named above, here it is:

```
CREATE DATABASE DBHR
AUTOMATIC STORAGE YES;

CREATE STOGROUP SGHR
ON 'C:' OVERHEAD 6.725
```

```
DEVICE READ RATE 100.0
DATA TAG NONE;

CREATE BUFFERPOOL BPHR
IMMEDIATE
ALL DBPARTITIONNUMS SIZE 1000
AUTOMATIC PAGESIZE 4096;

CREATE REGULAR TABLESPACE TSHR
IN DATABASE PARTITION GROUP IBMDEFAULTGROUP
PAGESIZE 4096
MANAGED BY AUTOMATIC STORAGE USING STOGROUP SGHR
AUTORESIZE YES
BUFFERPOOL BPHR
OVERHEAD INHERIT
TRANSFERRATE INHERIT
DROPPED TABLE RECOVERY
ON DATA TAG INHERIT;

CREATE SCHEMA HRSCHEMA
AUTHORIZATION robert;  ← This should be your DB2 id, whatever it is.
```

DML SQL Statements

Data Manipulation Language (DML) is at the core of working with relational databases. You need to be very comfortable with DML statements: INSERT, UPDATE, DELETE, MERGE and SELECT. We'll cover the syntax and use of each of these. For purposes of this section, let's plan and create a very simple table. Here are the columns and data types for our table which we will name EMPLOYEE.

Field Name	Type	Attributes
EMP_ID	INTEGER	NOT NULL, PRIMARY KEY
EMP_LAST_NAME	VARCHAR(30)	NOT NULL
EMP_FIRST_NAME	VARCHAR(20)	NOT NULL
EMP_SERVICE_YEARS	INTEGER	NOT NULL, DEFAULT IS ZERO
EMP_PROMOTION_DATE	DATE	

The table can be created with the following DDL:

```
CREATE TABLE HRSCHEMA.EMPLOYEE(
EMP_ID INT NOT NULL,
EMP_LAST_NAME VARCHAR(30) NOT NULL,
EMP_FIRST_NAME VARCHAR(20) NOT NULL,
EMP_SERVICE_YEARS INT NOT NULL WITH DEFAULT 0,
EMP_PROMOTION_DATE DATE,
PRIMARY KEY(EMP_ID))
IN TSHR;
```

INSERT Statement

The INSERT statement adds one or more rows to a table. There are two forms of the INSERT statement and you need to know the syntax of each of these.

1. Insert via values
2. Insert via select

Insert Via Values

There are actually two sub-forms of the insert by values. One form explicitly names the target fields and the other does not. Generally when inserting a record you explicitly name the target columns, followed by a VALUES clause that includes the actual values to apply to the columns in the new record. Let's use our `EMPLOYEE` table for this example:

```
INSERT INTO HRSCHEMA.EMPLOYEE
(EMP_ID,
 EMP_LAST_NAME,
 EMP_FIRST_NAME,
 EMP_SERVICE_YEARS,
 EMP_PROMOTION_DATE)

VALUES (3217,
'JOHNSON',
'EDWARD',
4,
'01/01/2017');

Updated 1 rows.
```

A second sub-form of the INSERT statement via values is to omit the target fields and simply provide the VALUES clause. You can do this only if your values clause includes values for ALL the columns in the correct positional order.

Here's an example of this second sub-form of insert via values for the `EMPLOYEE` table:

```
INSERT INTO HRSCHEMA.EMPLOYEE
VALUES (7459,
'STEWART',
'BETTY',
7,
'07/31/2016');

Updated 1 rows.
```

Another consideration for the INSERT statement concerns the use of the `DEFAULT` keyword. If you define a column with a default value using the `WITH DEFAULT` clause, you can assign that default value to a record by simply specifying the word `DEFAULT` instead of an actual value in the INSERT statement.

One final consideration concerns the use of the NULL value. If a column is not defined as NOT NULL (or if it is explicitly defined as NULL meaning NULL values are allowed), and you don't want to assign a value to that column, then you must specify NULL for the column in the values clause.

Note that `EMP_ID` is defined as a primary key on the table. If you try inserting a row for which the primary key already exists, you will receive a -803 error SQL code (meaning a record already exists with that key).

Here's an example of specifying the `DEFAULT` value for the `EMP_SERVICE_YEARS` column, and the NULL value for the `EMP_PROMOTION_DATE`.

```
INSERT INTO HRSCHEMA.EMPLOYEE
(EMP_ID,
EMP_LAST_NAME,
EMP_FIRST_NAME,
EMP_SERVICE_YEARS,
EMP_PROMOTION_DATE)

VALUES (9134,
'FRANKLIN',
'ROSEMARY',
DEFAULT,
NULL);

Updated 1 rows.
```

When you define a column using `WITH DEFAULT`, you do not necessarily have to specify an actual default value when you define the table. DB2 provides **implicit** default values for most data types and if you just specify `WITH DEFAULT` and no specific value, the implicit default value will be used.

In the `EMPLOYEE` table we specified `WITH DEFAULT 0` for the employee's service years. However, the implicit default value is also zero because the column is defined as INTEGER. So we could have simply specified `WITH DEFAULT` and it would have the same result.

We provided this information previously, but it is important enough to repeat it here. The following table provides the default values for the various data types.

Default Values for DB2 Data Types

For columns of	Type	Default
Numbers	SMALLINT, INTEGER, BIGINT, DECIMAL, NUMERIC, REAL, DOUBLE, DECFLOAT, or FLOAT	0
Fixed-length strings	CHAR or GRAPHIC BINARY	Blanks Hexadecimal zeros
Varying-length strings	VARCHAR, CLOB, VARGRAPHIC, DBCLOB, VARBINARY, or BLOB	Empty string
Dates	DATE	CURRENT DATE
Times	TIME	CURRENT TIME
Timestamps	TIMESTAMP	CURRENT TIMESTAMP
ROWIDs	ROWID	DB2-generated

Before moving on to the **Insert via Select** option, let's take a look at the data we have in the table so far.

```
SELECT
EMP_ID,
EMP_LAST_NAME,
EMP_FIRST_NAME,
EMP_SERVICE_YEARS,
EMP_PROMOTION_DATE
FROM HRSCHEMA.EMPLOYEE
ORDER BY EMP_ID;

EMP_ID       EMP_LAST_NAME  EMP_FIRST_NAME EMP_SERVICE_YEARS   EMP_PROMOTION_DATE
------       -------------  -------------- -----------------   ------------------
  3217       JOHNSON        EDWARD                         4   2017-01-01
  7459       STEWART        BETTY                          7   2016-07-31
  9134       FRANKLIN       ROSEMARY                       0   NULL
```

Insert via Select

You can use a SELECT query to extract data from one table and load it to another. You can even include literals or built in functions in the SELECT query in lieu of column names (if you need them). This is often useful for loading tables. Let's do an example.

Suppose you have an employee recognition request table named EMPRECOG. This table is used to generate/store recognition requests for employees who have been promoted during a certain time frame. HR will print a recognition certificate and deliver it to the employee. Once the request is fulfilled, the date completed will be entered by HR in a separate process.

The table specification is as follows:

Field Name	Type	Attributes
EMP_ID	INTEGER	NOT NULL
EMP_PROMOTION_DATE	DATE	NOT NULL
EMP_RECOG_RQST_DATE	DATE	NOT NULL WITH DEFAULT
EMP_RECOG_COMP_DATE	DATE	

The DDL to create the table is as follows:

```
CREATE TABLE HRSCHEMA.EMPRECOG(
EMP_ID INT NOT NULL,
EMP_PROMOTION_DATE DATE NOT NULL,
EMP_RECOG_RQST_DATE DATE
NOT NULL WITH DEFAULT,
EMP_RECOG_COMP_DATE DATE)
IN TSHR;
```

Your objective is to load this table with data from the `EMPLOYEE` table for any employee whose promotion date occurs during the current month. The selection criteria could be expressed as:

```
SELECT
EMP_ID,
EMP_PROMOTION_DATE
FROM HRSCHEMA.EMPLOYEE
WHERE MONTH(EMP_PROMOTION_DATE)
 = MONTH(CURRENT DATE);
```

To use this SQL in an INSERT statement on the `EMPRECOG` table, you would need to add a value for another column for the request date (`EMP_RECOG_RQST_DATE`). Let's use the `CURRENT DATE` function to insert today's date as the requested date. Now our select statement looks like this:

```
SELECT
EMP_ID,
EMP_PROMOTION_DATE,
CURRENT DATE AS RQST_DATE
FROM HRSCHEMA.EMPLOYEE
WHERE MONTH(EMP_PROMOTION_DATE)
    = MONTH(CURRENT DATE);
```

Assuming we are running the SQL on January 10, 2017 we should get the following results:

```
EMP_ID  EMP_PROMOTION_DATE  RQST_DATE
------  ------------------  ----------
  3217  2017-01-01          2017-01-10
```

Let's create the INSERT statement for the EMPRECOG table. Since our query does not include the EMP_RQST_COMP_DATE (assume that the **request complete** column will be populated by another HR process when the request is complete), we must specify the target column names we are populating. Otherwise we will get a mismatch between the number of columns we are loading and the number in the table.

Of course, in circumstances where you have values for all the table's columns, you needn't include the column names and you can just use the INSERT INTO and SELECT statement. But in many cases it is handy to include the target column names, even when you don't have to. It makes the DML more self-documenting and helpful for the next developer.

Here is our SQL:

```
INSERT INTO HRSCHEMA.EMPRECOG
(EMP_ID,
 EMP_PROMOTION_DATE,
 EMP_RECOG_RQST_DATE)
 SELECT
 EMP_ID,
 EMP_PROMOTION_DATE,
 CURRENT DATE AS RQST_DATE
 FROM HRSCHEMA.EMPLOYEE
 WHERE MONTH(EMP_PROMOTION_DATE)
  = MONTH(CURRENT DATE);
```

If you are following along and running the examples, you may notice it doesn't work if the real date is not a January 2017 date. You can make this one work by specifying the comparison date as 1/1/2017. So your query would be:

```
INSERT INTO HRSCHEMA.EMPRECOG
(EMP_ID,
 EMP_PROMOTION_DATE,
 EMP_RECOG_RQST_DATE)
 SELECT
 EMP_ID,
 EMP_PROMOTION_DATE,
 CURRENT DATE AS RQST_DATE
 FROM HRSCHEMA.EMPLOYEE
 WHERE MONTH(EMP_PROMOTION_DATE)
  = MONTH('01/01/2017');

Updated 1 rows.
```

After you run the SQL, query the EMPRECOG table, and you can see the result:

```
SELECT * FROM HRSCHEMA.EMPRECOG;

EMP_ID  EMP_PROMOTION_DATE    EMP_RECOG_RQST_DATE   EMP_RECOG_COMP_DATE
------  ------------------    -------------------   --------------------
  3217  2017-01-01            2017-01-10            NULL
```

The above is what we expect. Only one of the employees has a promotion date in January, 2017. This employee has been added to the **EMPRECOG** table with request date of January 10 and a NULL recognition completed date.

Insert via Values for Multiple Rows

In DB2 LUW you can insert multiple sets of values in the VALUES clause on a single INSERT statement. You simply need to delimit the sets of values by the open and close parentheses, and use a comma between each set.

Let's look at an example.

```
INSERT INTO HRSCHEMA.EMPLOYEE
(EMP_ID,
EMP_LAST_NAME,
EMP_FIRST_NAME,
EMP_SERVICE_YEARS,
EMP_PROMOTION_DATE)
VALUES (4720,
'SCHULTZ',
'TIM',
9,
'01/01/2017'),
(6288,
'WILLARD',
'JOE',
6,
'01/01/2016');

Updated 2 rows.
```

A brief query shows our current table data to be the following:

```
SELECT
EMP_ID,
EMP_LAST_NAME,
EMP_FIRST_NAME,
EMP_SERVICE_YEARS,
EMP_PROMOTION_DATE
FROM HRSCHEMA.EMPLOYEE
ORDER BY EMP_ID;

EMP_ID  EMP_LAST_NAME     EMP_FIRST_NAME   EMP_SERVICE_YEARS  EMP_PROMOTION_DATE
------  -------------     --------------   -----------------  ------------------
  3217  JOHNSON           EDWARD                           4  2017-01-01
  4720  SCHULTZ           TIM                              9  2017-01-01
```

```
6288  WILLARD       JOE                    6  2016-01-01
7459  STEWART       BETTY                  7  2016-01-01
9134  FRANKLIN      ROSEMARY               0  NULL
```

Note: You can also INSERT to an underlying table via a view. The syntax is exactly the same as for inserting to a table. This topic will be considered in a later chapter.

UPDATE Statement

The UPDATE statement is pretty straightforward. It changes one or more records based on specified conditions. There are two forms of the UPDATE statement:

1. The searched update
2. The positioned update

The positioned update is used with a result set and we'll deal with that when we get to chapter five. Meanwhile let's look at the searched update.
Searched Update

The searched update is performed on records that meet a certain search criteria using a WHERE search clause. The basic form and syntax you need to know for the searched update is:

```
UPDATE <TABLENAME>
SET COLUMN NAME = <VALUE>
WHERE <CONDITION>
```

For example, recall that we left the promotion date for employee 9134 with a NULL value. Now let's say we want to update the promotion date to October 1, 2016. We could use this SQL to do that:

```
UPDATE HRSCHEMA.EMPLOYEE
SET EMP_PROMOTION_DATE = '10/01/2016'
WHERE EMP_ID = 9134;

Updated 1 rows.
```

If you have more than one column to update, you must use a comma to separate the column names. For example, let's update both the promotion date and the first name of the employee. We'll make the first name Brianna and the promotion date 10/1/2016.

```
UPDATE HRSCHEMA.EMPLOYEE
SET EMP_PROMOTION_DATE = '10/01/2016',
    EMP_FIRST_NAME = 'BRIANNA'
WHERE EMP_ID = 9134;
```

```
Updated 1 rows.
```

Another sub-form of the UPDATE statement to be aware of is UPDATE without a WHERE clause. For example, to set the **EMP_RECOG_COMP_DATE** field to January 31, 2017 for every row in the EMPRECOG table, you could use this statement:

```
UPDATE HRSCHEMA.EMPRECOG
SET EMP_RECOG_COMP_DATE = '01/31/2017';

Updated 1 rows.
```

Obviously you should be very careful using this form of UPDATE, as it will apply the value you specify for that column to every row in the table. This is normally not what you want, but it could be useful in cases where you need to initialize one or more columns for all rows of a relatively small table.

Let's do another example where we'll update multiple rows based on a condition. We need to specially set up test data for our example, so if you are following along, execute the following query:

```
UPDATE HRSCHEMA.EMPLOYEE
SET EMP_LAST_NAME = LOWER(EMP_LAST_NAME)
WHERE
EMP_LAST_NAME IN ('JOHNSON', 'STEWART', 'FRANKLIN');

Updated 3 rows.
```

Now here is the current content of our EMPLOYEE table:

```
SELECT EMP_ID, EMP_LAST_NAME, EMP_FIRST_NAME
FROM HRSCHEMA.EMPLOYEE
ORDER BY EMP_ID;

EMP_ID  EMP_LAST_NAME    EMP_FIRST_NAME
------  -------------    --------------
  3217  johnson          EDWARD
  4720  SCHULTZ          TIM
  6288  WILLARD          JOE
  7459  stewart          BETTY
  9134  franklin         BRIANNA
```

As you can see we have some last names that are in lower case. Assume that we have decided we want to store all names in upper case. We want to correct the lowercase data. We will check all records in the EMPLOYEE table and if the last name contains lower case, we want to change it to upper case.

To accomplish our objective we need only run a query using the UPPER function. We'll also only specify those rows for which the last name contains lower case letters. Put another way, we only want to apply the update to those rows that actually need to be changed.

Here is our update query:

```
UPDATE HRSCHEMA.EMPLOYEE
SET EMP_LAST_NAME = UPPER(EMP_LAST_NAME)
WHERE EMP_LAST_NAME <> UPPER(EMP_LAST_NAME);

Updated 3 rows.
```

And here is the modified data in the table:

```
SELECT * FROM HRSCHEMA.EMPLOYEE
ORDER BY EMP_ID;

EMP_ID  EMP_LAST_NAME    EMP_FIRST_NAME
------  -------------    --------------
  3217  JOHNSON          EDWARD
  4720  SCHULTZ          TIM
  6288  WILLARD          JOE
  7459  STEWART          BETTY
  9134  FRANKLIN         BRIANNA
```

DELETE Statement

The DELETE statement is also pretty straightforward. It removes one or more records from the table based on specified conditions. As with the UPDATE statement, there are two forms of the DELETE statement:

1. The searched delete
2. The positioned delete

And as with the UPDATE statement, the positioned delete is used with a result set and we'll deal with that when we get to chapter five. Meanwhile let's look at the searched delete.

Searched DELETE

The searched delete is performed on records that meet a certain criteria, i.e., based on a WHERE clause. The basic form and syntax you need to remember for the searched DELETE is:

```
DELETE FROM <TABLENAME> WHERE <CONDITION>
```

For example, we might want to remove the record with employee id 9134. We could use this SQL to do that:

```
DELETE FROM HRSCHEMA.EMPLOYEE WHERE EMP_ID = 9134;

Updated 1 rows.
```

Let's add this employee back to the table so it will be available for a later example.

```
INSERT INTO HRSCHEMA.EMPLOYEE
(EMP_ID,
EMP_LAST_NAME,
EMP_FIRST_NAME,
EMP_SERVICE_YEARS,
EMP_PROMOTION_DATE)

VALUES (9134,
'FRANKLIN',
'BRIANNA',
DEFAULT,
NULL);

Updated 1 rows.
```

Now let's do another example of the searched delete. Suppose we want to delete any employee record which does not have a promotion date. Checking our data, we find a single row lacks a promotion date (sorry Brianna!).

```
SELECT EMP_ID,
EMP_LAST_NAME,
EMP_FIRST_NAME,
EMP_PROMOTION_DATE
FROM HRSCHEMA.EMPLOYEE;

EMP_ID EMP_LAST_NAME P_FIRST_NAME    EMP_PROMOTION_DATE
------ ------------- ------------    ------------------
  3217 JOHNSON        EDWARD         2017-01-01
  7459 STEWART        BETTY          2016-07-31
  9134 FRANKLIN       BRIANNA        NULL
  4720 SCHULTZ        TIM            2017-01-01
  6288 WILLARD        JOE            2016-01-01
```

Our delete SQL would look like this:

```
DELETE HRSCHEMA.EMPLOYEE
WHERE EMP_PROMOTION_DATE IS NULL;

Updated 1 rows.
```

A single row was deleted from the table, as we can confirm by querying `EMPLOYEE`:

```
SELECT EMP_ID, EMP_PROMOTION_DATE
FROM HRSCHEMA.EMPLOYEE
ORDER BY EMP_ID;

EMP_ID  EMP_PROMOTION_DATE
------  ------------------
3217    2017-01-01
4720    2017-01-01
6288    2016-01-01
7459    2016-07-31
```

Ok let's add Brianna back to the table.

```
INSERT INTO HRSCHEMA.EMPLOYEE
(EMP_ID,
EMP_LAST_NAME,
EMP_FIRST_NAME,
EMP_SERVICE_YEARS,
EMP_PROMOTION_DATE)

VALUES (9134,
'FRANKLIN',
'BRIANNA',
DEFAULT,
'10/01/2016');

Updated 1 rows.
```

Finally, another sub-form of the DELETE statement to be aware of is the DELETE without a WHERE clause. For example, to remove all records from the `EMPRECOG` table, use this statement:

```
DELETE FROM HRSCHEMA.EMPRECOG;
```

Be very careful using this form of DELETE, as it will remove every record from the target table. This is normally not what you want, but it could be useful in cases where you need to initialize a relatively small table to empty.

MERGE Statement

The MERGE statement updates a target table or view using specified input data. Rows that already exist in the target table are updated as specified by the input source, and rows that do not exist in the target are inserted using data from that same input source.

So what problem does the merge solve? It adds/updates records for a table from a data source when you don't know whether the row already exists in the table or not. An example could be if you are updating data in your table based on a flat file you receive from another system, department or even another company. Assuming the other system does not send you an action code (add, change or delete), you won't know whether to use the INSERT or UPDATE statement.

One way of handling this situation is to first try doing an INSERT and if you get a -803 SQL error code, then you know the record already exists. In that case you would need to do an UPDATE instead. Or you could first try doing an UPDATE and then if you received an SQLCODE +100, you would know the record does not exist and you would do an INSERT. This solution works, but it inevitably wastes some DB2 calls.

A more elegant solution is the MERGE statement. We'll look at an example of this below. You'll notice the example is a pretty long SQL statement, but don't be put off by that. The SQL is only slightly longer than the combined INSERT and UPDATE statements you would have needed to use otherwise.

Single Row Merge Using Values

Let's go back to our `EMPLOYEE` table for this example. Let's say we have the employee information for Deborah Jenkins. This information is being fed to us from another system which also supplied an `EMP_ID`, but we don't know whether that `EMP_ID` already exists in our `EMPLOYEE` table or not. So let's use the MERGE statement:

```
MERGE INTO HRSCHEMA.EMPLOYEE AS T
USING
(VALUES (1122,
'JENKINS',
'DEBORAH',
5,
NULL))
AS S
(EMP_ID,
 EMP_LAST_NAME,
 EMP_FIRST_NAME,
 EMP_SERVICE_YEARS,
 EMP_PROMOTION_DATE)
ON S.EMP_ID = T.EMP_ID

WHEN MATCHED
   THEN UPDATE
      SET T.EMP_LAST_NAME      = S.EMP_LAST_NAME,
          T.EMP_FIRST_NAME     = S.EMP_FIRST_NAME,
          T.EMP_SERVICE_YEARS  = S.EMP_SERVICE_YEARS,
          T.EMP_PROMOTION_DATE = S.EMP_PROMOTION_DATE
```

```
WHEN NOT MATCHED
   THEN INSERT
      VALUES (S.EMP_ID,
      S.EMP_LAST_NAME,
      S.EMP_FIRST_NAME,
      S.EMP_SERVICE_YEARS,
      S.EMP_PROMOTION_DATE);
```

Note that the existing EMPLOYEE table is given with a T qualifier and the new information is given with S as the qualifier (these qualifiers are arbitrary – you can use anything tag you want). We are matching the new information to the table based on employee id. When the specified employee id is matched to an employee id on the table, an update is performed using the S values, i.e., the new information. If it is not matched to an existing record, then an insert is performed – again based on the S values.

To see that our MERGE action was successful, let's take another look at our EMPLOYEE table.

```
SELECT
EMP_ID,
EMP_LAST_NAME,
EMP_FIRST_NAME,
EMP_PROMOTION_DATE
FROM HRSCHEMA.EMPLOYEE
ORDER BY EMP_ID;

EMP_ID EMP_LAST_NAME  EMP_FIRST_NAME         EMP_PROMOTION_DATE
------ -------------  --------------         ------------------
  1122 JENKINS        DEBORAH                NULL
  3217 JOHNSON        EDWARD                 2017-01-01
  4720 SCHULTZ        TIM                    2017-01-01
  6288 WILLARD        JOE                    2016-01-01
  7459 STEWART        BETTY                  2016-07-31
  9134 FRANKLIN       BRIANNA                2016-10-01
```

Merge With Multiple VALUES

As with the INSERT statement, you can MERGE multiple records by including more than one set of values in the VALUES CLAUSE. To prove it, let's create a new table EMP_PAY and it will include the base and bonus pay for each employee identified by employee id. We will also use this table later when we look at built in functions. Here are the columns we need to define.

Field Name	Type	Attributes
EMP_ID	INTEGER	NOT NULL
EMP_REGULAR_PAY	DECIMAL	NOT NULL
EMP_BONUS	DECIMAL	

The DDL would look like this:

```
CREATE TABLE HRSCHEMA.EMP_PAY(
EMP_ID INT NOT NULL,
EMP_REGULAR_PAY DECIMAL (8,2) NOT NULL,
EMP_BONUS_PAY DECIMAL   (8,2));
```

Next, let's add three records:

```
INSERT INTO HRSCHEMA.EMP_PAY
VALUES (3217, 80000.00, 4000),
       (7459, 80000.00, 4000),
       (9134, 70000.00, NULL);

Updated 3 rows.
```

Now the current data in the table is as follows:

```
SELECT * FROM HRSCHEMA.EMP_PAY;

EMP_ID  EMP_REGULAR_PAY  EMP_BONUS_PAY
------  ---------------  -------------
  3217         80000.00        4000.00
  7459         80000.00        4000.00
  9134         70000.00           NULL
```

Ok, let's create some update data for our MERGE query such that some of the employee ids are already on the table and some are not, so some records will be inserted and some will be updated.

Here's the data we want to use:

```
EMP_ID     EMP_REGULAR_PAY    EMP_BONUS_PAY
------     ---------------    -------------
3217           65000.00             5500.00
7459           85000.00             4500.00
9134           75000.00             2500.00
4720           80000.00             2500.00
6288           70000.00             2000.00
```

Looking at this data we know we will need to update three records that are already on the table, and we will add two that don't currently exist on the table (4720 and 6288).

Here is our merge query:

```
MERGE INTO HRSCHEMA.EMP_PAY AS TARGET
USING (VALUES(3217,
```

```
65000.00,
5500.00),
(7459,
85000.00,
4500.00),
(9134,
75000.00,
2500.00),
(4720,
80000.00,
2500.00),
(6288,
70000.00,
2000.00))

AS SOURCE(EMP_ID,
EMP_REGULAR_PAY,
EMP_BONUS_PAY)
ON TARGET.EMP_ID = SOURCE.EMP_ID

WHEN MATCHED THEN UPDATE
   SET TARGET.EMP_REGULAR_PAY
          = SOURCE.EMP_REGULAR_PAY,
       TARGET.EMP_BONUS_PAY
          = SOURCE.EMP_BONUS_PAY

WHEN NOT MATCHED THEN INSERT
  (EMP_ID,
   EMP_REGULAR_PAY,
   EMP_BONUS_PAY)
   VALUES
   (SOURCE.EMP_ID,
   SOURCE.EMP_REGULAR_PAY,
   SOURCE.EMP_BONUS_PAY);

  Updated 5 rows.
```

And now we can verify that the results were actually applied to the table.

```
SELECT *
FROM HRSCHEMA.EMP_PAY;

EMP_ID  EMP_REGULAR_PAY  EMP_BONUS_PAY
------  ---------------  -------------
  3217         65000.00        5500.00
  7459         85000.00        4500.00
  9134         75000.00        2500.00
  4720         80000.00        2500.00
  6288         70000.00        2000.00
```

The power of the MERGE statement is that you do not need to know whether a record already exists when you apply the data to the table. The program logic is simplified – there is no trial and error to determine whether or not the record exists. We'll look at a program example in a later chapter.

SELECT Statement

SELECT is the main statement you will use to retrieve data from a table or view. The basic syntax for the select statement is:

```
SELECT        <column names>
FROM          <table or view name>
WHERE         <condition>
ORDER BY      <column name or number to sort by>
```

Let's return to our EMPLOYEE table for an example:

```
SELECT EMP_ID, EMP_LAST_NAME, EMP_FIRST_NAME
FROM HRSCHEMA.EMPLOYEE
WHERE EMP_ID = 3217'

EMP_ID EMP_LAST_NAME    EMP_FIRST_NAME
------ -------------    --------------
  3217 JOHNSON           EDWARD
```

You can also change the column heading on the result set by specifying <column name> AS <literal>. For example:

```
SELECT EMP_ID AS "EMPLOYEE NUMBER",
EMP_LAST_NAME AS "EMPLOYEE LAST NAME",
EMP_FIRST_NAME AS "EMPLOYEE FIRST NAME"
FROM HRSCHEMA.EMPLOYEE
WHERE EMP_ID = 3217;

EMPLOYEE NUMBER  EMPLOYEE LAST NAME    EMPLOYEE FIRST NAME
---------------  ------------------    -------------------
           3217  JOHNSON               EDWARD
```

Now let's look at some clauses that will further qualify the rows that are returned.

WHERE CONDITION

There are quite a lot of options for the WHERE condition. In fact, you can use multiple where conditions by specifying AND and OR clauses. Be aware of the equality operators which are:

```
=            Equal to
<>           Not equal to
>            Greater than
```

```
>=          Greater than or equal to
<           Less than
<=          Less than or equal to
```

Let's look at some examples of WHERE conditions.

OR

```
SELECT EMP_ID, EMP_LAST_NAME, EMP_FIRST_NAME
FROM HRSCHEMA.EMPLOYEE
WHERE EMP_ID = 3217 OR EMP_ID = 9134;

EMP_ID  EMP_LAST_NAME          EMP_FIRST_NAME
------  -------------          --------------
  3217  JOHNSON                EDWARD
  9134  FRANKLIN               BRIANNA
```

AND

```
SELECT EMP_ID,
EMP_LAST_NAME,
EMP_FIRST_NAME,
EMP_PROMOTION_DATE
FROM HRSCHEMA.EMPLOYEE
WHERE (EMP_SERVICE_YEARS > 1)
  AND (EMP_PROMOTION_DATE > '12/31/2016');

EMP_ID  EMP_LAST_NAME          EMP_FIRST_NAME         EMP_PROMOTION_DATE
------  -------------          --------------         ------------------
  3217  JOHNSON                EDWARD                 2017-01-01
  4720  SCHULTZ                TIM                    2017-01-01
```

IN

You can specify that the column value must be present in a specified collection of values, either those you code in the SQL explicitly or a collection that is a result of a query. Let's look at an example of specifying specific `EMP_ID` values as a set.

```
SELECT EMP_ID,
EMP_LAST_NAME,
EMP_FIRST_NAME
FROM HRSCHEMA.EMPLOYEE
WHERE EMP_ID IN (3217, 9134);

   EMP_ID  EMP_LAST_NAME          EMP_FIRST_NAME
   ------  -------------          --------------
     3217  JOHNSON                EDWARD
     9134  FRANKLIN               BRIANNA
```

Now let's provide a listing of employees who are in the EMPLOYEE table but who are NOT in the EMP_PAY table yet. This example shows us two new techniques, use of the NOT keyword and use of a subselect query to create a collection result set. First, let's add a couple of records to the EMPLOYEE table:

```
INSERT INTO HRSCHEMA.EMPLOYEE
(EMP_ID,
EMP_LAST_NAME,
EMP_FIRST_NAME,
EMP_SERVICE_YEARS,
EMP_PROMOTION_DATE)

VALUES (3333,
'FORD',
'JAMES',
7,
'10/01/2015'),
(7777,
'HARRIS',
'ELISA',
2,
NULL);

Updated 2 rows.
```

Now let's run our mismatch query:

```
SELECT EMP_ID,
EMP_LAST_NAME,
EMP_FIRST_NAME
FROM HRSCHEMA.EMPLOYEE
WHERE EMP_ID
NOT IN (SELECT EMP_ID FROM HRSCHEMA.EMP_PAY);

 EMP_ID EMP_LAST_NAME  EMP_FIRST_NAME
 ------ -------------- --------------
   3333 FORD           JAMES
   1122 JENKINS        DEBORAH
   7777 HARRIS         ELISA
```

Our results show those records in the EMPLOYEE table that have no corresponding records in the EMP_PAY table.

Incidentally, you can also use the **EXCEPT** predicate to identify rows in one table that have no counterpart in the other. For example, suppose we want the employee ids of any employee who has not received a paycheck. You could quickly identify them with this SQL:

```
SELECT EMP_ID
FROM HRSCHEMA.EMPLOYEE EXCEPT (SELECT EMP_ID FROM HRSCHEMA.EMP_PAY);
```

```
EMP_ID
------
  1122
  3333
  7777
```

One limitation of the EXCEPT clause is that the columns in the two queries have to match exactly, so you could not bring back a column from EMPLOYEE that does not also exist in the EMP_PAY table. Still the EXCEPT is useful in some cases, especially where you need to identify discrepancies between tables using a single column.

BETWEEN

The BETWEEN clause allows you to specify a range of values inclusive of the start and end value you provide. Here's an example where we want to retrieve the employee id and pay rate for all employees whose pay rate is between 60,000 and 85,000 annually.

```
SELECT EMP_ID,
EMP_REGULAR_PAY
FROM HRSCHEMA.EMP_PAY
WHERE EMP_REGULAR_PAY
BETWEEN 60000 AND 85000;

EMP_ID  EMP_REGULAR_PAY
------  ---------------
  3217         65000.00
  7459         85000.00
  9134         75000.00
  4720         80000.00
  6288         70000.00
```

LIKE

You can use the LIKE predicate to select values that match a pattern. For example, let's choose all rows for which the last name begins with the letter B. The % character is used as a wild card for any string value or character. So in this case we are retrieving every record for which the EMP_FIRST_NAME starts with the letter B.

```
SELECT EMP_ID,
EMP_LAST_NAME,
EMP_FIRST_NAME
FROM HRSCHEMA.EMPLOYEE
WHERE EMP_FIRST_NAME LIKE 'B%';

EMP_ID  EMP_LAST_NAME           EMP_FIRST_NAME
------  -------------           --------------
  7459  STEWART                 BETTY
  9134  FRANKLIN                BRIANNA
```

DISTINCT

Use the DISTINCT operator when you want to eliminate duplicate values. To illustrate this, let's create a couple of new tables. The first is called EMP_PAY_CHECK and we will use to store a calculated bi-monthly pay amount for each employee based on their annual salary. The DDL to create EMP_PAY_CHECK is a s follows:

```
CREATE TABLE HRSCHEMA.EMP_PAY_CHECK(
EMP_ID INT NOT NULL,
EMP_REGULAR_PAY  DECIMAL (8,2) NOT NULL,
EMP_SEMIMTH_PAY DECIMAL (8,2) NOT NULL)
IN TSHR;
```

Now let's insert some data into the EMP_PAY_CHECK table by calculating a twice monthly pay check:

```
INSERT INTO HRSCHEMA.EMP_PAY_CHECK
(SELECT EMP_ID,
EMP_REGULAR_PAY,
EMP_REGULAR_PAY / 24 FROM HRSCHEMA.EMP_PAY);

Updated 5 rows.
```

Let's look at the results:

```
SELECT *
FROM HRSCHEMA.EMP_PAY_CHECK;

EMP_ID  EMP_REGULAR_PAY  EMP_SEMIMTH_PAY
------   --------------  ---------------
  3217         65000.00          2708.33
  7459         85000.00          3541.66
  9134         75000.00          3125.00
  4720         80000.00          3333.33
  6288         70000.00          2916.66
```

We now know how much each employee should make in their pay check. The next step is to create a history table of each pay check the employee receives. First we'll create the table and then we'll load it with data.

```
CREATE TABLE HRSCHEMA.EMP_PAY_HIST(
EMP_ID INT NOT NULL,
EMP_PAY_DATE  DATE NOT NULL,
EMP_PAY_AMT   DECIMAL (8,2) NOT NULL)
IN TSHR;
```

We can load the history table by creating pay checks for the first four pay periods of the year like this:

```
INSERT INTO HRSCHEMA.EMP_PAY_HIST
SELECT EMP_ID,
 '01/15/2017',
 EMP_SEMIMTH_PAY
 FROM HRSCHEMA.EMP_PAY_CHECK;

INSERT INTO HRSCHEMA.EMP_PAY_HIST
SELECT EMP_ID,
 '01/31/2017',
 EMP_SEMIMTH_PAY
 FROM HRSCHEMA.EMP_PAY_CHECK;

INSERT INTO HRSCHEMA.EMP_PAY_HIST
SELECT EMP_ID,
 '02/15/2017',
 EMP_SEMIMTH_PAY
 FROM HRSCHEMA.EMP_PAY_CHECK;

INSERT INTO HRSCHEMA.EMP_PAY_HIST
SELECT EMP_ID,
 '02/28/2017',
 EMP_SEMIMTH_PAY
 FROM HRSCHEMA.EMP_PAY_CHECK;
```

Now we can look at the history table content which is as follows:

```
SELECT * from HRSCHEMA.EMP_PAY_HIST;

EMP_ID  EMP_PAY_DATE  EMP_PAY_AMT
------  ------------  -----------
  3217  2017-01-15        2708.33
  7459  2017-01-15        3541.66
  9134  2017-01-15        3125.00
  4720  2017-01-15        3333.33
  6288  2017-01-15        2916.66
  3217  2017-01-31        2708.33
  7459  2017-01-31        3541.66
  9134  2017-01-31        3125.00
  4720  2017-01-31        3333.33
  6288  2017-01-31        2916.66
  3217  2017-02-15        2708.33
  7459  2017-02-15        3541.66
  9134  2017-02-15        3125.00
  4720  2017-02-15        3333.33
  6288  2017-02-15        2916.66
  3217  2017-02-28        2708.33
  7459  2017-02-28        3541.66
  9134  2017-02-28        3125.00
  4720  2017-02-28        3333.33
  6288  2017-02-28        2916.66
```

If you want a list of all employees who got a paycheck during the month of February, you would need to eliminate the duplicate entries because there are two for each employee. You could accomplish that with this SQL:

```
SELECT DISTINCT EMP_ID
FROM HRSCHEMA.EMP_PAY_HIST
WHERE MONTH(EMP_PAY_DATE) = '02' ;

EMP_ID
------
  3217
  4720
  6288
  7459
  9134
```

The DISTINCT operator ensures that only unique records are selected based on the columns you are returning. This is important because if you included additional columns in the results, any value that makes the record unique will also make it **NOT** a duplicate.

For example, let's add the payment date to our query and see the results:

```
SELECT DISTINCT EMP_ID, EMP_PAY_DATE
FROM HRSCHEMA.EMP_PAY_HIST
WHERE MONTH(EMP_PAY_DATE) = '02';


EMP_ID  EMP_PAY_DATE
------  ------------
  3217  2017-02-15
  3217  2017-02-28
  4720  2017-02-15
  4720  2017-02-28
  6288  2017-02-15
  6288  2017-02-28
  7459  2017-02-15
  7459  2017-02-28
  9134  2017-02-15
  9134  2017-02-28
```

Since the combination of the employee id and payment date makes each record unique, you'll get multiple rows for each employee. So you must be careful in using DISTINCT to ensure that the structure of your query is really what you want.

FETCH FIRST X ROWS ONLY

You can limit your result set by using the `FETCH FIRST X ROWS ONLY` clause. For example, suppose you just want the employee id and names of the first four records from the employee table. You can code it as follows:

```
SELECT EMP_ID,
EMP_LAST_NAME,
EMP_FIRST_NAME
FROM HRSCHEMA.EMPLOYEE
FETCH FIRST 4 ROWS ONLY;

EMP_ID   EMP_LAST_NAME        EMP_FIRST_NAME
------   -------------        --------------
  3217    JOHNSON              EDWARD
  7459    STEWART              BETTY
  3333    FORD                 JAMES
  4720    SCHULTZ              TIM
```

Keep in mind that when you order the results you may get different records. For example if you order by last name, you would get this result:

```
SELECT EMP_ID,
EMP_LAST_NAME,
EMP_FIRST_NAME
FROM HRSCHEMA.EMPLOYEE
ORDER BY EMP_LAST_NAME FETCH FIRST 4 ROWS ONLY;

EMP_ID       EMP_LAST_NAME  EMP_FIRST_NAME
------       -------------  --------------
  3333       FORD           JAMES
  9134       FRANKLIN       BRIANNA
  7777       HARRIS         ELISA
  1122       JENKINS        DEBORAH
```

SUBQUERY

A subquery is essentially a query within a query. Suppose for example we want to list the employee or employees who make the largest salary in the company. You can use a subquery to determine the maximum salary, and then use that value in the WHERE clause of your main query.

```
SELECT EMP_ID, EMP_REGULAR_PAY
FROM HRSCHEMA.EMP_PAY
WHERE EMP_REGULAR_PAY
   = (SELECT MAX(EMP_REGULAR_PAY)
        FROM HRSCHEMA.EMP_PAY);

EMP_ID  EMP_REGULAR_PAY
------  ---------------
  7459         85000.00
```

What if there is more than one employee who makes the highest salary? Let's bump two people up to 85000 (and 4500 bonus) and see.

```
UPDATE HRSCHEMA.EMP_PAY
SET EMP_REGULAR_PAY = 85000.00,
    EMP_BONUS_PAY = 4500
WHERE EMP_ID IN (4720,9134);

Updated 2 rows.
```

Here are the results:

```
SELECT * FROM HRSCHEMA.EMP_PAY;

EMP_ID  EMP_REGULAR_PAY  EMP_BONUS_PAY
------  ---------------  -------------
  3217         65000.00        5500.00
  7459         85000.00        4500.00
  9134         85000.00        4500.00
  4720         85000.00        4500.00
  6288         70000.00        2000.00
```

Now let's see if our subquery still works:

```
SELECT EMP_ID, EMP_REGULAR_PAY
FROM HRSCHEMA.EMP_PAY
WHERE EMP_REGULAR_PAY
   = (SELECT MAX(EMP_REGULAR_PAY)
        FROM HRSCHEMA.EMP_PAY);

EMP_ID  EMP_REGULAR_PAY
------  ---------------
  7459         85000.00
  9134         85000.00
  4720         85000.00
```

And in fact the query pulls all three of the employees who earn the top pay. Subqueries are very powerful in that any value you can produce via a subquery can be substituted into a main query as selection criteria.

GROUP BY

You can summarize data using the GROUP BY clause. For example, let's determine how many distinct employee salary rates there are and how many employees are paid those amounts.

```
SELECT EMP_REGULAR_PAY,
COUNT(*) AS "HOW MANY"
FROM HRSCHEMA.EMP_PAY
GROUP BY EMP_REGULAR_PAY;
```

```
EMP_REGULAR_PAY     HOW MANY
---------------     --------
       65000.00            1
       70000.00            1
       85000.00            3
```

ORDER BY

You can sort the display into ascending or descending sequence using the ORDER BY clause. To take the query we were just using for the group-by, let's present the data in descending sequence:

```
SELECT EMP_REGULAR_PAY,
COUNT(*) AS "HOW MANY"
FROM HRSCHEMA.EMP_PAY
GROUP BY EMP_REGULAR_PAY ORDER BY EMP_REGULAR_PAY DESC;
```

```
EMP_REGULAR_PAY     HOW MANY
---------------     --------
       85000.00            3
       70000.00            1
       65000.00            1
```

HAVING

You could also use the GROUP BY with a HAVING clause that limits the results to only those groups that meet another condition. Let's specify that the group must have more than one employee in it to be included in the results.

```
SELECT EMP_REGULAR_PAY, COUNT(*) AS "HOW MANY"
FROM HRSCHEMA.EMP_PAY
GROUP BY EMP_REGULAR_PAY HAVING COUNT(*) > 1
ORDER BY EMP_REGULAR_PAY DESC;
```

```
EMP_REGULAR_PAY     HOW MANY
---------------     --------
       85000.00            3
```

Or if you want pay rates that have only one employee you could specify the count 1.

```
SELECT EMP_REGULAR_PAY, COUNT(*) AS "HOW MANY"
FROM HRSCHEMA.EMP_PAY
GROUP BY EMP_REGULAR_PAY
HAVING COUNT(*) = 1
ORDER BY EMP_REGULAR_PAY DESC;
```

```
EMP_REGULAR_PAY     HOW MANY
---------------     --------
       70000.00            1
       65000.00            1
```

Before we move on, let's reset our two employees to whom we gave a temporary raise. Otherwise our EMP_PAY and EMP_PAY_CHECK tables will not be in sync.

```
UPDATE HRSCHEMA.EMP_PAY
SET EMP_REGULAR_PAY = 80000.00,
    EMP_BONUS_PAY = 2500
WHERE EMP_ID = 4720;

Updated 1 rows.

UPDATE HRSCHEMA.EMP_PAY
SET EMP_REGULAR_PAY = 75000.00,
    EMP_BONUS_PAY = 2500
WHERE EMP_ID = 9134;

Updated 1 rows.
```

Now our EMP_PAY table is restored:

```
SELECT * FROM HRSCHEMA.EMP_PAY;
EMP_ID  EMP_REGULAR_PAY  EMP_BONUS_PAY
------  ---------------  -------------
  3217         65000.00        5500.00
  7459         85000.00        4500.00
  9134         75000.00        2500.00
  4720         80000.00        2500.00
  6288         70000.00        2000.00
```

CASE Expressions

In some situations you may need to code rather complex conditional logic into your queries. Assume we have a requirement to report all employees according to seniority. We've invented the classifications ENTRY, ADVANCED and SENIOR. Report those who have less than a year service as ENTRY, employees who have a year or more service but less than 5 years as ADVANCED, and all employees with 5 years or more service as SENIOR. Here is a sample query that performs this using a CASE expression:

```
SELECT EMP_ID,
EMP_LAST_NAME,
EMP_FIRST_NAME,
CASE
   WHEN EMP_SERVICE_YEARS  < 1 THEN 'ENTRY'
   WHEN EMP_SERVICE_YEARS  < 5 THEN 'ADVANCED'
   ELSE 'SENIOR'
END CASE
FROM HRSCHEMA.EMPLOYEE;
```

```
EMP_ID  EMP_LAST_NAME          EMP_FIRST_NAME         CASE
------  -------------          --------------         --------
  3217  JOHNSON                EDWARD                 SENIOR
  7459  STEWART                BETTY                  SENIOR
  9134  FRANKLIN               BRIANNA                ENTRY
  4720  SCHULTZ                TIM                    SENIOR
  6288  WILLARD                JOE                    SENIOR
  3333  FORD                   JAMEs                  SENIOR
  7777  HARRIS                 ELISA                  ADVANCED
```

You'll notice that the column heading for the case result is CASE. If you want to use a more meaningful column heading, then instead of closing the CASE statement with END CASE, close it with END AS <some literal>. So if we want to call the result of the CASE expression an employee's "LEVEL", code it this way:

```
SELECT EMP_ID,
EMP_LAST_NAME,
EMP_FIRST_NAME,
CASE
   WHEN EMP_SERVICE_YEARS  < 1 THEN 'ENTRY'
   WHEN EMP_SERVICE_YEARS  < 5 THEN 'ADVANCED'
   ELSE 'SENIOR'
END AS LEVEL
FROM HRSCHEMA.EMPLOYEE ;

EMP_ID  EMP_LAST_NAME          EMP_FIRST_NAME         LEVEL
------  -------------          --------------         ---------
  3217  JOHNSON                EDWARD                 SENIOR
  7459  STEWART                BETTY                  SENIOR
  9134  FRANKLIN               BRIANNA                ENTRY
  4720  SCHULTZ                TIM                    SENIOR
  6288  WILLARD                JOE                    SENIOR
  3333  FORD                   JAMES                  SENIOR
  7777  HARRIS                 ELISA                  ADVANCED
```

JOINS

Now let's look at some cases where we need to extract data from more than one table. To do this we can use a join. Before we start running queries I want to add one row to the EMP_PAY_CHECK table. This is needed to make some of the joins work later, so bear with me.

```
INSERT INTO HRSCHEMA.EMP_PAY_CHECK
VALUES
(7033,
77000.00,
77000 / 24);

Updated 1 rows.
```

Now our EMP_PAY_CHECK table now has these rows.

```
SELECT * FROM HRSCHEMA.EMP_PAY_CHECK;

EMP_ID  EMP_REGULAR_PAY  EMP_SEMIMTH_PAY
------  ---------------  ---------------
  3217         65000.00          2708.33
  7459         85000.00          3541.66
  9134         75000.00          3125.00
  4720         80000.00          3333.33
  6288         70000.00          2916.66
  7033         77000.00          3208.00
```

Inner Joins

An inner join combines each row of one table with each row of the other table, keeping only the rows in which the join condition is true. You can join more than two tables but keep in mind that the more tables you join, the more record I/O is required and this could be a performance consideration. When I say a "performance consideration" I do not mean it is necessarily a problem. I mean it is one factor of many to keep in mind when designing an application process.

Let's do an example. Assume we want a report that includes employee id, first and last names and pay rate for each employee. To accomplish this we need data from both the EMPLOYEE and the EMP_PAY tables. We can match the tables on EMP_ID which is the column they have in common.

We can perform our join either implicitly or with the JOIN verb (explicitly). The first example will perform the join implicitly by specifying we will only include rows for which the EMP_ID in the EMPLOYEE table matches the EMP_ID in the EMP_PAY table.

```
SELECT A.EMP_ID,
A.EMP_LAST_NAME,
A.EMP_FIRST_NAME,
B.EMP_REGULAR_PAY
FROM HRSCHEMA.EMPLOYEE A, HRSCHEMA.EMP_PAY B
WHERE A.EMP_ID = B.EMP_ID
ORDER BY EMP_ID;

EMP_ID  EMP_LAST_NAME          EMP_FIRST_NAME          EMP_REGULAR_PAY
------  -------------          --------------          ---------------
  3217  JOHNSON                EDWARD                         65000.00
  4720  SCHULTZ                TIM                            80000.00
  6288  WILLARD                JOE                            70000.00
  7459  STEWART                BETTY                          85000.00
  9134  FRANKLIN               BRIANNA                        75000.00
```

Notice that in the SQL the column names are prefixed with a tag that is associated with the table being referenced. This is needed in all cases where the column being referenced exists in both tables with the same column name. If you do not specify the tag, you will get an error that your column name reference is ambiguous, i.e., DB2 does not know which table you are referencing when you refer to a column.

Moving on, you can use an explicit join by specifying the JOIN or INNER JOIN verbs. This is actually a best practice because it helps keep the query clearer for those developers who follow you, especially as your queries get more complex.

```
SELECT A.EMP_ID,
A.EMP_LAST_NAME,
A.EMP_FIRST_NAME,
B.EMP_REGULAR_PAY
FROM HRSCHEMA.EMPLOYEE A
INNER JOIN
HRSCHEMA.EMP_PAY B
ON A.EMP_ID = B.EMP_ID
ORDER BY EMP_ID;
```

```
EMP_ID  EMP_LAST_NAME           EMP_FIRST_NAME         EMP_REGULAR_PAY
------  -------------           --------------         ---------------
  3217  JOHNSON                 EDWARD                        65000.00
  4720  SCHULTZ                 TIM                           80000.00
  6288  WILLARD                 JOE                           70000.00
  7459  STEWART                 BETTY                         85000.00
  9134  FRANKLIN                BRIANNA                       75000.00
```

Finally let's do a join with three tables just to extend the concepts. We'll join the `EMPLOYEE`, `EMP_PAY` and `EMP_PAY_HIST` tables for pay date February 15 as follows:

```
SELECT A.EMP_ID,
A.EMP_LAST_NAME,
B.EMP_REGULAR_PAY,
C.EMP_PAY_AMT
FROM HRSCHEMA.EMPLOYEE A
   INNER JOIN
     HRSCHEMA.EMP_PAY  B ON A.EMP_ID = B.EMP_ID
   INNER JOIN
     HRSCHEMA.EMP_PAY_HIST C ON B.EMP_ID = C.EMP_ID
WHERE C.EMP_PAY_DATE = '2/15/2017';
```

```
EMP_ID  EMP_LAST_NAME           EMP_REGULAR_PAY  EMP_PAY_AMT
------  -------------           ---------------  -----------
  3217  JOHNSON                        65000.00      2708.33
  7459  STEWART                        85000.00      3541.66
  9134  FRANKLIN                       75000.00      3125.00
  4720  SCHULTZ                        80000.00      3333.33
  6288  WILLARD                        70000.00      2916.66
```

Outer Joins

Now let's move on to outer joins. There are three types of outer joins. A **left outer join** includes matching rows from both tables plus any rows from the first table (the LEFT table) that were not matched to the other table but otherwise satisfied the WHERE condition. A **right outer join** includes matching rows from both tables plus any rows from the second (the RIGHT) table that were not matched to the first table, but that otherwise satisfied the WHERE condition. A **full outer join** includes matching rows from both tables, plus those in either table that were not matched but which otherwise satisfied the WHERE condition. We'll look at examples of all three types of outer joins.

Left Outer Join

Let's try a left outer join to include matching rows from the EMPLOYEE and EMP_PAY tables, plus any rows in the EMPLOYEE table that might not be in the EMP_PAY table. In this case we are not using a WHERE clause because the table is very small and we want to see all the results. But keep in mind that we could use a WHERE clause.

```
SELECT A.EMP_ID,
A.EMP_LAST_NAME,
A.EMP_FIRST_NAME,
B.EMP_REGULAR_PAY
FROM HRSCHEMA.EMPLOYEE A
LEFT OUTER JOIN
HRSCHEMA.EMP_PAY B
ON A.EMP_ID = B.EMP_ID
ORDER BY EMP_ID;

EMP_ID       EMP_LAST_NAME  EMP_FIRST_NAME        EMP_REGULAR_PAY
------       -------------  --------------        ---------------
  1122       JENKINS        DEBORAH                          NULL
  3217       JOHNSON        EDWARD                       65000.00
  3333       FORD           JAMES                            NULL
  4720       SCHULTZ        TIM                          80000.00
  6288       WILLARD        JOE                          70000.00
  7459       STEWART        BETTY                        85000.00
  7777       HARRIS         ELISA                            NULL
  9134       FRANKLIN       BRIANNA                      75000.00
```

As you can see, we've included three employees who have not been assigned an annual salary yet. Deborah Jenkins, James Ford and Elisa Harris have NULL as their regular pay. The LEFT JOIN says we want all records in the first (left) table that satisfy the query even if there is no matching record in the other (right) table. That's why the query results included the three unmatched records.

Let's do another left join, and this time we'll join the EMPLOYEE table with the EMP_PAY_CHECK table. Like before, we want all records from EMPLOYEE and EMP_PAY_CHECK that match on

EMP_ID, plus any EMPLOYEE records that could not be matched to EMP_PAY_CHECK.

```
SELECT A.EMP_ID,
A.EMP_LAST_NAME,
A.EMP_FIRST_NAME,
B.EMP_SEMIMTH_PAY
FROM HRSCHEMA.EMPLOYEE A
LEFT OUTER JOIN
HRSCHEMA.EMP_PAY_CHECK B
ON A.EMP_ID = B.EMP_ID
ORDER BY EMP_ID;
```

```
EMP_ID        EMP_LAST_NAME  EMP_FIRST_NAME       EMP_SEMIMTH_PAY
------        -------------  --------------       ---------------
  1122        JENKINS        DEBORAH                         NULL
  3217        JOHNSON        EDWARD                       2708.33
  3333        FORD           JAMES                           NULL
  4720        SCHULTZ        TIM                          3333.33
  6288        WILLARD        JOE                          2916.66
  7459        STEWART        BETTY                        3541.66
  7777        HARRIS         ELISA                           NULL
  9134        FRANKLIN       BRIANNA                      3125.00
```

Again we find that three records in the EMPLOYEE table have no matching EMP_PAY_CHECK records. From a business standpoint that could be a problem unless the three are new hires who have not received their first pay check. We'll comment more on this condition shortly.

Right Outer Join

Meanwhile, now let us turn it around and do a right join. In this case we want all matching records in the EMPLOYEE and EMP_PAY_CHECK records plus any records in the EMP_PAY_CHECK table that were not matched to the EMPLOYEE table. We can also add a WHERE condition such that the EMP_SEMIMTH_PAY column has to be populated (cannot be NULL). Let's do that.

```
SELECT B.EMP_ID,
A.EMP_LAST_NAME,
A.EMP_FIRST_NAME,
B.EMP_SEMIMTH_PAY
FROM HRSCHEMA.EMPLOYEE A
   RIGHT OUTER JOIN
      HRSCHEMA.EMP_PAY_CHECK B
         ON A.EMP_ID = B.EMP_ID
WHERE EMP_SEMIMTH_PAY IS NOT NULL;
```

```
EMP_ID  EMP_LAST_NAME         EMP_FIRST_NAME        EMP_SEMIMTH_PAY
------  -------------         --------------        ---------------
  3217  JOHNSON               EDWARD                        2708.33
  4720  SCHULTZ               TIM                           3333.33
  6288  WILLARD               JOE                           2916.66
  7033  --------------------  --------------------          3208.00
```

```
7459  STEWART          BETTY             3541.66
9134  FRANKLIN         BRIANNA           3125.00
```

Now we have a case where there is a record in the EMP_PAY_CHECK table for employee 7033, but that same employee number is NOT in the EMPLOYEE table. That is something to research to find out why this condition exists.

But let's pause for a moment. You may be thinking that this is not a realistic example because any employee getting a paycheck would also **have** to be in the EMPLOYEE table, so this mismatch condition would never happen. I chose this example for a few reasons. One reason is to point out the importance of referential data integrity. The reason the above exception is even *possible* is because we haven't defined a referential relationship between these two tables (we covered referential constraints in chapter two). For now, just know that these things can and do happen when a system has not been designed with tight referential constraints in place.

A second reason I chose this example is to highlight outer joins as a useful tool in tracking down data discrepancies between tables (subqueries are another useful tool). Keep this example in mind when you are called on by your boss or your client to troubleshoot a data mismatch problem in a high pressure, time sensitive situation. You need all the tools you can get.

The third reason for choosing this example is that it very clearly demonstrates what a right join is – it includes all records from both tables that satisfy the join condition, plus any records in the "right" table that otherwise meet the WHERE condition (in this case that the EMP_SEMIMTH_PAY is populated).

Full Outer Join

Finally, let's do a full outer join to include both matched and unmatched records from both tables that meet the where condition. This will expose all the discrepancies we already uncovered with a single query.

```
SELECT A.EMP_ID,
  A.EMP_LAST_NAME,
  B.EMP_SEMIMTH_PAY
  FROM HRSCHEMA.EMPLOYEE A
     FULL OUTER JOIN
        HRSCHEMA.EMP_PAY_CHECK B
           ON A.EMP_ID = B.EMP_ID;

EMP_ID     EMP_LAST_NAME          EMP_SEMIMTH_PAY
------     -------------          ---------------
  3217     JOHNSON                        2708.33
  7459     STEWART                        3541.66
  4720     SCHULTZ                        3333.33
  6288     WILLARD                        2916.66
```

```
9134    FRANKLIN                    3125.00
NULL    NULL                        3208.00
3333    FORD                           NULL
7777    HARRIS                         NULL
1122    JENKINS                        NULL
```

So with the FULL OUTER join we have identified the missing EMPLOYEE record, as well as the three EMP_PAY_CHECK records that are missing. Again these examples are intended both to explain the difference between the join types, and also to lend support to troubleshooting efforts where data integrity is involved.

One final comment. The outer join examples we've given so far point to potential issues with the data, and these joins are in fact helpful in diagnosing such problems. But there many cases where an entry in one table does not necessarily imply an entry in another.

For example, suppose we have an EMP_SPOUSE table that exists to administer company benefits. A person who is single has no spouse and presumably does not have an entry in the EMP_SPOUSE table. When querying for all persons covered by company benefits, an **inner** join between EMPLOYEE and EMP_SPOUSE would incorrectly exclude any employee who doesn't have a spouse. So you'd need a LEFT JOIN using EMPLOYEE and EMP_SPOUSE to return all insured employees plus their spouses. What I am saying is: your data model will govern what type of join is needed, so be very familiar with it.

UNION and INTERSECT

Another way to combine the results from two or more tables (or in some complex cases, to combine different result sets from a single table) is to use the UNION and INTERSECT statements. In some cases this can be preferable to doing a join.

Union

The UNION predicate combines the result sets from multiple SELECT queries. To understand how this might be useful, let's look at three examples. First, let's say we have two companies that have merged to form a third company. We have two tables EMP_COMPA and EMP_COMPB that we have structured with an EMP_ID, EMP_LAST_NAME and EMP_FIRST_NAME. We are going to structure a third table EMPLOYEE_NEW which will combine all the employees from both companies, and it will auto-generate new employee ids.

The DDL for the new table looks like this:

```
CREATE TABLE HRSCHEMA.EMPLOYEE_NEW(
EMP_ID INT GENERATED ALWAYS AS IDENTITY,
EMP_OLD_ID INTEGER,
EMP_LAST_NAME VARCHAR(30) NOT NULL,
```

```
EMP_FIRST_NAME VARCHAR(20) NOT NULL)
IN TSHR;
```

Now we can load the table using a UNION as follows:

```
INSERT INTO
HRSCHEMA.EMPLOYEE_NEW

SELECT EMP_ID,
EMP_LAST_NAME,
EMP_FIRST_NAME
FROM HRSCHEMA.EMP_COMPA
UNION
SELECT EMP_ID,
EMP_LAST_NAME,
EMP_FIRST_NAME
FROM HRSCHEMA.EMP_COMPB;
```

This will load the new table with data from both the old tables, and the new employee numbers will be auto-generated. Notice also we also keep the old employee numbers for cross reference if needed.

When using a UNION, the number of columns and data types must be the same for each SELECT statement. But the column names need not be the same. The UNION operation looks at the columns in the queries by position, not by name.

Let's look at two other examples of UNION actions. First, recall that earlier we used a full outer join to return all employee ids, including those that exist in one table but not the other.

```
SELECT A.EMP_ID,
B.EMP_ID,
A.EMP_LAST_NAME,
B.EMP_SEMIMTH_PAY
FROM HRSCHEMA.EMPLOYEE A
   FULL OUTER JOIN
      HRSCHEMA.EMP_PAY_CHECK B
         ON A.EMP_ID = B.EMP_ID;
```

If we just needed a unique list of employee id numbers from the `EMPLOYEE` and `EMP_PAY_CHECK` tables, we could use a UNION:

```
SELECT EMP_ID
FROM HRSCHEMA.EMPLOYEE
UNION
SELECT EMP_ID
FROM HRSCHEMA.EMP_PAY_CHECK;
```

```
EMP_ID
------
  1122
  3217
  3333
  4720
  6288
  7033
  7459
  7777
  9134
```

If you are wondering why we didn't get duplicate employee numbers in our list, it is because the UNION statement automatically eliminates duplicates. If for some reason you need to retain the duplicates, you would need to specify UNION ALL.

One final example will show how handy the UNION predicate is. Suppose that you want to query the `EMPLOYEE` table to get a list of all employee names for an upcoming company party. But you also have a contractor who (by business rules) cannot be in the `EMPLOYEE` table. You still want to include the contractor's name in the result set for whom to invite to the party. Let's say you want to identify the contractor with a pseudo-employee-id of 9999, and the contractor's name is Janet Ko.

You could code the query as follows:

```
SELECT EMP_ID,
EMP_LAST_NAME,
EMP_FIRST_NAME
FROM HRSCHEMA.EMPLOYEE
UNION
SELECT 9999,
'KO',
'JANET'
FROM SYSIBM.SYSDUMMY1;

EMP_ID EMP_LAST_NAME  EMP_FIRST_NAME
------ -------------  --------------
  1122 JENKINS        DEBORAH
  3217 JOHNSON        EDWARD
  3333 FORD           JAMES
  4720 SCHULTZ        TIM
  6288 WILLARD        JOE
  7459 STEWART        BETTY
  7777 HARRIS         ELISA
  9134 FRANKLIN       BRIANNA
  9999 KO             JANET
```

Now you have listed all the employees plus your contractor friend Janet on your query results.

This is a useful technique when you have a "mostly" table driven system that also has some exceptions to the business rules. Sometimes a system has one-off situations that simply don't justify full blown changes to the system design. UNION can help in these cases.

Intersect

The INTERSECT predicate returns a combined result set that consists of all of the rows existing in **both** result sets. In one of the earlier UNION examples, we wanted all employee ids as long as they existed in **either** the `EMPLOYEE` table or the `EMP_PAY_CHECK` table.

```
SELECT EMP_ID
FROM HRSCHEMA.EMPLOYEE
UNION
SELECT EMP_ID
FROM HRSCHEMA.EMP_PAY_CHECK;

EMP_ID
------
  1122
  3217
  3333
  4720
  6288
  7033
  7459
  7777
  9134
```

Now let's say we only want a list of employee ids that appear in both tables. The INTERSECT will accomplish that for us and we only need to change that one word in the query:

```
SELECT EMP_ID
FROM HRSCHEMA.EMPLOYEE
INTERSECT
SELECT EMP_ID
FROM HRSCHEMA.EMP_PAY_CHECK;

EMP_ID
------
  3217
  4720
  6288
  7459
  9134
```

Common Table Expression

A common table expression is a result set that you can create and then reference in a query as though it were a table. It sometimes makes coding easier. For example, suppose we need to work with an aggregated total pay for each employee. Recall that our table named `EMP_PAY_`

HIST includes these fields:

```
(EMP_ID INTEGER NOT NULL,
EMP_PAY_DATE DATE NOT NULL,
EMP_PAY_AMT DECIMAL (8,2) NOT NULL);
```

Assume further that we have created the following SQL that includes aggregated totals for the employees' pay. And we want to join this aggregated total data with the EMPLOYEE table to produce a final result set. Here's our SQL:

```
WITH EMP_PAY_SUM (EMP_ID, EMP_PAY_TOTAL) AS
(SELECT EMP_ID,
SUM(EMP_PAY_AMT)
AS EMP_PAY_TOTAL
FROM EMP_PAY_HIST
GROUP BY EMP_ID)

SELECT B.EMP_ID,
A.EMP_LAST_NAME,
A.EMP_FIRST_NAME,
B.EMP_PAY_TOTAL
FROM EMPLOYEE A
INNER JOIN
EMP_PAY_SUM B
ON A.EMP_ID = B.EMP_ID;
```

What we've done is to create a temporary result set named EMP_PAY_SUM that can be queried by SQL as if it were a table. This helps break down the data requirement into two pieces, one of which summarizes the pay data and the other of which adds columns from other tables to create a joined result set.

The above example may not seem like much because you could have as easily combined the two queries into one. But as your data stores get more numerous, and your queries and joins grow more complex, you may find that common table expressions can simplify queries both for you and for the developer that follows you.

Here's the result of our common table expression and the query against it.

```
WITH EMP_PAY_SUM (EMP_ID, EMP_PAY_TOTAL) AS
(SELECT EMP_ID,
SUM(EMP_PAY_AMT)
AS EMP_PAY_TOTAL
FROM HRSCHEMA.EMP_PAY_HIST
GROUP BY EMP_ID)

SELECT B.EMP_ID,
```

```
A.EMP_LAST_NAME,
A.EMP_FIRST_NAME, B.EMP_PAY_TOTAL
FROM HRSCHEMA.EMPLOYEE A
INNER JOIN
EMP_PAY_SUM B
ON A.EMP_ID = B.EMP_ID;

EMP_ID EMP_LAST_NAME  EMP_FIRST_NAME      EMP_PAY_TOTAL
------ -------------  --------------      -------------
  3217 JOHNSON        EDWARD                   10833.32
  7459 STEWART        BETTY                    14166.64
  4720 SCHULTZ        TIM                      13333.32
  6288 WILLARD        JOE                      11666.64
  9134 FRANKLIN       BRIANNA                  12500.00
```

This concludes our discussion of the SELECT statement.

XML

XML is a highly used standard for exchanging self-describing data files or documents. Even if you work in a shop that does not use the DB2 XML data type or XML functions, it is good to know how to use these. A complete tutorial on XML is well beyond the scope of this book. We'll review some XML basics, but if you have little or no experience with XML, I strongly suggest that you purchase some books to acquire this knowledge. The following are a few that can help fill in the basics:

XML in a Nutshell, Third Edition 3rd Edition by Elliotte Rusty Harold (ISBN 978-0596007645)

XSLT 2.0 and XPath 2.0 Programmer's Reference by Michael Kay (ISBN: 978-0470192740)

XQuery: Search Across a Variety of XML Data by Priscilla Walmsley
(ISBN: ISBN-13: 978-1491915103)

Basic XML Concepts

You may know that XML stands for Extensible Markup Language. XML technology is cross-platform and independent of machine and software. It provides a structure that consists of both data and data element tags, and so it describes the data in both human readable and machine readable format. The tag names for the elements are defined by the developer/user of the data.

XML Structure

XML has a tree type structure that is required to begin with a root element and then it expands to branches which are called nodes. To continue in the EMPLOYEE domain, let's take a simple XML example with an employee profile as the root. We'll include the employee id, the address and birth date. The XML document might look like this:

```
<?xml version="1.0" encoding="UTF-8"?>
<EMP_PROFILE>
        <EMP_ID>4175</EMP_ID>
        <EMP_ADDRESS>
                <STREET>6161 MARGARET LANE</STREET>
                <CITY>ERINDALE</CITY>
                <STATE>AR</STATE>
                <ZIP_CODE>72653</ZIP_CODE>
        </EMP_ADDRESS>
        <BIRTH_DATE>07/14/1991</BIRTH_DATE>
</EMP_PROFILE>
```

XML documents frequently begin with a declaration that includes the XML version and the encoding scheme of the document. In our example, we are using XML version 1.0 which is still very common. This declaration is optional but it's a best practice to include it.

Notice after the version specification that we continue with the tag name EMP_PROFILE enclosed between <> symbols. The employee profile element ends with /EMP_PROFILE enclosed between <> symbols. Similarly each sub-element is tagged and enclosed and the value (if any) appears between the opening and closing of the element.

XML documents must have a single root element, i.e., one element that is the root of all other elements. If you want more than one `EMP_PROFILE` in a single document, then you would need a higher level element to contain the profiles. For example you could have a DEPARTMENT element that contains employee profiles, and a COMPANY element that contains DEPARTMENTS.

All elements must have a closing tag. Elements that are not populated can be represented by an opening and closing with nothing in between. For example, if an employee's birthday is not known, it can be represented by either <BIRTH_DATE></BIRTH_DATE> or you can use the short hand form <BIRTH_DATE/>.

The example document includes elements such as the employee id, address and birth date. The address is broken down into a street name, city, state and zip code. Comments can be included in an XML document by following the following format:

```
<!--  This is a sample comment  -->
```

By default, white space is preserved in XML documents.

Ok, so we've given you a drive-thru version of XML. We have almost enough information to move on to how to manipulate XML data in DB2. Before we get to that, let's briefly look at two XML-related technologies that we will need.

XML Related Technologies

XPath

The extensible path language (XPath) is used to locate and extract information from an XML document using "path" expressions through the XML nodes. For example, in the case of the employee XML document we created earlier, you could locate and return a zip code value by specifying the path.

Recall this structure:

```
<EMP_PROFILE>
        <EMP_ID>4175</EMP_ID>
        <EMP_ADDRESS>
                <STREET>6161 MARGARET LANE</STREET>
                <CITY>ERINDALE</CITY>
                <STATE>AR</STATE>
                <ZIP_CODE>72653</ZIP_CODE>
        </EMP_ADDRESS>
        <BIRTH_DATE>07/14/1991</BIRTH_DATE>
</EMP_PROFILE>
```

In this example, the employee profile nodes with zip code 72653 can be identified using the following path:

```
/EMP_PROFILE/ADDRESS[ZIP_CODE=72653]
```

The XPath expression for all employees who live in Texas as follows:

```
/EMP_PROFILE/ADDRESS[STATE="TX"]
```

XQuery

XQuery enables us to query XML data using XPath expressions. It is similar to how we query relational data using SQL, but of course the syntax is different. Here's an example of pulling the employee id of every employee who lives at a zip code greater than 90000 from an XML document named **employees.xml**.

```
for $x in doc("employees.xml")employee/profile/address/zipcode
where $x/zipcode>90000
order by $x/zipcode
return $x/empid
```

In DB2 you run an XQuery using the built-in function **XMLQUERY**. We'll show you some examples using XMLQUERY shortly.

DB2 Support for XML

DB2 LUW includes an XML data type and many built-in DB2 functions to validate, traverse and manipulate XML data. The DB2 XML data type can store well-formed XML documents in their hierarchical form and retrieve entire documents or portions of documents.

You can execute DML operations such as inserting, updating and deleting XML documents. You can index and create triggers on XML columns. Finally, you can extract data items from an XML document and then store those values in columns of relational tables using the SQL XMLTABLE built-in function.

XML Examples

XML for the EMPLOYEE table

Suppose that we need to implement a new interface with our employee benefits providers who use XML as the data exchange format. This could give us a reason to store our detailed employee information in an XML structure within the `EMPLOYEE` table. For our purposes, we will add a column named `EMP_PROFILE` to the `EMPLOYEE` table and make it an XML column. Here's the DDL:

```
ALTER TABLE HRSCHEMA.EMPLOYEE
ADD COLUMN EMP_PROFILE XML;
```

We could also establish an XML schema to validate our data structure, but for the moment we'll just deal with the basic SQL operations. As long as the XML is well formed, DB2 will accept it without a schema to validate against.

Let's assume we are going to add a record to the `EMPLOYEE` table for employee Fred Turnbull who has employee id 4175, has 1 year if service and was promoted on 12/1/2016.

Here's the XML document structure we want for storing his employee profile, and the data is included as follows:

```
<EMP_PROFILE>
        <EMP_ID>4175</EMP_ID>
        <EMP_ADDRESS>
                <STREET>6161 MARGARET LANE</STREET>
                <CITY>ERINDALE</CITY>
                <STATE>AR</STATE>
                <ZIP_CODE>72653</ZIP_CODE>
        </EMP_ADDRESS>
        <BIRTH_DATE>07/14/1991</BIRTH_DATE>
</EMP_PROFILE>
```

INSERT With XML

Now we can insert the new record as follows:

```
INSERT INTO HRSCHEMA.EMPLOYEE
(EMP_ID,
 EMP_LAST_NAME,
 EMP_FIRST_NAME,
 EMP_SERVICE_YEARS,
 EMP_PROMOTION_DATE,
 EMP_PROFILE)
VALUES (4175,
'TURNBULL',
'FRED',
1,
'12/01/2016',
'
<EMP_PROFILE>
        <EMP_ID>4175</EMP_ID>
        <EMP_ADDRESS>
                <STREET>6161 MARGARET LANE</STREET>
                <CITY>ERINDALE</CITY>
                <STATE>AR</STATE>
                <ZIP_CODE>72653</ZIP_CODE>
        </EMP_ADDRESS>
        <BIRTH_DATE>07/14/1991</BIRTH_DATE>
</EMP_PROFILE>
');

Updated 1 rows.
```

SELECT With XML

You can do a SELECT on an XML column and display the content of the record. Since the XML data is stored as one long string, it may be difficult to read in its entirety without reformatting. Let's do a query to see what we have.

```
SELECT EMP_ID, EMP_PROFILE FROM HRSCHEMA.EMPLOYEE
WHERE EMP_ID = 4175;

 EMP_ID        EMP_PROFILE
 ------        -------------------------------------------------------------

 4175          <EMP_PROFILE><EMP_ID>4175</EMP_ID><EMP_ADDRESS><STREET>6161  MARGARET  LANE</
               STREET><CITY>ERINDALE</CITY><STATE>AR</STATE><ZIP_CODE>72653</ZIP_CODE></
               EMP_ADDRESS><BIRTH_DATE>07/14/1991</BIRTH_DATE></EMP_PROFILE>
```

UPDATE With XML

To update an XML column you can use standard SQL if you want to update the entire content of the column. This SQL will do it:

```
UPDATE HRSCHEMA.EMPLOYEE
SET EMP_PROFILE
 =     '<EMP_PROFILE>
        <EMP_ID>3217</EMP_ID>
        <EMP_ADDRESS>
               <STREET>2913 PATE DR</STREET>
               <CITY>FORT WORTH</CITY>
               <STATE>TX</STATE>
               <ZIP_CODE>76105</ZIP_CODE>
        </EMP_ADDRESS>
        <BIRTH_DATE>03/15/1952</BIRTH_DATE>
        </EMP_PROFILE> '

WHERE EMP_ID = 3217;

Updated 1 rows.
```

DELETE With XML

If you wish to delete the entire `EMP_PROFILE` document, you can set it to NULL as follows:

```
UPDATE HRSCHEMA.EMPLOYEE
SET EMP_PROFILE = NULL
WHERE EMP_ID = 3217;

Updated 1 rows.

SELECT EMP_ID, EMP_PROFILE FROM HRSCHEMA.EMPLOYEE
WHERE EMP_ID = 3217;

 EMP_ID        EMP_PROFILE
 ------        -----------
   3217        NULL
```

As you can see, the `EMP_PROFILE` column has been set to NULL. At this point, only one row in the `EMPLOYEE` table has the `EMP_PROFILE` populated.

```
SELECT EMP_ID, EMP_PROFILE FROM HRSCHEMA.EMPLOYEE;

 EMP_ID EMP_PROFILE
 ------      ----------------------------------------------------------
   3217         NULL
   7459         NULL
   3333         NULL
   4720         NULL
   6288         NULL
   1122         NULL
   9134         NULL
```

```
7777          NULL
4175          <EMP_PROFILE><EMP_ID>4175</EMP_ID><EMP_ADDRESS><STREET>6161       MARGARET
              LANE</STREET><CITY>ERINDALE</CITY><STATE>AR</STATE><ZIP_CODE>72653</ZIP_
              CODE></EMP_ADDRESS><BIRTH_DATE>07/14/1991</BIRTH_DATE></EMP_PROFILE>
```

Let's go ahead and add the XML data back to this record so we can use it later for other XML queries.

```
UPDATE HRSCHEMA.EMPLOYEE
SET EMP_PROFILE
 = '<EMP_PROFILE>
       <EMP_ID>3217</EMP_ID>
       <EMP_ADDRESS>
              <STREET>2913 PATE DR</STREET>
              <CITY>FORT WORTH</CITY>
              <STATE>TX</STATE>
              <ZIP_CODE>76105</ZIP_CODE>
       </EMP_ADDRESS>
       <BIRTH_DATE>03/15/1952</BIRTH_DATE>
    </EMP_PROFILE>
   '
WHERE EMP_ID = 3217;

Updated 1 rows.
```

Also, let's update one more record so we have a bit more data to work with.

```
UPDATE HRSCHEMA.EMPLOYEE
SET EMP_PROFILE
 = '<EMP_PROFILE>
  <EMP_ID>7459</EMP_ID>
  <EMP_ADDRESS>
       <STREET>6742 OAK ST</STREET>
       <CITY>DALLAS</CITY>
       <STATE>TX</STATE>
       <ZIP_CODE>75277</ZIP_CODE>
  </EMP_ADDRESS>
  <BIRTH_DATE>09/22/1963</BIRTH_DATE>
</EMP_PROFILE>
   '
WHERE EMP_ID = 7459;

Updated 1 rows.
```

XML BUILTIN FUNCTIONS

XMLQUERY

XMLQUERY is the DB2 builtin function that enables you to run XQuery. Here is an example of using XMLQUERY to retrieve an XML element from the `EMP_PROFILE` element. In this case we will select the zip code for employee 4175.

```
SELECT XMLQUERY
('$d/EMP_PROFILE/EMP_ADDRESS/ZIP_CODE'
PASSING EMP_PROFILE as "d") AS ZIP
FROM HRSCHEMA.EMPLOYEE
WHERE EMP_ID = 4175;

ZIP
-------------------------
<ZIP_CODE>72653</ZIP_CODE>
```

Notice that the data is returned in XML format. If you don't want the data returned with its XML structure, simply add the XQuery text() function at the end of the return string, as below:

```
SELECT XMLQUERY
('$d/EMP_PROFILE/EMP_ADDRESS/ZIP_CODE/text()'
PASSING EMP_PROFILE as "d") AS ZIP
FROM HRSCHEMA.EMPLOYEE
WHERE EMP_ID = 4175;
```

The result of this query will not include the XML format.

```
ZIP
-----
72653
```

XMLEXISTS

The XMLEXISTS predicate specifies an XQuery expression. If the XQuery expression returns an empty sequence, the value of the XMLEXISTS predicate is false. Otherwise, XMLEXISTS returns true and those rows matching the XMLEXISTS value of true are returned.

XMLEXISTS enables us to specify rows based on the XML content which is often what you want to do. Suppose you want to return the first and last names of all employees who live in the state of Texas? This query with XMLEXISTS would accomplish it:

```
SELECT EMP_LAST_NAME, EMP_FIRST_NAME
FROM HRSCHEMA.EMPLOYEE
WHERE
XMLEXISTS('$info/EMP_PROFILE[EMP_ADDRESS/STATE/text()="TX"]'
PASSING EMP_PROFILE AS "info");

EMP_LAST_NAME                  EMP_FIRST_NAME
-------------                  --------------
JOHNSON                        EDWARD
STEWART                        BETTY
```

You can also use XMLEXISTS with update and delete functions.

XMLSERIALIZE

The XMLSERIALIZE function returns a serialized XML value of the specified data type that is generated from the first argument. You can use this function to generate an XML structure from relational data. Here's an example.

```
SELECT E.EMP_ID,
XMLSERIALIZE(XMLELEMENT ( NAME "EMP_FULL_NAME",
   E.EMP_FIRST_NAME || ' ' || E.EMP_LAST_NAME)
                AS CLOB(100)) AS "RESULT"
      FROM HRSCHEMA.EMPLOYEE E;
```

```
 EMP_ID        RESULT
 ------        ---------------------------------------------
   3217        <EMP_FULL_NAME>EDWARD JOHNSON</EMP_FULL_NAME>
   7459        <EMP_FULL_NAME>BETTY STEWART</EMP_FULL_NAME>
   3333        <EMP_FULL_NAME>JAMES FORD</EMP_FULL_NAME>
   4720        <EMP_FULL_NAME>TIM SCHULTZ</EMP_FULL_NAME>
   6288        <EMP_FULL_NAME>JOE WILLARD</EMP_FULL_NAME>
   1122        <EMP_FULL_NAME>DEBORAH JENKINS</EMP_FULL_NAME>
   9134        <EMP_FULL_NAME>BRIANNA FRANKLIN</EMP_FULL_NAME>
   7777        <EMP_FULL_NAME>ELISA HARRIS</EMP_FULL_NAME>
   4175        <EMP_FULL_NAME>FRED TURNBULL</EMP_FULL_NAME>
```

XMLTABLE

This function can be used to convert XML data to relational data. You can then use it for traditional SQL such as in joins. To use XMLTABLE you must specify the relational column names you want to use. Then you point these column names to the XML content using path expressions. For this example we'll pull address information from the profile:

```
SELECT X.*
FROM HRSCHEMA.EMPLOYEE,
XMLTABLE ('$x/EMP_PROFILE'
           PASSING EMP_PROFILE as "x"
   COLUMNS
      EMP_ID  INTEGER     PATH 'EMP_ID',
      STREET  VARCHAR(20) PATH 'EMP_ADDRESS/STREET',
      CITY    VARCHAR(20) PATH 'EMP_ADDRESS/CITY',
      STATE   VARCHAR(02) PATH 'EMP_ADDRESS/STATE',
      ZIP     VARCHAR(10) PATH 'EMP_ADDRESS/ZIP_CODE')
             AS X;
```

```
 EMP_ID STREET             CITY       STATE ZIP
 ------ ------------------ ---------- ----- -----
   3217 2913 PATE DR       FORT WORTH TX    76105
   7459 6742 OAK ST        DALLAS     TX    75277
   4175 6161 MARGARET LANE ERINDALE   AR    72653
```

SPECIAL REGISTERS

Special registers allow you to access detailed information about the DB2 instance settings as well as certain session information. CURRENT DATE is an example of a special register that is often used in programming (see example below). The following are SQL examples of some commonly used special registers. I suggest that you focus on these.

CURRENT DATE

CURRENT DATE specifies a date that is based on a reading of the time-of-day clock when the SQL statement is executed at the current server. This is often used in application programs to establish the processing date.

```
SELECT CURRENT DATE
FROM SYSIBM.SYSDUMMY1;

1
----------
2017-09-18
```

CURRENT DEGREE

CURRENT DEGREE specifies the degree of parallelism for the execution of queries that are dynamically prepared by the application process. A value of "ANY" enables parallel processing. A value of 1 prohibits parallel processing. You can query for the value of the CURRENT DEGREE as follows:

```
SELECT CURRENT DEGREE
FROM SYSIBM.SYSDUMMY1;

  1
  -----
  1
```

CURRENT MEMBER

CURRENT MEMBER specifies the member name of a current DB2 data sharing member on which a statement is executing. The value of CURRENT MEMBER is a character string. If your DB2 instance is not part of a data sharing architecture, this value will be blank.

CURRENT SCHEMA

CURRENT SCHEMA specifies the schema name used to qualify unqualified database object references in dynamically prepared SQL statements.

```
SELECT CURRENT SCHEMA
FROM SYSIBM.SYSDUMMY1;
```

```
1
--------
ROBERT
```

CURRENT SERVER

CURRENT SERVER specifies the location name of the current server.

```
SELECT CURRENT SERVER
FROM SYSIBM.SYSDUMMY1;

1
----
DBHR
```

CURRENT TEMPORAL BUSINESS_TIME

The CURRENT TEMPORAL BUSINESS_TIME special register specifies a TIMESTAMP(12) value that is used in the default BUSINESS_TIME period specification for references to application-period temporal tables.

CURRENT TEMPORAL SYSTEM_TIME

The CURRENT TEMPORAL SYSTEM_TIME special register specifies a TIMESTAMP(12) value that is used in the default SYSTEM_TIME period specification for references to system-period temporal tables.

CURRENT TIME

The CURRENT TIME special register specifies a time that is based on a reading of the time-of-day clock when the SQL statement is executed at the current server.

```
SELECT CURRENT TIME
FROM SYSIBM.SYSDUMMY1;

1
--------
10:41:04
```

CURRENT TIMESTAMP

The CURRENT TIMESTAMP special register specifies a timestamp based on the time-of-day clock at the current server.

```
SELECT CURRENT TIMESTAMP
FROM SYSIBM.SYSDUMMY1;
----------------------------
2017-01-13-10.12.51.778225
```

SESSION_USER

`SESSION_USER` specifies the primary authorization ID of the process.

```
SELECT SESSION_USER
FROM SYSIBM.SYSDUMMY1;

1
--------
ROBERT
```

NOTE: You can use all special registers in a user-defined function or a stored procedure. However, you can modify only some of the special registers. The following are the special registers that can be modified:

```
CURRENT DEFAULT TRANSFORM GROUP
CURRENT DEGREE
CURRENT EXPLAIN MODE
CURRENT EXPLAIN SNAPSHOT
CURRENT FEDERATED ASYNCHRONY
CURRENT IMPLICIT XMLPARSE OPTION
CURRENT ISOLATION
CURRENT LOCALE LC_MESSAGES
CURRENT LOCALE LC_TIME
CURRENT LOCK TIMEOUT
CURRENT MAINTAINED TABLE TYPES FOR OPTIMIZATION
CURRENT MDC ROLLOUT MODE
CURRENT OPTIMIZATION PROFILE
CURRENT PACKAGE PATH
CURRENT PATH
CURRENT QUERY OPTIMIZATION
CURRENT REFRESH AGE
CURRENT SCHEMA
CURRENT SQL_CCFLAGS
CURRENT TEMPORAL BUSINESS_TIME
CURRENT TEMPORAL SYSTEM_TIME
SESSION_USER
USER
```

BUILT-IN FUNCTIONS

Built-in functions can be used in SQL statements to return a result based on an argument. These functions are great productivity tools because they can replace custom coded functionality in an application program. Whether your role is application developer, DBA or business services professional, the DB2 built-in functions can save you a great deal of time and effort if you know what they are and how to use them.

There are three types of builtin functions:

1. Aggregate
2. Scalar
3. Table

We'll look at examples of each of these types, and then we'll provide a complete list for your study.

AGGREGATE Functions

An aggregate function receives a set of values for each argument (such as the values of a column) and returns a single-value result for the set of input values. These are especially useful in data analytics. Here are some examples of commonly used aggregate functions.

AVERAGE

The average function returns the average of a set of numbers. Using our EMP_PAY table, you could get the average REGULAR_PAY for your employees like this:

```
SELECT AVG(EMP_REGULAR_PAY)
FROM HRSCHEMA.EMP_PAY;

1
---------------------------------
75000.000000000000000000000000
```

COUNT

The COUNT function returns the number of rows or values in a set of rows or values. Suppose you want to know how many employees you have. You could use this SQL to find out:

```
SELECT COUNT(*) AS EMP_CNT
FROM HRSCHEMA.EMPLOYEE;

EMP_CNT
-------
      9
```

MAX

The MAX function returns the maximum value in a set of values.

MIN

The MIN function returns the minimum value in a set of values.

In the next two examples, we use the MAX and MIN functions to determine the highest and lowest paid employees:

```
SELECT MAX(EMP_REGULAR_PAY)
FROM HRSCHEMA.EMP_PAY;

 1
 --------
 85000.00
```

Now if we want to know both the maximum salary and the employee who earns it, it is a bit more complex, but not much. We just need to add a subquery.

```
SELECT EMP_ID, EMP_REGULAR_PAY
FROM HRSCHEMA.EMP_PAY
WHERE EMP_REGULAR_PAY =
(SELECT MAX(EMP_REGULAR_PAY) FROM HRSCHEMA.EMP_PAY);

 EMP_ID       EMP_REGULAR_PAY
 ------       ---------------
   7459       85000.00
```

Similarly, we can find the minimum using the MIN function.

```
SELECT MIN(EMP_REGULAR_PAY)
FROM HRSCHEMA.EMP_PAY;

 1
 --------
 65000.00

SELECT EMP_ID, EMP_REGULAR_PAY
FROM HRSCHEMA.EMP_PAY
WHERE EMP_REGULAR_PAY =
(SELECT MIN(EMP_REGULAR_PAY) FROM HRSCHEMA.EMP_PAY);


EMP_ID EMP_REGULAR_PAY
------ ---------------
  3217         65000.00
```

SUM

The SUM function returns the sum of a set of numbers. Suppose you need to know what your base payroll will be for the year. You could find out with this SQL:

```
SELECT SUM(EMP_REGULAR_PAY)
FROM HRSCHEMA.EMP_PAY;
```

```
1
---------
375000.00
```

SCALAR Functions

A scalar function can be used wherever an expression can be used. It is often used to calculate a value or to influence the result of a query. Again we'll provide some examples.

COALESCE

The COALESCE function returns the value of the first non-null expression. It is normally used to assign some alternate value when a NULL value is encountered that would otherwise cause an entire record containing the NULL to be excluded from the results. For example, consider the EMP_PAY table with data as follows:

```
SELECT *
FROM HRSCHEMA.EMP_PAY;
EMP_ID  EMP_REGULAR_PAY          EMP_BONUS_PAY
------  ---------------          -------------
  3217         65000.00                        5500.00
  7459         85000.00                        4500.00
  9134         75000.00                        2500.00
  6288         70000.00                        2000.00
  4720         80000.00                        2500.00
```

To demonstrate how COALESCE works, let's first change the bonus pay amount for employee 9134 to NULL.

```
UPDATE HRSCHEMA.EMP_PAY
SET EMP_BONUS_PAY = NULL
WHERE EMP_ID = 9134;

Updated 1 rows.
```

Now our data looks like this:

```
SELECT *
FROM HRSCHEMA.EMP_PAY;

EMP_ID  EMP_REGULAR_PAY         EMP_BONUS_PAY
------  ---------------         -------------
  3217         65000.00              5500.00
  7459         85000.00              4500.00
  9134         75000.00                 NULL
  6288         70000.00              2000.00
  4720         80000.00              2500.00
```

Ok, here's the example. Let's find the average bonus pay in the EMP_PAY table.

```
SELECT AVG(EMP_BONUS_PAY)
AS AVERAGE_BONUS
FROM HRSCHEMA.EMP_PAY;

AVERAGE_BONUS
------------------------------
3625.0000000000000000000000000
```

If we look at this result and do a bit of arithmetic, we can see there is a problem here! The problem is that the average bonus is not 3625, it is 2900 (total 14,500 bonus pay divided by five employees). The problem here is that one of the employee records has NULL in the EMP_BONUS_PAY column. Consequently this record was excluded from the calculated average because NULL is not a numeric value and therefore cannot be included in a numeric computation.

Assuming that you do want to include this record in your results to get the correct average, what we need is to convert the NULL to numeric value zero. You can do this using the COALESCE function.

```
SELECT AVG(COALESCE(EMP_BONUS_PAY,0))
AS AVERAGE_BONUS
FROM HRSCHEMA.EMP_PAY;

AVERAGE_BONUS
-------------------------
2900.00000000000000000000
```

The above says calculate the average EMP_BONUS_PAY using the first non-null value of EMP_BONUS_PAY or zero. Since employee 9134 has a NULL value in the EMP_BONUS_PAY field, DB2 skips over that value and substitutes a zero instead. Zero is a numeric value, so this record can now be included in the computation of the average. This gives the correct average which is 2900.

Before we move on let's reset the bonus pay on our employee 9134 so that it can be used correctly for other queries later in the text book.

```
UPDATE HRSCHEMA.EMP_PAY SET EMP_BONUS_PAY = 2500.00
WHERE EMP_ID = 9134;

Updated 1 rows.
```

You can use COALESCE anytime you need to include a record that would otherwise be excluded due to a NULL value. Converting the NULL to an actual value will ensure the record can be included in the results.

CONCAT

The CONCAT function combines two or more strings. Suppose for example you want to list each employee's first and last names from the EMPLOYEE table. You could so it with this SQL:

```
SELECT
CONCAT(CONCAT(EMP_FIRST_NAME,' '),EMP_LAST_NAME)
AS EMP_FULL_NAME
FROM HRSCHEMA.EMPLOYEE;

EMP_FULL_NAME
----------------
EDWARD JOHNSON
BETTY STEWART
JAMES FORD
TIM SCHULTZ
JOE WILLARD
DEBORAH JENKINS
BRIANNA FRANKLIN
ELISA HARRIS
FRED TURNBULL
```

LCASE

The LCASE function returns a string in which all the characters are converted to lowercase characters. I can't think of many good applications for this, but here is an example of formatting the last name of each employee to lower case. Note: this function does not change any value on the table, it is only formatting the value for presentation.

```
SELECT EMP_ID, LCASE(EMP_LAST_NAME)
FROM HRSCHEMA.EMPLOYEE;

EMP_ID       2
------       --------
  3217       johnson
  7459       stewart
  3333       ford
  4720       schultz
  6288       willard
  1122       jenkins
  9134       franklin
  7777       harris
  4175       turnbull
```

LEFT

The LEFT function returns a string that consists of the specified number of leftmost bytes of the specified string units. Suppose you have an application that needs the first four letters of the last name (my pharmacy does this as part of the automated prescription filling process). You could accomplish that with this SQL:

```
SELECT EMP_ID, LEFT(EMP_LAST_NAME,4) AS NAME_FIRST_4
FROM HRSCHEMA.EMPLOYEE;

 EMP_ID        NAME_FIRST_4
 ------        ------------
   3217        JOHN
   7459        STEW
   3333        FORD
   4720        SCHU
   6288        WILL
   1122        JENK
   9134        FRAN
   7777        HARR
   4175        TURN
```

MAX

The MAX function returns the maximum value in a set of values. For example if we wanted to know the largest base pay for our `EMP_PAY` table, we could use this SQL:

```
SELECT MAX(EMP_REGULAR_PAY)
AS HIGHEST_PAY
FROM HRSCHEMA.EMP_PAY;

HIGHEST_PAY
-----------
   85000.00
```

MIN

The MIN scalar function returns the minimum value in a set of values. For example if we wanted to know the largest base pay for our `EMP_PAY` table, we could use this SQL:

```
SELECT MIN(EMP_REGULAR_PAY)
AS LOWEST_PAY
FROM HRSCHEMA.EMP_PAY;

LOWEST_PAY
----------
  65000.00
```

MONTH

The MONTH function returns the month part of a date value. We used this one earlier to compare the month of the employee's promotion to the current month. Assume that the query is run on January 19.

```
SELECT
EMP_ID,
EMP_PROMOTION_DATE,
CURRENT DATE AS RQST_DATE
FROM HRSCHEMA.EMPLOYEE
WHERE MONTH(EMP_PROMOTION_DATE)
 = MONTH(CURRENT DATE);

EMP_ID       EMP_PROMOTION_DATE    RQST_DATE
 ------       ------------------    ----------
   3217              2017-01-01    01/19/2017
   4720              2017-01-01    01/19/2017
   6288              2016-01-01    01/19/2017
```

REPEAT

The REPEAT function returns a character string that is composed of an argument that is repeated a specified number of times. Suppose for example that you wanted to display 10 asterisks as a literal field on a report. You could specify it this way:

```
SELECT
EMP_ID,
REPEAT('*',10) AS "FILLER LITERAL",
EMP_SERVICE_YEARS
FROM HRSCHEMA.EMPLOYEE;

EMP_ID   FILLER LITERAL     EMP_SERVICE_YEARS
------   --------------     -----------------
  3217        **********                    4
  7459        **********                    7
  3333        **********                    7
  4720        **********                    9
  6288        **********                    6
  1122        **********                    5
  9134        **********                    0
  7777        **********                    2
  4175        **********                    1
```

SPACE

The SPACE function returns a character string that consists of the number of blanks that the argument specifies. You could use this in place of the quotation literals (especially when you want a lot of spaces). The example I'll give uses the SPACE function instead of having to concatenate an empty string using quotation marks.

```
SELECT
CONCAT(CONCAT(EMP_FIRST_NAME,SPACE(1)),
EMP_LAST_NAME)
AS EMP_FULL_NAME
```

```
FROM HRSCHEMA.EMPLOYEE;

EMP_FULL_NAME
----------------
EDWARD JOHNSON
BETTY STEWART
JAMES FORD
TIM SCHULTZ
JOE WILLARD
DEBORAH JENKINS
BRIANNA FRANKLIN
ELISA HARRIS
FRED TURNBULL
```

SUBSTR

The SUBSTR function returns a substring of a string. Let's use the earlier example of retrieving the first four letters of the last name via the LEFT function. You could also accomplish that with this SQL:

```
SELECT EMP_ID, SUBSTR(EMP_LAST_NAME,1,4)
FROM HRSCHEMA.EMPLOYEE;

EMP_ID       2
------       ----
  3217       JOHN
  7459       STEW
  3333       FORD
  4720       SCHU
  6288       WILL
  1122       JENK
  9134       FRAN
  7777       HARR
  4175       TURN
```

The 1,4 means starting in position one for a length of four. Of course, you could use a different starting position. An example that might make more sense is reformatting the current date. For example:

```
SELECT CURRENT DATE,
SUBSTR(CHAR(CURRENT DATE),6,2)
|| '/'
||SUBSTR(CHAR(CURRENT DATE),9,2)
|| '/'
|| SUBSTR(CHAR(CURRENT DATE),1,4)
AS REFORMED_DATE FROM SYSIBM.SYSDUMMY1;

 1             REFORMED_DATE
 ----------    -------------
 2017-09-18          09/18/2017
```

UCASE

The UCASE function returns a string in which all the characters are converted to uppercase characters. Here is an example of changing the last name of each employee to upper case. First we will have to covert the uppercase EMP_LAST_NAME values to lowercase. We can do that using the LOWER function. Let's do this for a single row:

```
UPDATE HRSCHEMA.EMPLOYEE
SET EMP_LAST_NAME
= LOWER(EMP_LAST_NAME)
WHERE EMP_ID = 3217;

Updated 1 rows.
```

We can verify that the data did in fact get changed to lower case.

```
SELECT EMP_LAST_NAME
FROM HRSCHEMA.EMPLOYEE
WHERE EMP_ID = 3217;


EMP_LAST_NAME
-------------
johnson
```

Now let's use the UCASE function to have the EMP_LAST_NAME display as upper case.

```
SELECT EMP_ID, UCASE(EMP_LAST_NAME)
FROM HRSCHEMA.EMPLOYEE
WHERE EMP_ID = 3217;

EMP_ID      2
------      -------
  3217      JOHNSON
```

Note that the SELECT query did not change any data on the table. We have simply reformatted the data for presentation. Now let's actually convert the data on the record back to upper case:

```
UPDATE HRSCHEMA.EMPLOYEE
SET EMP_LAST_NAME = UPPER(EMP_LAST_NAME)
WHERE EMP_ID = 3217;

Updated 1 rows.
```

And we'll verify that it reverted back to uppercase:

```
SELECT EMP_LAST_NAME
FROM HRSCHEMA.EMPLOYEE
WHERE EMP_ID = 3217;

EMP_LAST_NAME
-------------
JOHNSON
```

YEAR

The YEAR function returns the year part of a value that is a character or graphic string. The value must be a valid string representation of a date or timestamp.

```
SELECT CURRENT DATE AS TODAYS_DATE,
YEAR(CURRENT DATE) AS CURRENT_YEAR
FROM SYSIBM.SYSDUMMY1;

TODAYS_DATE   CURRENT_YEAR
-----------   ------------
2017-09-18    2017
```

Chapter Three Questions

1. Suppose you issue this INSERT statement:

```
UPDATE HRSCHEMA.EMPLOYEE
SET EMP_PROFILE
=(
'<?xml version="1.0"?>
<EMP_PROFILE>
<EMP_ID>4175</EMP_ID>
<EMP_ADDRESS>
<STREET>6161 MARGARET LANE</STREET>
<CITY>ERINDALE</CITY>
<STATE>AR</STATE>
<ZIP_CODE>72653</ZIP_CODE>
</EMP_ADDRESS>
<BIRTH_DATE>07/14/1991</BIRTH_DATE>
<EMP_PROFILE>
' )
WHERE EMP_ID = 4175;
```

What will the result be?

   a. An error – the specified XML version is incorrect
   b. An error – the document is not well formed
   c. The record will be inserted with a warning that it has not been validated
   d. The record will be inserted successfully

2. Which of the following is NOT a special register in DB2 11?

   a. CURRENT DEGREE
   b. CURRENT RULES
   c. CURRENT ISOLATION
   d. All of the above are valid DB2 special registers.

3. Which built-in routine returns the value of the first non-null expression?

   a. COALESCE
   b. ABS
   c. CEILING
   d. MIN

4. If you want the aggregate total for a set of values, which function would you use?

    a. COUNT
    b. SUM
    c. ABS
    d. CEIL

5. If you want to return the first 3 characters of a 10-character column, which function would you use?

    a. SUBSTR
    b. LTRIM
    c. DIGITS
    d. ABS

6. Assuming all referenced onjects are valid, what will the result of this statement be?

```
DELETE FROM EMPLOYEE;
```

    a. The statement will fail because there is no WHERE clause.
    b. The statement will fail because you must specify DELETE * .
    c. The statement will succeed and all rows in the table will be deleted.
    d. The statement will run but no rows in the table will be deleted.

Chapter Three Exercises

1. Write a query to display the last and first names of all employees in the `EMPLOYEE` table. Display the names in alphabetic order by `EMP_LAST_NAME`.

2. Write a query to change the first name of Edward Johnson (employee 3217) to Eddie.

3. Write a query to produce the number of employees in the `EMPLOYEE` table.

CHAPTER FOUR: DATA CONTROL LANGUAGE

You don't need to be an expert on DB2 security unless your company puts you in charge of security. However, it is important to understand the basics of how DB2 security works.

Authorities and privileges are the domain of DBAs and system administrators. As such I won't be going into these. However, as an application developer you may need to know the GRANT and REVOKE statements, as well as something about roles and trusted contexts.

Data Control Language Statements

Data Control Language (DCL) is used to grant and revoke privileges. You basically just need to know the syntax of these statements, plus a few options.

GRANT

The basic syntax of the GRANT statement is:

```
GRANT <PRIVILEGE> ON <OBJECT> TO <USER OR GROUP>
```

Let's look at some examples for table privileges. Suppose you want to grant the SELECT privilege on table `EMPLOYEE` to user USER01. You would issue the following DCL:

```
GRANT SELECT ON HRSCHEMA.EMPLOYEE TO USER01;
```

If you wanted to grant multiple privileges (such as SELECT, INSERT, UPDATE, DELETE), you would use this DCL:

```
GRANT SELECT, INSERT, UPDATE, DELETE
ON HRSCHEMA.EMPLOYEE TO USER01;
```

If you wanted to grant the SELECT privilege to all users, you would grant the privilege to PUBLIC:

```
GRANT SELECT ON HRSCHEMA.EMPLOYEE TO PUBLIC;
```

Keep in mind that in a well secured environment, granting privileges to PUBLIC is usually not what you want, and it is not a best practice. Create appropriate profiles and/or roles, and grant access based on those.

If you want to grant all table privileges to a user or group, use this syntax:

```
GRANT ALL ON HRSCHEMA.EMPLOYEE TO USER01;
```

Finally, if you want the recipient of the grant to be able to grant the same privilege to others, use the `WITH GRANT OPTION` clause:

```
GRANT SELECT ON HRSCHEMA.EMPLOYEE
TO USER01
WITH GRANT OPTION;
```

Although the basic structure of a GRANT is the same for most objects, the privileges and keywords are somewhat different.

- Collection
- Database
- Distinct type
- Function or stored procedure
- Package
- Plan
- Schema
- Sequence
- System
- Table or view
- Distinct type, array type, or JAR file
- Variable
- Use

We'll show an example of each of the above after we look at the REVOKE statement.

REVOKE

The revoke statement removes a privilege from a user or role. There are basically two forms of the REVOKE. One removes the specified privilege only from the specified user or role. However if REVOKE is used with the `BY` <userid/role> and the `INCLUDING DEPENDENT PRIVILEGES` clause, then in addition to revoking the access from the user/role specified in the

FROM clause, all access that was explicitly granted to other users by the specified user/role will also be automatically revoked.

The basic syntax of the REVOKE statement is:

```
REVOKE <PRIVILEGE> FROM <user/role>
```

For example, you could revoke the INSERT, UPDATE and DELETE privileges on `EMPLOYEE` from user `USER01` and role `HRUSER` as follows:

```
REVOKE INSERT, UPDATE, DELETE
ON HRSCHEMA.EMPLOYEE
FROM USER USER01, ROLE HRUSER;
```

Or you could revoke all table privileges on `EMPLOYEE` from PUBLIC as follows:

```
REVOKE ALL
ON HRSCHEMA.EMPLOYEE
FROM PUBLIC;
```

DCL Examples

COLLECTION

```
Syntax is:    GRANT CREATE ON COLLECTION <collection-id> TO <USER/ROLE>

              GRANT PACKADM ON COLLECTION <collection-id> TO <USER/ROLE>

Example:      GRANT CREATE ON COLLECTION HRDATA TO HRUSER01
```

DATABASE

```
Syntax is:    GRANT [authority or privilege] ON DATABASE <database name>
              TO <USER/ROLE>

Example:      GRANT CREATETAB ON DATABASE HRDB TO HRUSER01
```

FUNCTION/STORED PROCEDURE

```
Syntax is:    GRANT EXECUTE ON FUNCTION  <FUNCTION>  TO <USER/ROLE>

              GRANT EXECUTE ON PROCEDURE <PROCEDURE> TO <USER/ROLE>

Examples:     GRANT EXECUTE ON FUNCTION EMP_SALARY TO HRUSER09;

              GRANT EXECUTE ON PROCEDURE GET_EMP_INFO TO ROLE HRUSERS;
```

PACKAGE

```
Syntax is:    GRANT <privilege> ON PACKAGE <package name> TO <USER/ROLE>

Example:      GRANT EXECUTE ON PACKAGE COBEMP1 TO ROLE HRUSERS;
```

PLAN

```
Syntax is:    GRANT <privilege> ON PLAN <plan name> TO <USER/ROLE>

Example:      GRANT EXECUTE ON PLAN COBEMP1 TO ROLE HRUSERS;

              GRANT BIND,EXECUTE ON PLAN COBEMP4 TO DBA001;
```

SCHEMA

```
Syntax is:    GRANT <privilege> ON SCHEMA <schema name> TO <USER/ROLE>

Example:      GRANT CREATEIN ON SCHEMA HRDATA TO ROLE HR_DEVELOPERS;
```

SEQUENCE

```
Syntax is:    GRANT <privilege> ON SEQUENCE <sequence name> TO <USER/ROLE>

Example:      GRANT USAGE ON SEQUENCE EMPL_ID__SEQ TO PUBLIC;
```

TABLE OR VIEW

```
Syntax is:    GRANT <privilege> ON TABLE/VIEW <table/view name> TO <USER/ROLE>

Example:      GRANT ALTER ON TABLE EMPLOYEE TO DBA001

              GRANT SELECT,INSERT,UPDATE,DELETE ON EMPLOYEE TO USER01
```

USE

```
Syntax is:    GRANT USAGE ON OBJECT <Object name> TO <USER/ROLE>

Example:      GRANT USAGE ON SEQUENCE EMPL_ID__SEQ TO PUBLIC;
```

VIEWS

A view is a classic way to restrict a subset of data columns to a specific set of users who are allowed to see or manipulate those columns. If you are going to use views for security you must make sure that users do not have direct access to the base tables, i.e. that they can only access the data via view(s). Otherwise they could circumvent the view restrictions by accessing the base table.

Let's look back at our EMPLOYEE table. Suppose we add a column for the employee's Social Security number. That is obviously a very private piece of information that not everyone should see. Our business rule will be that users HRUSER01, HRUSER02 and HRUSER99 are the only ones who should be able to view Social Security numbers. All other users and/or groups are not allowed to see the content of this column, but they can access all the other columns.

We might implement this as follows:

```
ALTER TABLE HRSCHEMA.EMPLOYEE
ADD COLUMN EMP_SSN CHAR(09);
```

Now let's update the value of this column for employee 3217:

```
UPDATE HRSCHEMA.EMPLOYEE
SET EMP_SSN = '238297536'
WHERE EMP_ID = 3217;
```

Now let's create two views, one of which includes the EMP_SSN column. The other view will not include the EMP_SSN column.

```
CREATE VIEW HRSCHEMA.EMPLOYEE_ALL
AS SELECT
EMP_ID,
EMP_LAST_NAME,
EMP_FIRST_NAME,
EMP_SERVICE_YEARS,
EMP_PROMOTION_DATE,
EMP_PROFILE
FROM HRSCHEMA.EMPLOYEE;

CREATE VIEW HRSCHEMA.EMPLOYEE_HR
AS SELECT
EMP_ID,
EMP_LAST_NAME,
EMP_FIRST_NAME,
EMP_SERVICE_YEARS,
EMP_PROMOTION_DATE,
EMP_PROFILE,
EMP_SSN
FROM HRSCHEMA.EMPLOYEE;
```

Finally, issue the appropriate grants.

```
GRANT SELECT ON HRSCHEMA.EMPLOYEE_ALL TO PUBLIC;

GRANT SELECT on HRSCHEMA.EMPLOYEE_HR
TO HRUSER01, HRUSER02, HRUSER99;
```

At this point, assuming we are only accessing data through views, the three HR users are the only users able to access the EMP_SSN column. Other users cannot access the EMP_SSN column because it is not included in the EMPLOYEE_ALL view that they have access to. To prove this:

```
SELECT EMP_ID, EMP_SSN
FROM HRSCHEMA.EMPLOYEE_ALL
WHERE EMP_ID = 3217;

"EMP_SSN" is not valid in the context where it is used.. SQLCODE=-206,
 SQLSTATE=42703, DRIVER=4.18.60
```

If you are one of the HR users, you will be able to access the EMP_SSN column via the other view, EMPLOYEE_HR:

```
SELECT EMP_ID, EMP_SSN
FROM HRSCHEMA.EMPLOYEE_HR
WHERE EMP_ID = 3217;

EMP_ID        EMP_SSN
------        ---------
3217          238297536
```

Roles

A role is a database object upon which database privileges can be granted and revoked. Often roles are created to simplify maintenance of privileges for a group of people (users, developers, administrators, etc).

Let's do an example. Assume we want to create a role HRUSER to which we will grant privileges to the role as needed. We will then assign the role to system users as needed. Here's how to create the HRUSER role:

```
CREATE ROLE HRUSER;
```

That's it. Very simple. Next let's grant SELECT access on our special view EMPLOYEE_HR to the HRUSER role that we just created.

```
GRANT SELECT ON HRSCHEMA.EMPLOYEE_HR TO HRUSER;
```

Finally, we can grant the HRUSER role to specific userids for the HR personnel:

```
GRANT ROLE HRUSER TO HRUSER01, HRUSER02, HRUSER99;
```

Now everyone in the `HRUSER` group has SELECT access to the `EMPLOYEE_HR` view, and that role is assigned to `HRUSER01, HRUSER02` and `HRUSER99`. Powerfully, we can add or revoke other privileges for the HRUSER role at any time, and the change takes place immediately for all users who have been granted that role.

Trusted Contexts

A *trusted context* is a database object that defines a secure connection based on a system authorization id and connection attributes such as an IP address. Trusted contexts are typically used to establish a trust relationship between DB2 and some middleware product such as a web service or an application. A trusted context is associated with a single system authorization ID.

The point of a trusted context is to enable an entity such as a web service to establish a connection to the database that can be used by that service without the service users themselves having to logon to the DB2 server. It also prevents the connection from being used from any location except the specified IP address. Finally, the authorized id can be granted appropriate privileges to operate on the database via use of a role or roles.

Now that we understand the basics, let's do an example. Earlier we created a role named HRUSER that allowed user ids assigned that role to view the Social Security numbers of employees. We can further require that the `HRUSER` access DB2 using a trusted context from a work location. Let's say we want the users to logon via a service application. Assume that the work server IP for the application is `202.033.178.267`, and that we have a DB2 authorization id `HRADM` to be used by our service.

To define and implement our trusted context, we need to grant the role `HRUSER` to the system authorization id that the service app will be using to login to DB2 with, specifically `HRADM`. We can do that with this GRANT statement:

```
GRANT ROLE HRUSER TO HRADM;
```

Now we can grant create the trusted context for our application, specifying the IP address, authorization logon id and a default role of `HRUSER`.

```
CREATE TRUSTED CONTEXT HRCTXT

BASED UPON CONNECTION USING SYSTEM AUTHID HRADM
```

```
ATTRIBUTES (ADDRESS '192.168.1.98')

DEFAULT ROLE HRUSER

ENABLE
```

At this point, the service application can logon to DB2 with the HRADM id and use the privileges granted to the role HRUSER, provided that the connection is made from the specified server (IP address or domain name) defined in the trusted context, which in this case is 192.168.1.98.

NOTE: If you provide an invalid IP address, the CREATE statement will fail.

Other examples of trusted contexts:

```
CREATE TRUSTED CONTEXT GGX1
BASED UPON CONNECTION USING SYSTEM AUTHID DBA001
ATTRIBUTES (ADDRESS '10.41.667.389',
            ENCRYPTION 'LOW')
DEFAULT ROLE GGXROLE
ENABLE
WITH USE FOR BILL, ROBERT ROLE ROLE1 WITH AUTHENTICATION;
```

```
CREATE TRUSTED CONTEXT CTX2

 BASED UPON CONNECTION USING SYSTEM AUTHID DBA002

 ATTRIBUTES (JOBNAME 'ABCPROD')

 DEFAULT ROLE ABCROLE WITH ROLE AS OBJECT OWNER AND QUALIFIER

 ENABLE

 WITH USE FOR SALLY;
```

Chapter Four Questions

1. Assume a table where certain columns contain sensitive data and you don't want all users to see these columns. Some other columns in the table must be made accessible to all users. What type of object could you create to solve this problem?

    a. INDEX
    b. SEQUENCE
    c. VIEW
    d. TRIGGER

2. To grant a privilege to all users of the database, grant the privilege to whom?

    a. ALL
    b. PUBLIC
    c. ANY
    d. DOMAIN

3. Tara wants to grant CONTROL of table TBL1 to Bill, and also allow Bill to grant the same privilege to other users. What clause should Tara use on the GRANT statement?

    a. WITH CONTROL OPTION
    b. WITH GRANT OPTION
    c. WITH USE OPTION
    d. WITH REVOKE OPTION

Chapter Four Exercises

1. Write a DCL statement to grant SELECT access on table `HRSCHEMA.EMPLOYEE` to users HR001 and HR002.

2. Write a DCL statement to revoke SELECT access on table `HRSCHEMA.EMPLOYEE` from user HR002.

3. Write a DCL statement to grant SELECT, INSERT, UPDATE and DELETE access on table `HRSCHEMA.EMP_PAY` to user `HRMGR01`.

CHAPTER FIVE: USING DB2 IN APPLICATION PROGRAMS

Java DB2 Programming

In this section we will look at how to work with DB2 using the Java programming language. Specifically we will learn how to connect to a DB2 database, and how to retrieve and update data from a table.

Sample Java DB2 Program

To work with DB2 in the Java programming language, you must:

1. Establish a Connection
2. Construct either a Statement or a PreparedStatement
3. Execute the Statement or Prepared Statement
4. Construct a ResultSet object to either read or update the query data
5. Handle any Errors

Let's start with a simple example where we will query the `EMPLOYEE` database for the first and last name of employee 3217. For this first example we will show creating the Java program in Eclipse Java Oxygen. For subsequent programs we will simply show the source code.

First, open Eclipse Java Oxygen and change to the Java perspective (if not already there). Also click on **Window --> Show View --> Package Explorer**

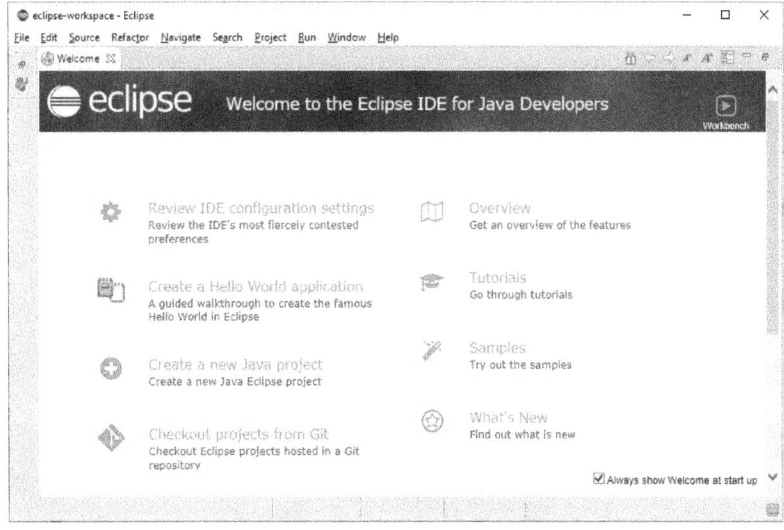

Now click on **File -->New -->Java Project.** Enter "employee" in the Project Name field and click on **Next.**

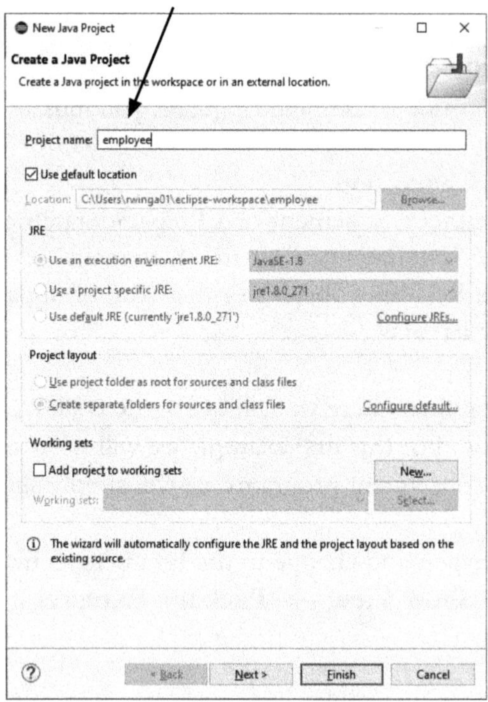

From this panel, click on the **Libraries** tab.

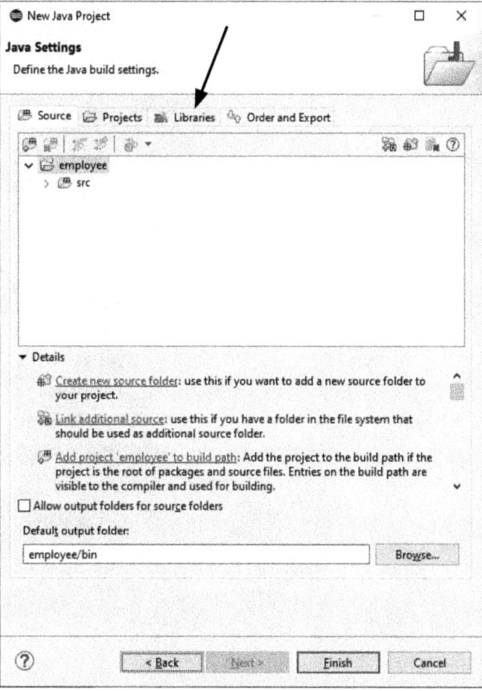

On the next panel you will need to point to the appropriate DB2 connection driver which is in file DB2jcc.jar. Click on **Libraries.** Then click on **Add External JARs**.

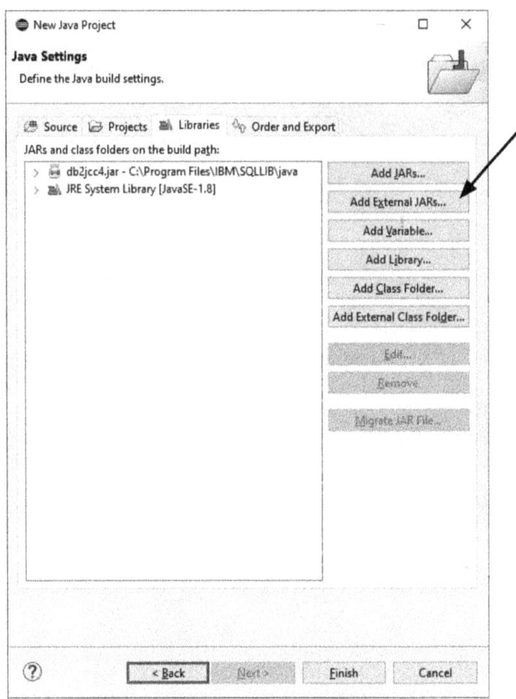

From this panel, navigate to the DB2 directory that contains the db2jcc.jar file. Typically that will be located in Program Files/IBM/SQLLIB/JAVA on your root directory.

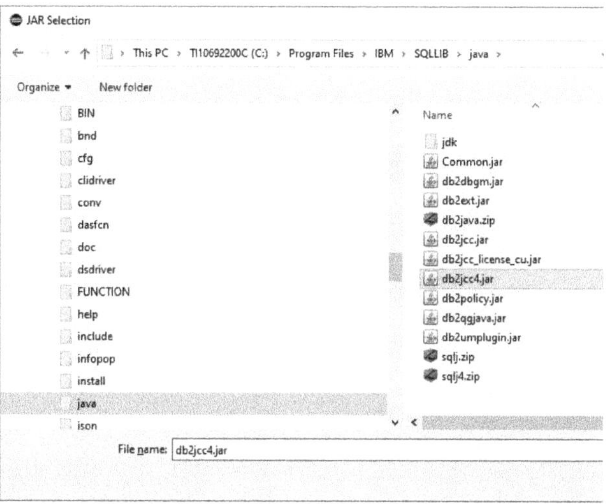

Select the db2jcc.jar file and then click **Open.** This will return you to the **Java Settings** panel. As you can see, the db2jcc.jar file is now added to your project and you should have no problems using the DB2 driver to make a connection. Click **Apply and Close.**

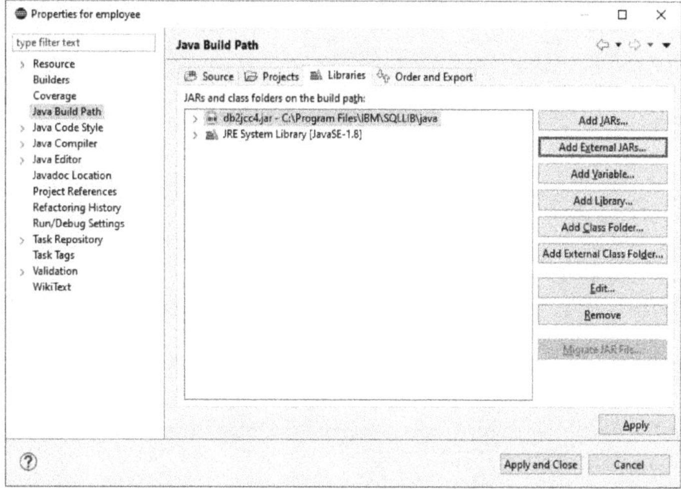

In the Project Explorer, select employee, right click and select **New -->Package**. Type employee, then click **Finish.**

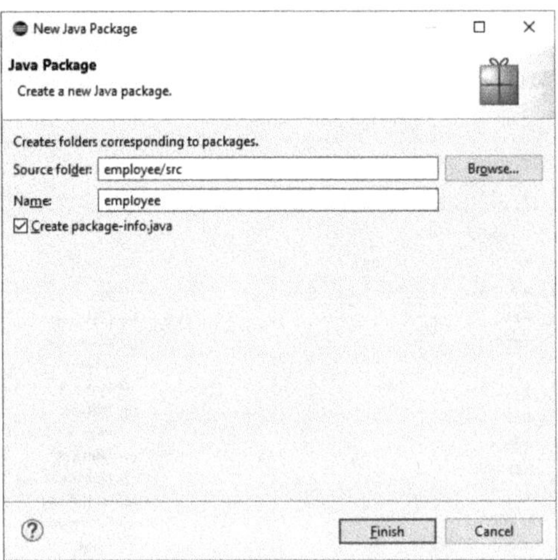

You should now see this structure in your Project Explorer. Right click on the employee package and select **New --> Class**

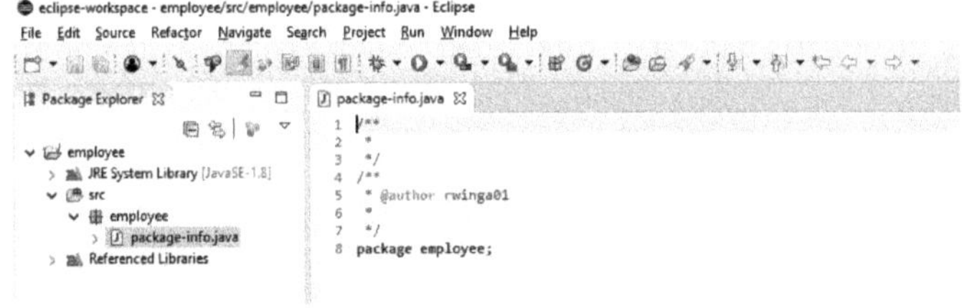

You'll see this panel. Let's name our class **EmpInfo**. Also check the box to generate a stub for a main method. Then click **Finish.**

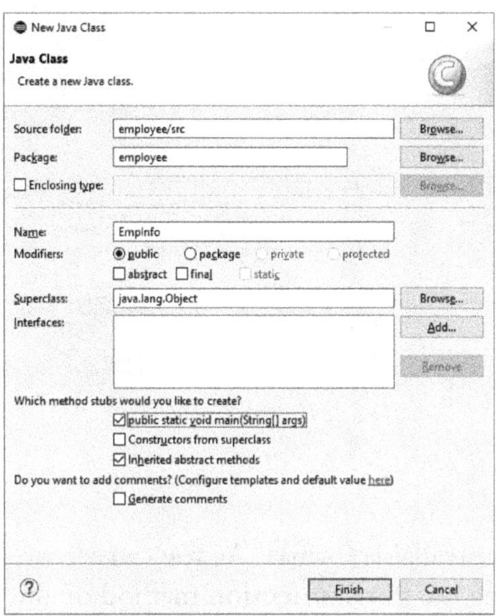

Your structure should look like this, and your new class should appear in the right portion of the screen.

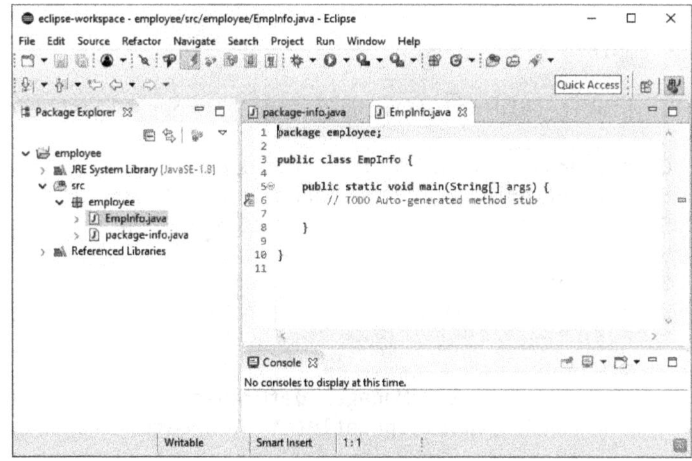

Now let's code the connection. I'm going to give you a screen shot for Eclipse on the next page, and then I am going to show you the code in a listing. I think you'll find the latter easier to read.

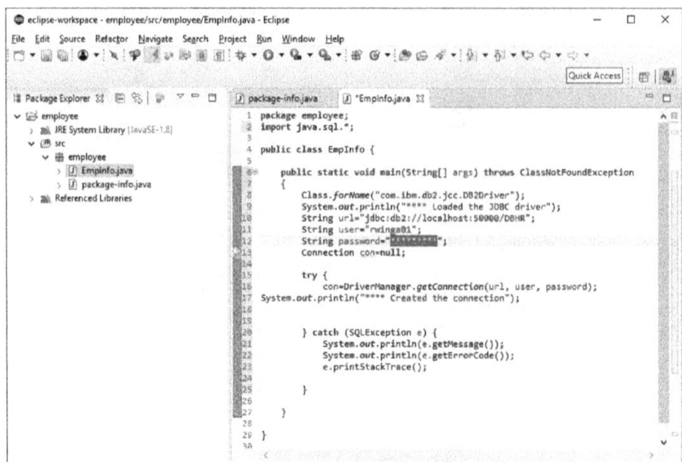

And here is the code in more readable format. As you can see we set up a few variables for the connection, and we are using the **getConnection** method of the connection object to establish a connection. We also coded an error handler to intercept any SQL errors we encounter. Note that I've used my login id and then asterisks for my password (you'll need to supply your own login credentials of course).

```
package employee;
import java.sql.*;

public class EmpInfo {

        public static void main(String[] args) throws ClassNotFoundException
        {
            Class.forName("com.ibm.db2.jcc.DB2Driver");
            System.out.println("**** Loaded the JDBC driver");
            String url="jdbc:db2://localhost:50000/DBHR";
            String user="rwinga01";
            String password="*********";
            Connection con=null;

               try {
                      con=DriverManager.getConnection(url, user, password);
                      System.out.println("**** Created the connection");


               } catch (SQLException e) {
                      System.out.println(e.getMessage());
                      System.out.println(e.getErrorCode());
                      e.printStackTrace();
```

```
                }
            }
    }
```

Let's test our code by running the program. Click on the green forward arrow at the top of the screen. And then check the Console window tab for the result.

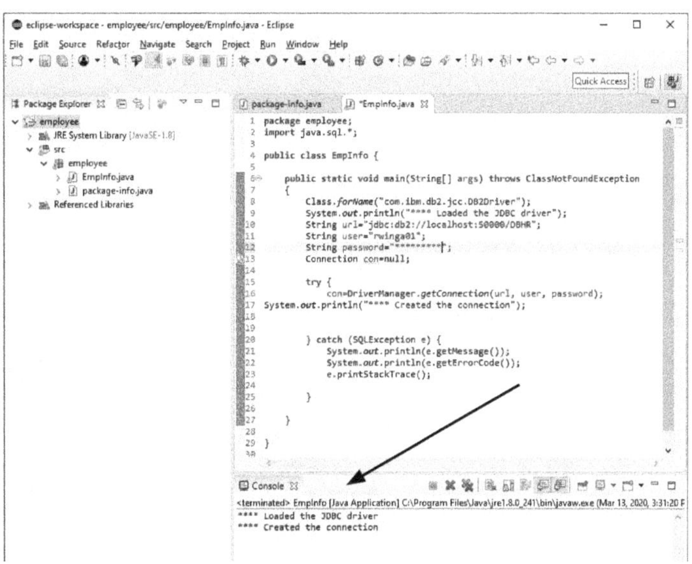

As you can see, our code execution was successful - we created the connection. Of course, that's just the first part of our task. Now we need some SQL to retrieve our `EMPLOYEE` information. To do that, we must create a String object to contain our query, a Statement object to execute the query and a ResultSet object to contain the results from the query.

Here is our current version of the program code with these additional features. Notice also that as we cycle through the result set, we are going to display the employee's last and first name in the console.

```
package employee;
import java.sql.*;

public class EmpInfo {

        public static void main(String[] args) throws ClassNotFoundException
        {
            Class.forName("com.ibm.db2.jcc.DB2Driver");
            System.out.println("**** Loaded the JDBC driver");
            String url="jdbc:db2://localhost:50000/DBHR";
```

```
        String user="rwinga01";
        String password="*********";
        Connection con=null;

          try {
               con=DriverManager.getConnection(url, user, password);
               System.out.println("**** Created the connection");

               String query = "SELECT EMP_LAST_NAME,"
                              + "EMP_FIRST_NAME "
                              + "FROM HRSCHEMA.EMPLOYEE "
                              + "WHERE EMP_ID = 3217";

               Statement stmt;
               stmt = con.createStatement();
               ResultSet rs=stmt.executeQuery(query);

               while(rs.next())
               {
                  String strLastName  = rs.getString(1);
                  String strFirstName = rs.getString(2);
                  System.out.println("Last name is  " + strLastName);
                  System.out.println("First name is " + strFirstName);
               }

          } catch (SQLException e) {
               System.out.println(e.getMessage());
               System.out.println(e.getErrorCode());
               e.printStackTrace();

          }

    }

}
```

Finally, let's look at the console to get the results:

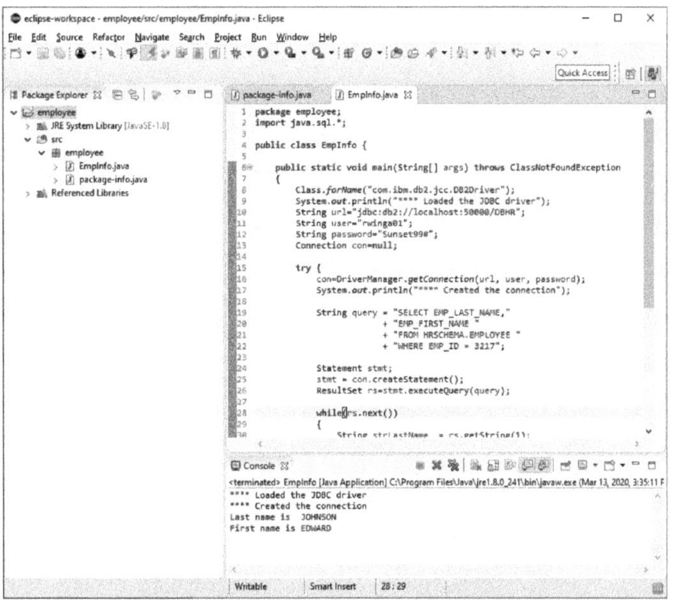

I've also cut and pasted the results here to make it easier to read:

```
**** Loaded the JDBC driver
**** Created the connection
Last name is  JOHNSON
First name is EDWARD
```

To review our example, what the Java code does is to:

1. Create a connection, connect to the `HRDB` database
2. Create a query string to read the `EMPLOYEE` table
3. Execute the query using a Statement object
4. Read the query results using a ResultSet object
5. Load the result set data to string variables
6. Display the data using the println method

Granted this program does not do much, but it did successfully connect to DB2 and retrieved data from it. That's what we wanted to accomplish!

Java DB2 INSERT Program

For the insert program, we only need to change a few things that we used in the query program. Of course we must change the SQL to an INSERT statement. Also, instead of creating a result set, we only need to execute the query with the **executeUpdate** method of the Statement object. Here is our listing.

```
package employee;
import java.sql.*;
public class JAVEMP1 {

       public static void main(String[] args) throws ClassNotFoundException
       {
           Class.forName("com.ibm.db2.jcc.DB2Driver");
           System.out.println("**** Loaded the JDBC driver");
           String url="jdbc:db2://localhost:50000/DBHR";
           String user= "rwinga01";
           String password="*********";

           Connection con=null;

              try {
                     con=DriverManager.getConnection(url, user, password);
                     System.out.println("**** Created the connection");

                     String query = "INSERT INTO HRSCHEMA.EMPLOYEE "
                                  + "(EMP_ID, "
                                  + "EMP_LAST_NAME, "
                                  + "EMP_FIRST_NAME, "
                                  + "EMP_SERVICE_YEARS, "
                                  + "EMP_PROMOTION_DATE) "
                                  + "VALUES (1111, "
                                  + "'VEREEN', "
                                  + "'CHARLES', "
                                  + "12, "
                                  + "'01/01/2017') " ;

                     Statement stmt;
                     stmt = con.createStatement();
                     stmt.executeUpdate(query);
                     System.out.println("Successful INSERT of employee "
                     + 1111);

              } catch (SQLException e) {
                     System.out.println(e.getMessage());
                     System.out.println(e.getErrorCode());
                     e.printStackTrace();

              }

       }
```

```
}
```

When you run this program, you should receive this output in the console window:

```
**** Loaded the JDBC driver
**** Created the connection
Successful INSERT of employee 1111
```

And we can verify the INSERT by checking the table.

```
SELECT * FROM
HRSCHEMA.EMPLOYEE
WHERE EMP_ID = 1111;

EMP_ID EMP_LAST_NAME EMP_FIRST_NAME EMP_SERVICE_YEARS   EMP_PROMOTION_DATE       EMP_PROFILE
------ ------------- -------------- -----------------   ------------------       -----------
  1111 VEREEN        CHARLES                       12   2017-01-01               NULL
```

Note: you could have also omitted the target columns and simply coded the INSERT statement as follows:

```
String query = "INSERT INTO HRSCHEMA.EMPLOYEE "
             + "VALUES (1111, "
             + "'VEREEN', "
             + "'CHARLES', "
             + "12, "
             + "'01/01/2017') " ;
```

Before we move on, let's check the error handling code. Let's try to add the same record again, which should result in an SQL Exception. And in fact we do trap an exception which is a -803 SQL code which means a record already exists with the same key. Here is the result shown in the console:

```
**** Loaded the JDBC driver
**** Created the connection
DB2 SQL Error: SQLCODE=-803, SQLSTATE=23505, SQLERRMC=1;HRSCHEMA.EMPLOYEE, DRIVER=3.72.24
-803
com.ibm.db2.jcc.am.SqlIntegrityConstraintViolationException: DB2 SQL Error: SQLCODE=-803, SQLSTATE=23505, SQLERRMC=1;HRSCHEMA.EMPLOYEE, DRIVER=3.72.24
	at com.ibm.db2.jcc.am.hd.a(hd.java:809)
	at com.ibm.db2.jcc.am.hd.a(hd.java:66)
	at com.ibm.db2.jcc.am.hd.a(hd.java:140)
	at com.ibm.db2.jcc.am.ip.c(ip.java:2788)
	at com.ibm.db2.jcc.am.ip.d(ip.java:2776)
	at com.ibm.db2.jcc.am.ip.b(ip.java:2154)
	at com.ibm.db2.jcc.t4.bb.j(bb.java:233)
	at com.ibm.db2.jcc.t4.bb.c(bb.java:48)
	at com.ibm.db2.jcc.t4.p.b(p.java:38)
```

```
        at com.ibm.db2.jcc.t4.vb.h(vb.java:124)
        at com.ibm.db2.jcc.am.ip.ib(ip.java:2149)
        at com.ibm.db2.jcc.am.ip.a(ip.java:3322)
        at com.ibm.db2.jcc.am.ip.c(ip.java:762)
        at com.ibm.db2.jcc.am.ip.executeUpdate(ip.java:741)
        at employee.JAVEMP1.main(JAVEMP1.java:37)
```

As you can see, we got a -803 SQL code which means a violation of the primary key. To get more information about the error, you could look up the SQL code. But I recommend you get this information automatically by adding a property setting to the connection string which is as follows: **retrieveMessagesFromServerOnGetMessage=true;**

So our connection string setting becomes:

```
jdbc:db2://localhost:50000/DBHR:retrieveMessagesFromServerOnGetMessage=true;"
```

Now let's run the program again, and this time our output is more descriptive:

```
**** Loaded the JDBC driver
**** Created the connection
One or more values in the INSERT statement, UPDATE statement, or foreign key up-
date caused by a DELETE statement are not valid because the primary key, unique
constraint or unique index identified by "1" constrains table "HRSCHEMA.EMPLOYEE"
from having duplicate values for the index key.. SQLCODE=-803, SQLSTATE=23505,
DRIVER=3.72.24
-803
com.ibm.db2.jcc.am.SqlIntegrityConstraintViolationException: One or more values
in the INSERT statement, UPDATE statement, or foreign key update caused by a DE-
LETE statement are not valid because the primary key, unique constraint or unique
index identified by "1" constrains table "HRSCHEMA.EMPLOYEE" from having duplicate
values for the index key.. SQLCODE=-803, SQLSTATE=23505, DRIVER=3.72.24
        at com.ibm.db2.jcc.am.hd.a(hd.java:809)
        at com.ibm.db2.jcc.am.hd.a(hd.java:66)
        at com.ibm.db2.jcc.am.hd.a(hd.java:140)
        at com.ibm.db2.jcc.am.ip.c(ip.java:2788)
        at com.ibm.db2.jcc.am.ip.d(ip.java:2776)
        at com.ibm.db2.jcc.am.ip.b(ip.java:2154)
        at com.ibm.db2.jcc.t4.bb.j(bb.java:233)
        at com.ibm.db2.jcc.t4.bb.c(bb.java:48)
        at com.ibm.db2.jcc.t4.p.b(p.java:38)
        at com.ibm.db2.jcc.t4.vb.h(vb.java:124)
        at com.ibm.db2.jcc.am.ip.ib(ip.java:2149)
        at com.ibm.db2.jcc.am.ip.a(ip.java:3322)
        at com.ibm.db2.jcc.am.ip.c(ip.java:762)
        at com.ibm.db2.jcc.am.ip.executeUpdate(ip.java:741)
        at employee.JAVEMP1.main(JAVEMP1.java:37)
```

This of course saves you the trouble of looking up the error code description. I recommend that you make this technique a coding standard.

Java DB2 INSERT Program with Parameter Markers

When processing multiple records you can use parameter markers with the VALUES clause. A parameter marker is a character (typically the question mark character "?") that serves as a placeholder for an unknown value that will be passed to the SQL using the various set methods of the statement object.

Let's add another record and this time we'll build the SQL using parameter markers. We'll supply the values at run time using the set methods of our Statement object (such as setInt, setString, etc).

```
package employee;
import java.sql.*;
public class JAVEMP1A {

       public static void main(String[] args) throws ClassNotFoundException
       {
           Class.forName("com.ibm.db2.jcc.DB2Driver");
           System.out.println("**** Loaded the JDBC driver");
           String    url="jdbc:db2://localhost:50000/DBHR:retrieveMessagesFromServerOn-
           GetMessage=true;";
           String user="rwinga01";
           String password="*********";

           Connection con=null;

              try {
                     con=DriverManager.getConnection(url, user,
                         password);
                     System.out.println("**** Created the connection");
                     String query = "INSERT INTO HRSCHEMA.EMPLOYEE "
                                   + "VALUES (?, "
                                   + "?, "
                                   + "?, "
                                   + "?, "
                                   + "?, "
                                   + "NULL) " ;
                     PreparedStatement stmt;
                     stmt = con.prepareStatement(query);

                     stmt.setInt(1,1112);
                     stmt.setString(2,"YATES");
                     stmt.setString(3,"JANENE");
                     stmt.setInt(4,7);
                     stmt.setString(5,"01/01/2015");

                     stmt.executeUpdate();

                     System.out.println("Successful INSERT of employee " + 1112);
```

```
        } catch (SQLException e) {
              System.out.println(e.getMessage());
              System.out.println(e.getErrorCode());
              e.printStackTrace();

        }

    }

}
```

And again we successfully inserted an employee record:

```
**** Loaded the JDBC driver
**** Created the connection
Successful INSERT of employee 1112
```

The PreparedStatement object is especially useful when you have multiple records to process and you want to use the same SQL statement and just reset the parameter values. In some respects this works like host variables in embedded SQL programs.

Java DB2 UPDATE Program

Now let's look at a process we performed earlier for changing lowercase characters in the surname to upper case.

```
UPDATE HRSCHEMA.EMPLOYEE
SET EMP_LAST_NAME = LOWER(EMP_LAST_NAME)
WHERE
EMP_LAST_NAME IN ('JOHNSON', 'STEWART', 'FRANKLIN');

SELECT EMP_LAST_NAME FROM HRSCHEMA.EMPLOYEE
WHERE EMP_LAST_NAME <> UPPER(EMP_LAST_NAME);


EMP_LAST_NAME
-------------
johnson
stewart
franklin
```

Now here is our Java program to change lower case to uppercase. We're creating a ResultSet object that contains the records we need to modify. We walk through the result set and move the surname to a string object. We modify the string using the Java **toUpperCase()** string function. Then we use the ResultSet **updateString**() method to replace the `EMP_LAST_NAME` column of the result set, and then we use **updateRow()** to apply the modified record to the table. Finally, we finalize our update records using the **commit()** method of the Connection object.

```
package employee;
import java.sql.*;
public class JAVEMP2 {

public static void main(String[] args) throws ClassNotFoundException
        {
            Class.forName("com.ibm.db2.jcc.DB2Driver");
            System.out.println("**** Loaded the JDBC driver");
            String url
               ="jdbc:db2://localhost:50000/DBHR:retrieveMessagesFromServerOnGetMes-
               sage=true;";
            String user="rwinga01";
            String password="*********";

            Connection con=null;
               try {
                      con=DriverManager.getConnection(url,
                                   user,
                                   password);
                      System.out.println("**** Created the connection");
                      String query = "SELECT EMP_ID, EMP_LAST_NAME "
                      + "FROM HRSCHEMA.EMPLOYEE "
                      + "WHERE EMP_LAST_NAME <> UPPER(EMP_LAST_NAME)" ;

                      Integer intEmpNo;
                      String strLastName;
                      Statement stmt;
                      stmt
                      = con.createStatement(ResultSet.TYPE_SCROLL_SENSITIVE,
                                   ResultSet.CONCUR_UPDATABLE);
                      ResultSet rs=stmt.executeQuery(query);
                      while(rs.next())
                      {
                          intEmpNo     = rs.getInt(1);
                          strLastName  = rs.getString(2);
                          System.out.println("Employee   " + intEmpNo
                          + " BEFORE Last name is: " + strLastName);
                          strLastName  = strLastName.toUpperCase();
                          rs.updateString(2, strLastName);
                          rs.updateRow();
                          System.out.println("Employee " + intEmpNo
                          + "AFTER  Last name is: " + strLastName);
                      }
                      con.commit();

               } catch (SQLException e) {
                      System.out.println(e.getMessage());
                      System.out.println(e.getErrorCode());
                      e.printStackTrace();

               }
```

```
    }
}
```

And this is our output:

```
**** Loaded the JDBC driver
**** Created the connection
Employee 3217 BEFORE Last name is: johnson
Employee 3217 AFTER  Last name is: JOHNSON
Employee 7459 BEFORE Last name is: stewart
Employee 7459 AFTER  Last name is: STEWART
Employee 9134 BEFORE Last name is: franklin
Employee 9134 AFTER  Last name is: FRANKLIN
```

Finally, let's verify that the data was set back to upper case.

```
SELECT EMP_LAST_NAME FROM HRSCHEMA.EMPLOYEE
WHERE EMP_LAST_NAME IN ('JOHNSON', 'STEWART', 'FRANKLIN');

EMP_LAST_NAME
-------------
JOHNSON
STEWART
FRANKLIN
```

The method of using a positioned update based on a ResultSet is something you will use often, particularly when you do not know your result set content beforehand, and anytime you need to examine the content of the record before you perform the update or delete action.

I'll leave the delete and merge statements for chapter exercises. I think you get the general idea of how to build SQL statements and run them in Java using the Connection object, the Statement (or PreparedStatement) object, and the ResultSet object.

.NET DB2 Programming

In this section we will look at how to work with DB2 using the c# .NET programming language. Specifically we will learn how to connect to a DB2 database, and how to retrieve and update data from a table.

Sample .NET DB2 Program

Basic RDMS objects and methods

To work with DB2 using .NET is similar to using Java. The object names and methods are somewhat different though.

1. Establish a Connection
2. Construct a command
3. Construct a DataReader object if you are reading data (otherwise you can use the command object to execute a non-query action).
4. Construct a DataAdaptor object and a DataSet object if you are updating data.
5. Handle any Errors.

There are three types of DB2 data providers using .NET:

1. DB2 .NET Data Provider
2. OleDB .NET Data Provider
3. ODBC .NET Data Provider

We'll mostly use the DB2 .NET data provider, but for our example program we'll show versions using each of these providers. Let's start with the same example we used with Java where we query the `EMPLOYEE` database for the first and last name of an employee. For this first example we will show creating the .NET c# program in Visual Studio 2017 Community edition. For subsequent programs we will simply display the c# source code as I think it is easier for you to read.

Begin by opening Visual Studio 2017. You will see the **Get Started** screen. Click on **Continue without code**.

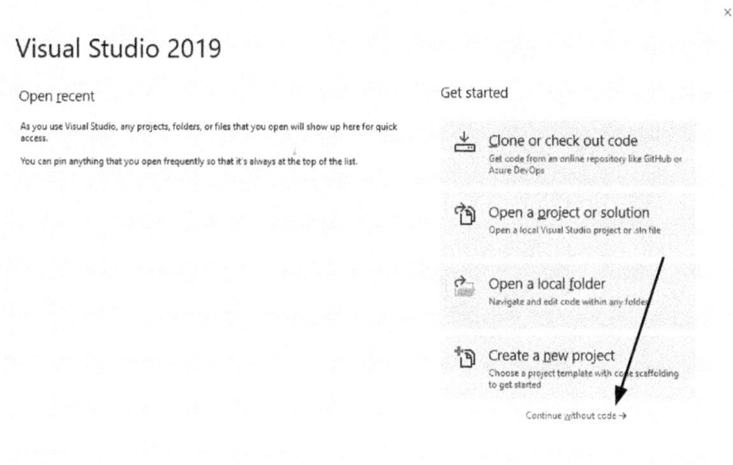

Now you should be in the main IDE. Select **File --> New --> Project.**

Select the type of project. In our case we will select a C# Console App (.NET framework). Then click **Next.**

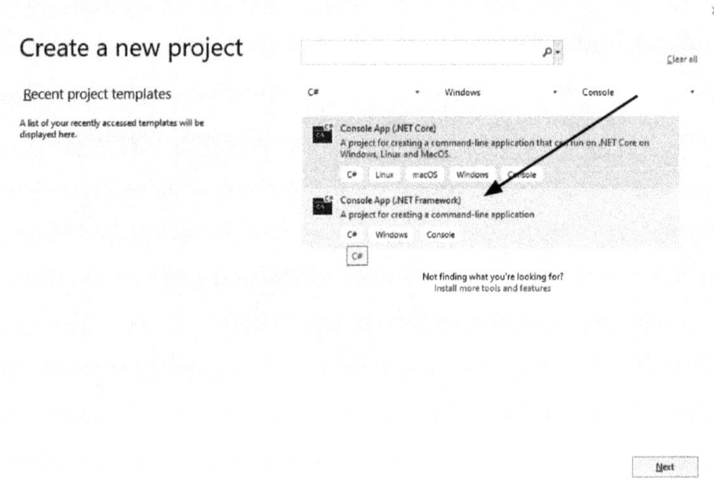

Type the name of your first project into the **Name** field. In my case I will name it `NETEMP0`. Then click **Create**.

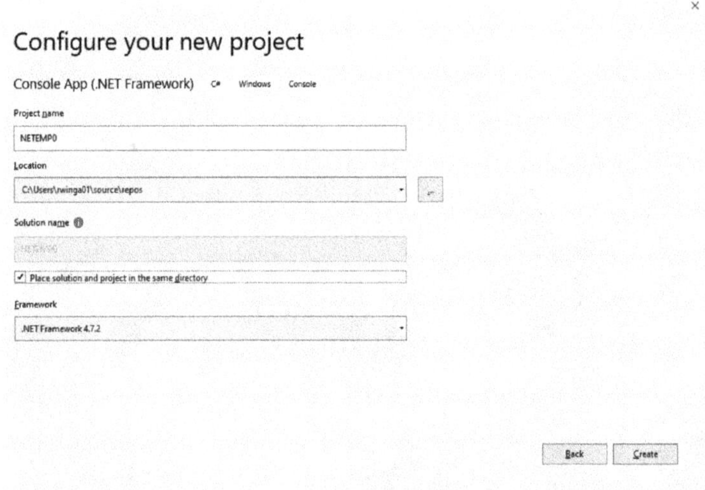

Now you will see this window with a program framework already filled in. First, change the program name in the solution explorer by right clicking it and selecting **rename** – let's use the name `NETEMP0`. When prompted to rename all references to the program to `NETEMP0`, click on **Yes**.

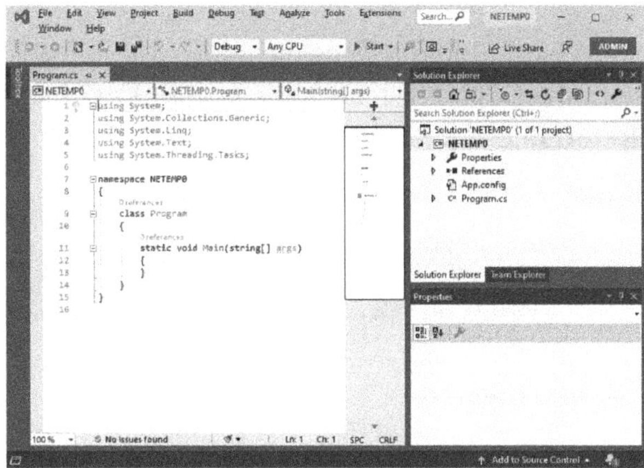

Now all program references are to NETEMP0.

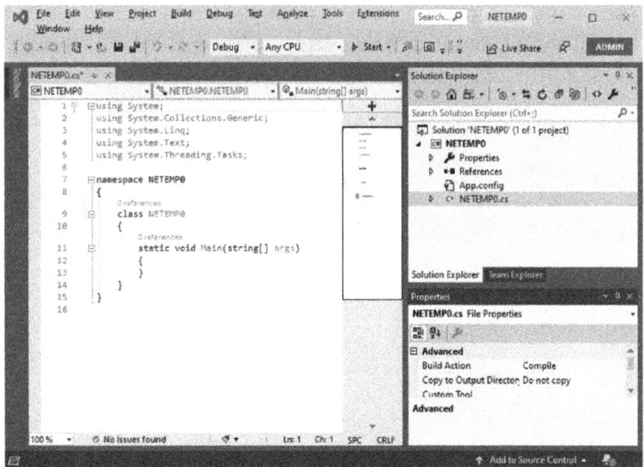

You'll notice the various "Using" entries at the beginning of the program. These are basically package names. We want to use the DB2 .NET data provider, so we must include this line of code:

```
Using IBM.Data.DB2;
```

However, when we add it, Visual Studio tells us the namespace IBM is unknown. To fix this, we must add a reference to the DB2 .NET DLL file in your DB2 installation directory. Click on **Project ->> Add Reference**

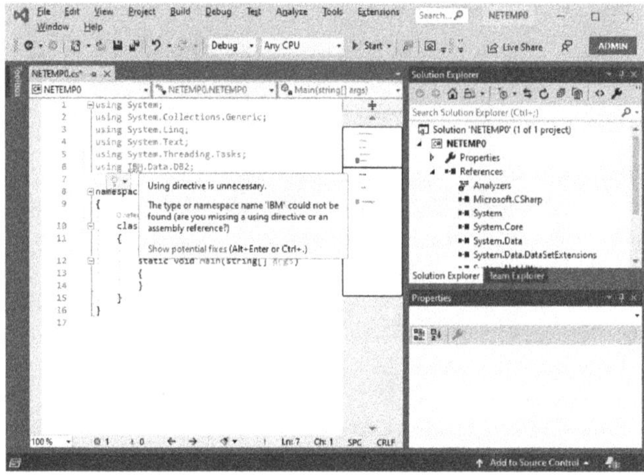

When the Reference Manager window opens, click on **Browse.**

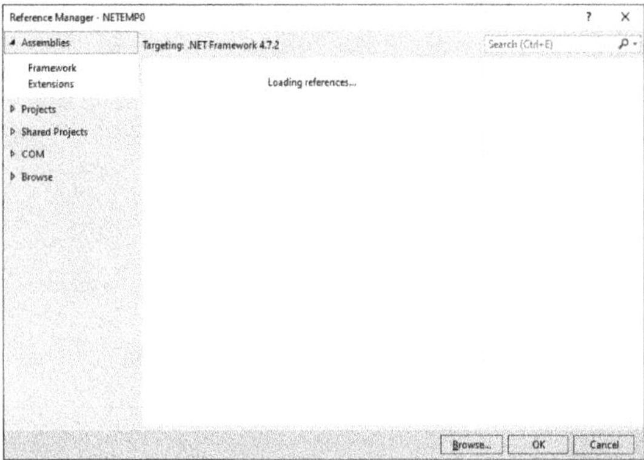

Navigate to your DB2 installation directory which is usually Program Files/IBM/SQLLIB. Go one further step to the BIN/NETF40 directory where you will see IBM.Data.DB2.dll. Click on this file and then click **Add.**

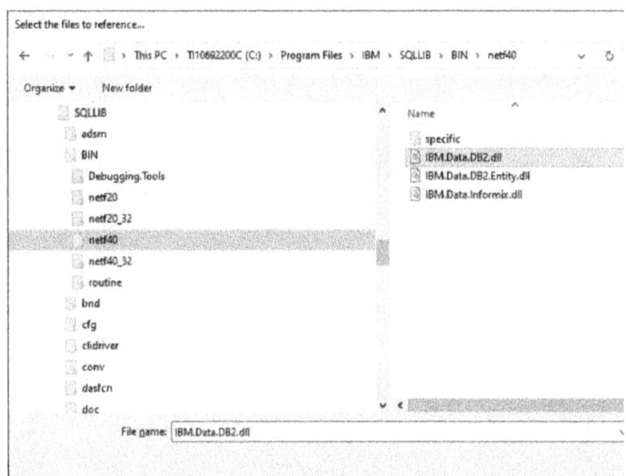

You will now see that the DLL has been added to the project. Click **OK.**

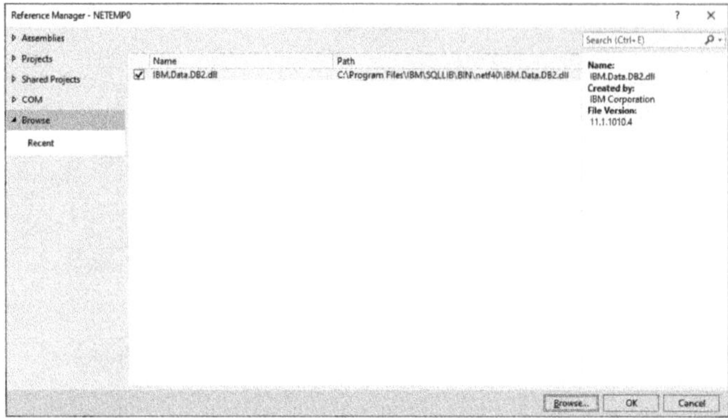

Now you will see there is no longer an error in the program listing.

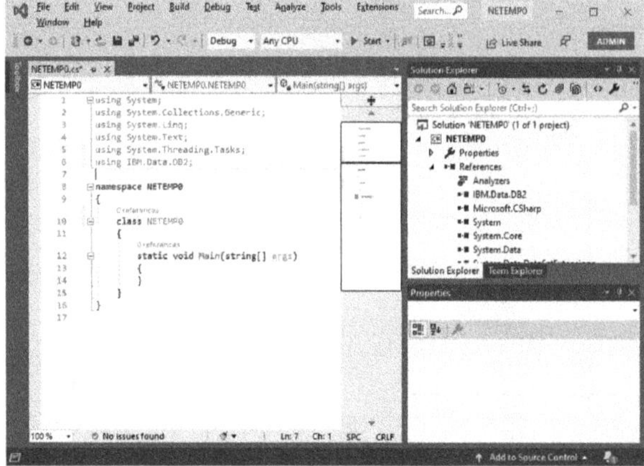

We need to make one more configuration change to the project, which is to change the output type from console application to windows application. If we don't do that, our output will appear in a black console window and disappear after a couple of seconds. We want our output to appear in the Output window in the IDE, and to persist so we can look at it!

So click on **Project – NETEMP0 Properties**. Then change the Output Type from Console Application to Windows Application.

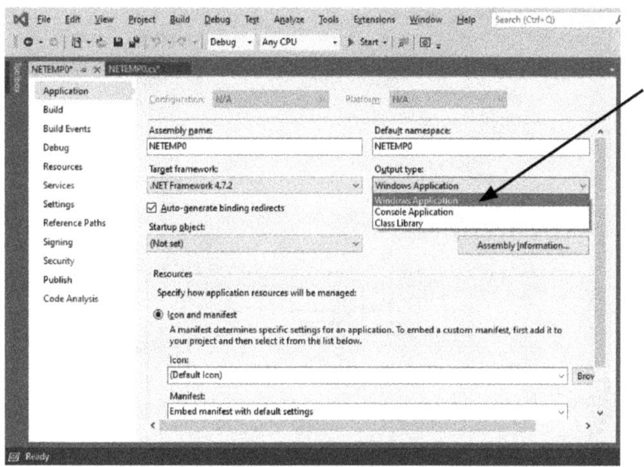

One other change that you may need is the Platform Target. If your version of windows is 64-bit, click on Build, and then change the Platform Target to **x64,** and then close this window.

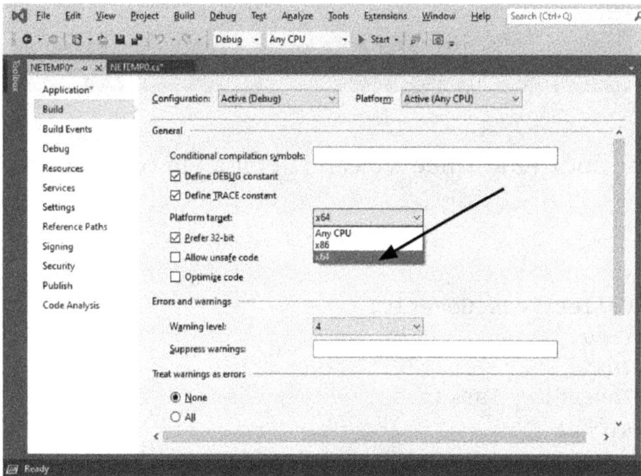

Now let's add the appropriate c# code to create our connection, command and data reader objects. Also we will add code to run the query and display the last and first name of the employee with employee id 3217.

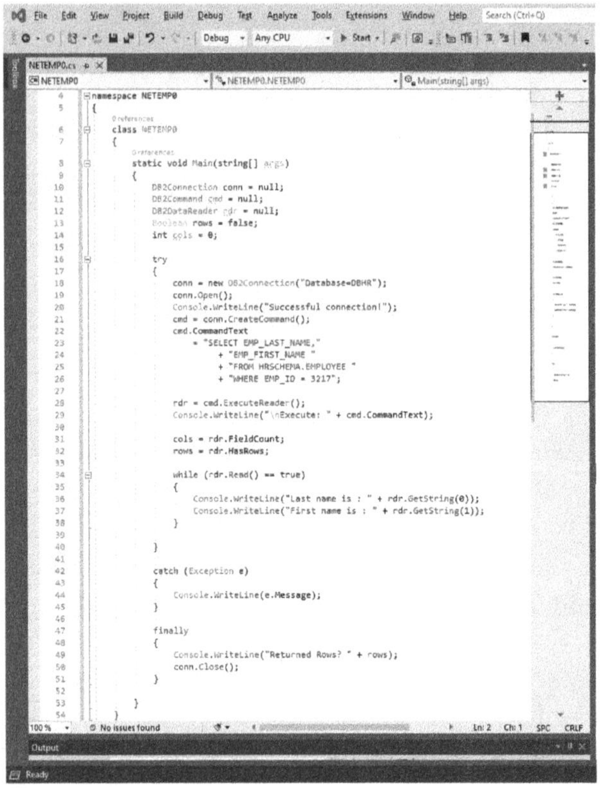

To more easily read the code (and since we can't get it all in one window anyway), I'll give you the entire listing here.

```
using System;
using System.Collections.Generic;
using System.Linq;
using System.Text;
using System.Threading.Tasks;
using IBM.Data.DB2;

namespace NETEMP0
{
    class NETEMP0
    {
        static void Main(string[] args)
        {
```

```
        DB2Connection conn = null;
        DB2Command cmd = null;
        DB2DataReader rdr = null;
        Boolean rows = false;
        int cols = 0;

        try
        {
            conn = new DB2Connection("Database=DBHR");
            conn.Open();
            Console.WriteLine("Successful connection!");
            cmd = conn.CreateCommand();
            cmd.CommandText
                = "SELECT EMP_LAST_NAME,"
                     + "EMP_FIRST_NAME "
                     + "FROM HRSCHEMA.EMPLOYEE "
                     + "WHERE EMP_ID = 3217";

            rdr = cmd.ExecuteReader();
            Console.WriteLine("\nExecute: " + cmd.CommandText);

            cols = rdr.FieldCount;
            rows = rdr.HasRows;

            while (rdr.Read() == true)
            {
                Console.WriteLine("Last name is : " + rdr.GetString(0));
                Console.WriteLine("First name is : " + rdr.GetString(1));
            }

        }

        catch (Exception e)
        {
            Console.WriteLine(e.Message);
        }

        finally
        {
            Console.WriteLine("Returned Rows? " + rows);
            conn.Close();
        }


    }

}


}
```

Now we can execute this program by clicking on the **Start** button (the forward green arrow at the top of the window which also reads "Start". You will see the result of the program execution in the Output window at the bottom:

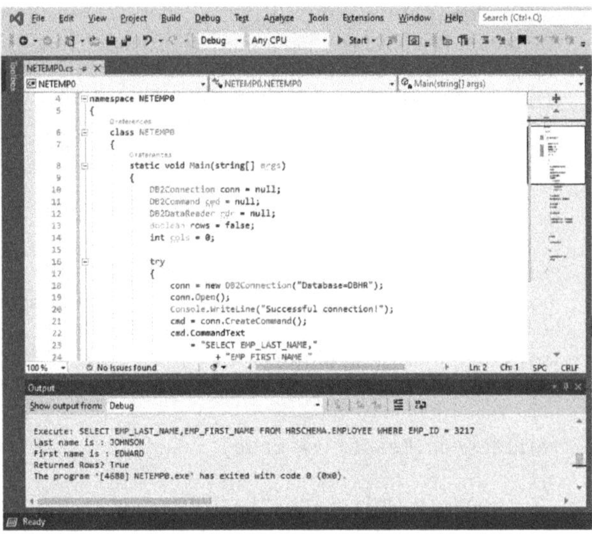

Congratulations, you have just run a .NET program to access and return data from a DB2 database!

Here's the output in easier-to-read format:

```
Execute: SELECT EMP_LAST_NAME,EMP_FIRST_NAME FROM HRSCHEMA.EMPLOYEE WHERE EMP_ID
= 3217
Last name is : JOHNSON
First name is : EDWARD
Returned Rows? True
```

The OLE DB and ODBC interfaces are a bit different in terms of the objects, but more or less the same in terms of the code.

Sample .NET OLE DB Program

For the most part you'll see those objects previously prefixed "DB2" replaced here with objects prefixed with OleDB. Note that in the connection, you must specify `Provider=IBMDADB2`.

```
using System;
using System.Collections.Generic;
using System.Linq;
using System.Text;
using System.Threading.Tasks;
using System.Data.OleDb;

namespace NETOLEDB
{
    class NETOLEDB
    {
        static void Main(string[] args)
        {
            OleDbConnection conn = null;
            OleDbCommand cmd = null;
            OleDbDataReader rdr = null;
            Boolean rows = false;
            int cols = 0;

            try
            {
                conn = new OleDbConnection("Provider=IBMDADB2;DSN=DBHR");
                conn.Open();
                Console.WriteLine("Successful connection!");
                cmd = conn.CreateCommand();
                cmd.CommandText
                    = "SELECT EMP_LAST_NAME,"
                        + "EMP_FIRST_NAME "
                        + "FROM HRSCHEMA.EMPLOYEE "
                        + "WHERE EMP_ID = 3217";

                rdr = cmd.ExecuteReader();
                Console.WriteLine("\nExecute: " + cmd.CommandText);
                cols = rdr.FieldCount;
                rows = rdr.HasRows;

                while (rdr.Read() == true)
                {
                    Console.WriteLine("Last name is : " + rdr.GetString(0));
                    Console.WriteLine("First name is : " + rdr.GetString(1));
                }
                rdr.Close();
            }

            catch (Exception e)
            {
                Console.WriteLine(e.Message);
```

```
                }

                finally
                {
                    Console.WriteLine("Returned Rows? " + rows);

                    conn.Close();
                }


            }

        }


    }
```

The output is:

```
Execute: SELECT EMP_LAST_NAME,EMP_FIRST_NAME FROM HRSCHEMA.EMPLOYEE WHERE EMP_ID
= 3217
Last name is : JOHNSON
First name is : EDWARD
Returned Rows? True
The program '[9700] NETEMPOLEDB.exe' has exited with code 0 (0x0).
```

Sample .NET ODBC Program

For the most part you'll see those objects previously prefixed "DB2" replaced here with objects prefixed with ODBC. Note that you must have registered the DB2 table you are reading as an ODBC data source (typically via **Control Panel --> Administrative Tools --> ODBC Data Sources**). Otherwise you cannot use the ODBC data provider.

```
using System;
using System.Collections.Generic;
using System.Linq;
using System.Text;
using System.Threading.Tasks;
using System.Data.Odbc;

namespace NETODBC
{
    class NETODBC
        // Use ODBC connector to retrieve data
    {
        static void Main(string[] args)
        {
            OdbcConnection conn = null;
            OdbcCommand cmd = null;
            OdbcDataReader rdr = null;
            Boolean rows = false;
            int cols = 0;

            try
            {
                conn = new OdbcConnection("DSN=DBHR");
                conn.Open();
                Console.WriteLine("Successful connection!");
                cmd = conn.CreateCommand();
                cmd.CommandText
                    = "SELECT EMP_LAST_NAME,"
                         + "EMP_FIRST_NAME "
                         + "FROM HRSCHEMA.EMPLOYEE "
                         + "WHERE EMP_ID = 3217";

                rdr = cmd.ExecuteReader();
                Console.WriteLine("\nExecute: " + cmd.CommandText);
                cols = rdr.FieldCount;
                rows = rdr.HasRows;

                while (rdr.Read() == true)
                {
                    Console.WriteLine("Last name is : " + rdr.GetString(0));
                    Console.WriteLine("First name is : " + rdr.GetString(1));
                }
                rdr.Close();
            }
```

```
                catch (Exception e)
                {
                    Console.WriteLine(e.Message);
                }

                finally
                {
                    Console.WriteLine("Returned Rows? " + rows);

                    conn.Close();
                }

            }

        }

    }
```

Here's the output:

```
Execute: SELECT EMP_LAST_NAME,EMP_FIRST_NAME FROM HRSCHEMA.EMPLOYEE WHERE EMP_ID = 3217
Last name is : JOHNSON
First name is : EDWARD
Returned Rows? True
```

.NET DB2 INSERT Program

Before we construct our C# INSERT program, let's first delete the records we added using our Java program so that our .NET examples will work correctly when we add these same records.

```
DELETE FROM HRSCHEMA.EMPLOYEE
WHERE EMP_ID IN (1111,1112);

Updated 2 rows.
```

Now let's construct our INSERT program to add these two rows back. You can use the Visual Studio IDE to create your version of the program. Here I'll just supply the code listing. Let's create solution NETEMP1 and class NETEMP1. Remember to change your application type from Console Application to Windows Application, and to add the DB2 reference.

We'll use the DB2Connection object along with a DB2Command object. That's all we need. Here is my example code:

```
using System;
using System.Collections.Generic;
using System.Linq;
using System.Text;
using System.Threading.Tasks;
using IBM.Data.DB2;

/* Program to connect to DB2 and insert a row    */
namespace NETEMP1
{
    class NETEMP1
    {
        static void Main(string[] args)
        {
            DB2Connection conn = null;
            DB2Command cmd = null;

            try
            {
                conn = new DB2Connection("Database=DBHR");
                conn.Open();
                Console.WriteLine("Successful connection!");
                cmd = conn.CreateCommand();
                cmd.CommandText
                    = "INSERT INTO HRSCHEMA.EMPLOYEE "
                    + "(EMP_ID, "
                    + "EMP_LAST_NAME, "
                    + "EMP_FIRST_NAME, "
                    + "EMP_SERVICE_YEARS, "
```

```
                + "EMP_PROMOTION_DATE) "
                + "VALUES (1111, "
                + "'VEREEN', "
                + "'CHARLES', "
                + "12, "
                + "'01/01/2017') ";

            int rowsAffected = cmd.ExecuteNonQuery();
            Console.WriteLine("\n Inserted Rows: " + rowsAffected + " \n");


    /* Now add a second record */


            cmd.CommandText
                = "INSERT INTO HRSCHEMA.EMPLOYEE "
                + "(EMP_ID, "
                + "EMP_LAST_NAME, "
                + "EMP_FIRST_NAME, "
                + "EMP_SERVICE_YEARS, "
                + "EMP_PROMOTION_DATE) "
                + "VALUES (1112, "
                + "'YATES', "
                + "'JANENE', "
                + "7, "
                + "'01/01/2015') ";

            rowsAffected = cmd.ExecuteNonQuery();
            Console.WriteLine("\n Inserted Rows: " + rowsAffected + " \n");

        }

        catch (Exception e)
        {
            Console.WriteLine(e.Message);
        }

        finally
        {
            conn.Close();
        }

        DB2Connection connection = new DB2Connection("Database=HRDB");
    }
  }
}
```

Now let's run it, and here is the result:

```
Successful connection!
Inserted Rows: 1
Inserted Rows: 1
```

And we can verify the result by querying the table.

```
SELECT * FROM HRSCHEMA.EMPLOYEE
WHERE EMP_ID IN(1111,1112);
```

```
EMP_ID    EMP_LAST_NAME    EMP_FIRST_NAME    EMP_SERVICE_YEARS  EMP_PROMOTION_DATE  EMP_PROFILE
------    -------------    --------------    -----------------  ------------------  -----------
  1111    VEREEN            CHARLES                         12  2017-01-01          NULL
  1112    YATES             JANENE                           7   2015-01-01          NULL
```

This is what we expect, so let's move on to the parameter based insert.

.NET DB2 INSERT Program with Parameters

Creating parameters for SQL statements in .NET is somewhat different than in Java but the principle is roughly the same. We'll create program `NETEMP1A` for this. You can use question marks for the parameter markers or you could use variable names preceded by the @ symbol. We'll use question marks. Then .NET requires you to add parameters by name to the Command object's Parameter collection. First let's change our query to use the ? character as parameter markers.

```
cmd.CommandText  = "INSERT INTO HRSCHEMA.EMPLOYEE "
                 + "(EMP_ID, "
                 + "EMP_LAST_NAME, "
                 + "EMP_FIRST_NAME, "
                 + "EMP_SERVICE_YEARS, "
                 + "EMP_PROMOTION_DATE) "
                 + "VALUES (?, "
                 + "?, "
                 + "?, "
                 + "?, "
                 + "?);" ;
```

Next we use the Add method of the Parameters collection of the Command object to specify parameters and values. In this case we are adding employee 1113 whose name is Rita Duggan with 5 years of service and a promotion date of 1/1/2016.

```
cmd.Parameters.Add(new DB2Parameter("@empid", 1113));
cmd.Parameters.Add(new DB2Parameter("@lname", "DUGGAN"));
cmd.Parameters.Add(new DB2Parameter("@fname", "RITA"));
cmd.Parameters.Add(new DB2Parameter("@yrsofservice", 5));
cmd.Parameters.Add(new DB2Parameter("@promdate",
    DateTime.Parse("01/01/2016")));
```

And now we could then execute the query as before.

```
int rowsAffected = cmd.ExecuteNonQuery();
Console.WriteLine("\n Inserted Rows: " + rowsAffected + " \n");
```

But before we run the program, let's add some additional code to change the parameter values and then run the same query again. This is where the value of using parameters comes in. We only need to modify the Value setting of each of our parameters and then perform the command again – no other modification is required. In this case we are adding employee 1114 who is Phyllis Miller with 11 years of service and promotion date 1/1/2017.

```
cmd.Parameters["@empid"].Value = 1114;
cmd.Parameters["@lname"].Value = "MILLER";
cmd.Parameters["@fname"].Value = "PHYLLIS";
cmd.Parameters["@yrsofservice"].Value = 11;
```

```
cmd.Parameters["@promdate"].Value = DateTime.Parse("01/01/2017");

rowsAffected = cmd.ExecuteNonQuery();
Console.WriteLine("\n Inserted Rows: " + rowsAffected + " \n");
```

Now we are ready to build the entire modified program and run it to add the two new records.

```
using System;
using System.Collections.Generic;
using System.Linq;
using System.Text;
using System.Threading.Tasks;
using IBM.Data.DB2;
using IBM.Data.DB2Types;

/* Program to connect to DB2 and insert a row    */
namespace NETEMP1A
{
    class NETEMP1A
    {
        static void Main(string[] args)
        {
            DB2Connection conn = null;
            DB2Command cmd = null;

            try
            {
                conn = new DB2Connection("Database=DBHR");
                conn.Open();
                Console.WriteLine("Successful connection!");
                cmd = conn.CreateCommand();


                cmd.CommandText
                    = "INSERT INTO HRSCHEMA.EMPLOYEE "
                    + "(EMP_ID, "
                    + "EMP_LAST_NAME, "
                    + "EMP_FIRST_NAME, "
                    + "EMP_SERVICE_YEARS, "
                    + "EMP_PROMOTION_DATE) "
                    + "VALUES (?, "
                    + "?, "
                    + "?, "
                    + "?, "
                    + "?);" ;
                /*  Now we add parameters to the Command object */

                cmd.Parameters.Add(new DB2Parameter("@empid", 1113));
                cmd.Parameters.Add(new DB2Parameter("@lname", "DUGGAN"));
                cmd.Parameters.Add(new DB2Parameter("@fname", "RITA"));
                cmd.Parameters.Add(new DB2Parameter("@yrsofservice", 5));
```

```
                    cmd.Parameters.Add(new DB2Parameter("@promdate",
                    DateTime.Parse("01/01/2016")));

                    /* Execute the query */

                    int rowsAffected = cmd.ExecuteNonQuery();
                    Console.WriteLine("\n Inserted Rows: "
                          + rowsAffected + " \n");

                    /* Update the parameter values and do another insert */

                    cmd.Parameters["@empid"].Value = 1114;
                    cmd.Parameters["@lname"].Value = "MILLER";
                    cmd.Parameters["@fname"].Value = "PHYLLIS";
                    cmd.Parameters["@yrsofservice"].Value = 11;
                    cmd.Parameters["@promdate"].Value
                          = DateTime.Parse("01/01/2017");

                    rowsAffected = cmd.ExecuteNonQuery();
                    Console.WriteLine("\n Inserted Rows: " + rowsAffected + "
                      \n");
                }

                catch (Exception e)
                {
                    Console.WriteLine(e.Message);
                }

                finally
                {
                    conn.Close();
                }

                DB2Connection connection = new DB2Connection("Database=HRDB");
            }

        }

    }
```

And here is the result:

```
Successful connection!
Inserted Rows: 1
Inserted Rows: 1
```

And we can verify that the records were added by querying the EMPLOYEE table.

```
SELECT * FROM HRSCHEMA.EMPLOYEE
WHERE EMP_ID IN(1113,1114);
```

```
EMP_ID EMP_LAST_NAME EMP_FIRST_NAME EMP_SERVICE_YEARS EMP_PROMOTION_DATE EMP_PROFILE
------ ------------- -------------- ----------------- ------------------ -----------
  1113 DUGGAN        RITA                           5  2016-01-01               NULL
  1114 MILLER        PHYLLIS                       11  2017-01-01               NULL
```

The value of using parameters is that you do not need to rebuild the INSERT statement each time you run the query. You need only change the value of the parameters and then execute the command again.. Obviously this model has utility when you are looping through a file of input records, or possibly processing data from another DB2 table or other data source.

OK, now let's move on to an update program.

NET DB2 UPDATE Program

This program will also correct the case on `EMP_LAST_NAME` as we did with Java, but this is a bit different in .NET. First, let's change all the `EMP_LAST_NAME` values to lower case.

```
UPDATE HRSCHEMA.EMPLOYEE
SET EMP_LAST_NAME = LOWER(EMP_LAST_NAME);
```

And we can demonstrate that our data is in fact in lower case:

```
SELECT EMP_LAST_NAME FROM HRSCHEMA.EMPLOYEE

EMP_LAST_NAME
-------------
johnson
stewart
franklin
schultz
willard
jenkins
ford
harris
turnbull
vereen
yates
duggan
miller
```

In .NET there are several ways to perform DB2 updates such as using an update query with the Command object. Since we want to capture our changes and report them in the console, we're going to use two other objects, which are a DataAdapter and a DataSet.

The DataAdaptor object is used to retrieve data from the `EMPLOYEE` table and load it to a DataSet. The DataSet is a disconnected copy of the records – we could think of it as a result set. We make changes to the records in the DataSet and then we use the **update()** method of the DataAdaptor to apply the changes in the Dataset to the table.

Here is our code:

```
using System;
using System.Collections.Generic;
using System.Linq;
using System.Text;
using System.Threading.Tasks;
using System.Data;
using IBM.Data.DB2;
using IBM.Data.DB2Types;
```

```
namespace NETEMP2
{
    class NETEMP2
    {
        static void Main(string[] args)
        {
            Console.WriteLine("Program NETEMP2 begins successfully");
            DB2Connection conn = new DB2Connection("Database=DBHR");
            DataSet EmployeesDataSet = new DataSet();
            DB2DataAdapter da;
            DB2CommandBuilder cmdBuilder = null;

            try
                {
                conn.Open();
                Console.WriteLine("Successful connection!");

                da = new DB2DataAdapter("SELECT EMP_ID, EMP_LAST_NAME "
                            + "FROM HRSCHEMA.EMPLOYEE "
                            + "WHERE EMP_LAST_NAME <> UPPER(EMP_LAST_NAME)",
                               conn);

                cmdBuilder = new DB2CommandBuilder(da);

                da.Fill(EmployeesDataSet, "Employees");

                foreach (DataTable DT in EmployeesDataSet.Tables)
                {
                    foreach (DataRow DR in DT.Rows)
                    {
                        string lName = DR[1].ToString();
                        Console.WriteLine("BEFORE surname is " + lName);

                        lName = lName.ToUpper();
                        DR[1] = lName;

                        Console.WriteLine("AFTER surname is " + lName);

                    }

                    // Update the table using values in the dataset

                    da.UpdateCommand = cmdBuilder.GetUpdateCommand();
                    da.Update(EmployeesDataSet, "Employees");
                }

                }

             catch (Exception e)
             {
                    Console.WriteLine(e.Message);
             }
```

```
                finally
                {
                    conn.Close();
                    Console.WriteLine("Program NETEMP2 concluded successfully");

                }


            }

        }

    }
```

And here is the result:

```
Program NETEMP2 begins successfully
BEFORE surname is johnson
AFTER surname is JOHNSON
BEFORE surname is stewart
AFTER surname is STEWART
BEFORE surname is franklin
AFTER surname is FRANKLIN
BEFORE surname is schultz
AFTER surname is SCHULTZ
BEFORE surname is willard
AFTER surname is WILLARD
BEFORE surname is jenkins
AFTER surname is JENKINS
BEFORE surname is ford
AFTER surname is FORD
BEFORE surname is harris
AFTER surname is HARRIS
BEFORE surname is turnbull
AFTER surname is TURNBULL
BEFORE surname is vereen
AFTER surname is VEREEN
BEFORE surname is yates
AFTER surname is YATES
BEFORE surname is duggan
AFTER surname is DUGGAN
BEFORE surname is miller
AFTER surname is MILLER
Program NETEMP2 concluded successfully
```

Error Handling

SQLCODES

When you encounter an SQL exception DB2 makes the error SQLCODE and a brief message available to you. DB2 sets the SQLCODE after each SQL statement. The value of the SQLCODE can be interpreted generally as follows:

- If SQLCODE = 0, execution was successful.
- If SQLCODE > 0, execution was successful with a warning.
- If SQLCODE < 0, execution was not successful.
- SQLCODE = 100, "no data" was found.

Common Error SQLCODES

SQLCODE	ERROR
-117	THE NUMBER OF VALUES ASSIGNED IS NOT THE SAME AS THE NUMBER OF SPECIFIED OR IMPLIED COLUMNS
-180	THE DATE, TIME, OR TIMESTAMP VALUE value IS INVALID
-181	THE STRING REPRESENTATION OF A DATETIME VALUE IS NOT A VALID DATETIME VALUE
-203	A REFERENCE TO COLUMN column-name IS AMBIGUOUS
-206	Object-name IS NOT VALID IN THE CONTEXT WHERE IT IS USED
-305	THE NULL VALUE CANNOT BE ASSIGNED TO OUTPUT HOST VARIABLE NUMBER position-number BECAUSE NO INDICATOR VARIABLE IS SPECIFIED
-501	THE CURSOR IDENTIFIED IN A FETCH OR CLOSE STATEMENT IS NOT OPEN
-502	THE CURSOR IDENTIFIED IN AN OPEN STATEMENT IS ALREADY OPEN

-803	AN INSERTED OR UPDATED VALUE IS INVALID BECAUSE THE INDEX IN INDEX SPACE indexspace-name CONSTRAINS COLUMNS OF THE TABLE SO NO TWO ROWS CAN CONTAIN DUPLICATE VALUES IN THOSE COLUMNS. RID OF EXISTING ROW IS X record-id
-805	DBRM OR PACKAGE NAME location-name.collection-id.dbrm-name.consistency-token NOT FOUND IN PLAN plan-name. REASON reason-code
-811	THE RESULT OF AN EMBEDDED SELECT STATEMENT OR A SUBSELECT IN THE SET CLAUSE OF AN UPDATE STATEMENT IS A TABLE OF MORE THAN ONE ROW, OR THE RESULT OF A SUBQUERY OF A BASIC PREDICATE IS MORE THAN ONE VALUE
-818	THE PRECOMPILER-GENERATED TIMESTAMP x IN THE LOAD MODULE IS DIFFERENT FROM THE BIND TIMESTAMP y BUILT FROM THE DBRM z
-904	UNSUCCESSFUL EXECUTION CAUSED BY AN UNAVAILABLE RESOURCE. REASON reason-code, TYPE OF RESOURCE resource-type, AND RESOURCE NAME resource-name.
-911	THE CURRENT UNIT OF WORK HAS BEEN ROLLED BACK DUE TO DEADLOCK OR TIMEOUT. REASON reason-code, TYPE OF RESOURCE resource-type, AND RESOURCE NAME resource-name
-913	UNSUCCESSFUL EXECUTION CAUSED BY DEADLOCK OR TIMEOUT. REASON CODE reason-code, TYPE OF RESOURCE resource-type, AND RESOURCE NAME resource-name.
-922	AUTHORIZATION FAILURE: error-type ERROR. REASON reason-code.

Dynamic versus Static SQL

Static SQL

Static SQL statements are embedded within an application program that is written in a traditional programming language such as COBOL. The statement is prepared before the program is executed, and the executable statement persists after the program ends. The JDBC and .NET interfaces use dynamic SQL. The Java SQLJ interface uses static SQL. Although we do not cover SQLJ in this text book, we'll briefly discuss the differences between static and dynamic SQL.

When you use static SQL, you cannot change the form of SQL statements unless you make changes to the program. So you must know before run time what SQL statements your application needs to use. However, you can increase the flexibility of those statements by using host variables. So for example you could write an SQL that retrieves employee information for all employees with X years of service where the X becomes a host variable that you load at run time. Using static SQL and host variables is more secure than using dynamic SQL.

Dynamic SQL

Unlike static SQL which is prepared before the program runs, with dynamic SQL, DB2 prepares and executes the SQL statements at run time as part of the program's execution. Dynamic SQL is a good choice when you do not know the format of an SQL statement before you write or run a program. An example might be a user interface that allows a web application to submit SQL statements to a background Java program for execution. In this case, you wouldn't know the structure of the statement the client submits until run time.

Applications that use dynamic SQL create an SQL statement in the form of a character string. A typical dynamic SQL application takes the following steps:

1. Translates the input data into an SQL statement.
2. Prepares the SQL statement to execute and acquires a description of the result table (if any).
3. Obtains, for SELECT statements, enough main storage to contain retrieved data.
4. Executes the statement or fetches the rows of data.
5. Processes the returned information.
6. Handles SQL return codes.

Chapter Five Questions

1. If you want to still reference data using a cursor after you issue a COMMIT, which clause would you use when you declare the cursor?

    a. WITH HOLD
    b. WITH RETAIN
    c. WITH STAY
    d. WITH REOPEN

2. Which of the following is not a valid Java object type?

    a. ResultSet
    b. Connection
    c. Statement
    d. Dataset

3. Which of the following is not a valid .NET object type?

    a. DB2Stmt
    b. DB2Connection
    c. DB2Command
    d. All of the above are valid .NET objects.

Chapter Five Exercises

For each of the following exercises, you can use either Java or .NET. Or you can do both.

1. Write a Java or .NET program that creates a result set of all employees who have 5 years or more of service from the `HRSCHEMA.EMPLOYEE` table. Include logic to display any rows that are returned. Display the employee number, last name and first name of these employees.

2. Write a Java or .NET program that tries to insert a record that already exists into the `HRSCHEMA.EMPLOYEE` table. Your program should capture the error, and your error routine should detail the cause of the error.

3. Write a Java or .NET program to delete employee number 1114.

4. Write a Java or .NET program to merge the following employee information into the `HRSCHEMA.EMPLOYEE` table.

Employee Number	: 1114
Employee Last Name	: MILLER
Employee First Name	: PHILLIS
Employee Years of Service	: 11
Employee Promotion Date	: 01/01/2017

Employee Number	: 1115
Employee Last Name	: JENSON
Employee First Name	: PAUL
Employee Years of Service	: 8
Employee Promotion Date	: 01/01/2016

CHAPTER SIX: DATA CONCURRENCY

Isolation Levels & Bind Release Options

Isolation level means the degree to which a DB2 application's activities are isolated from the operations of other DB2 applications. The isolation level for a package is specified when the package is bound, although you can override the package isolation level in an SQL statement. There are four isolation levels: Repeatable Read, Read Stability, Cursor Stability and Uncommitted Read.

ISOLATIONS LEVELS

Repeatable Read (RR)

Repeatable Read ensures that a query issued multiple times within the same unit of work will produce the exact same results. It does this by locking all rows that could affect the result. It does not permit any adds/changes/deletes to the table that could affect the result.

Read Stability (RS)

Read Stability locks for the duration of the transaction those rows that are returned by a query, but it allows additional rows to be added to the table.

Cursor Stability (CS)

Cursor Stability only locks the row that the cursor is placed on (and any rows it has updated during the unit of work). This is the default isolation level if no other is specified.

Uncommitted Read (UR)

Uncommitted Read permits reading of uncommitted changes which may never be applied to the database. It does not lock any rows at all unless the row(s) is updated during the unit of work.

An IBM recommended best practice prefers isolation levels in this order:

1. Cursor stability (CS)
2. Uncommitted read (UR)
3. Read stability (RS)
4. Repeatable read (RR)

Of course the chosen isolation level depends on the scenario. We'll look at specific scenarios now.

Isolation Levels for Specific Situations

When your environment is basically read-only (such as with data warehouse environments), use UR **(UNCOMMITTED READ)** because it incurs the least overhead.

If you want to maximize data concurrency without seeing uncommitted data, use the CS (**CURSOR STABILITY**) isolation level. CS only locks the row where the cursor is placed (and any other rows which have been changed since the last commit point), thus maximizing concurrency compared to RR or RS.

If you want no existing rows that were retrieved to be changed by other processes during your unit of work, but you don't mind if new rows are inserted, use RS (**READ STABILITY**).

Finally if you must lock all rows that satisfy the query and also not permit any new rows to be added that could change the result of the query, use RR (**REPEATABLE READ**).

Based on the above, if we wanted to order the isolation levels from most to least impact on performance, the order would be:

1. REPEATABLE READ (RR)
2. READ STABILITY (RS)
3. CURSOR STABILITY (CS)
4. UNCOMMITTED READ (UR)

How to Specify/Override Isolation Level

In Java the DB2 isolation level is an attribute of the DB2 connection. You use the SetTransactionIsolation method of the connection object and provide a value that corresponds to the desired isolation level. For example if you wanted to specify Cursor Stability for the isolation level and con is the connection variable, you would code:

```
con.setTransactionIsolation(con.TRANSACTION_READ_COMMITTED);
```

The values for the various isolation levels are as follows:

ANSI ISOLATION LEVEL	DB2 ISOLATION LEVEL	INTEGER VALUE
TRANSACTION_SERIALIZABLE	Repeatable read	8
TRANSACTION_REPEATABLE_READ[1]	Read stability	4
TRANSACTION_READ_COMMITTED	Cursor stability	2
TRANSACTION_READ_UNCOMMITTED	Uncommitted read	1

Given the above table, you could also specify the integer value for the isolation level rather than the named constant. For example, to specify the cursor stability isolation level, you could code:

```
con.setTransactionIsolation(2);
```

However, it is much better to use the named constant. It makes your program code more self-documenting - not only for you, but for the next programmer.

In .NET you can specify isolation level by adding the IsolationLevel keyword to the connection string:

```
DB2Connection conn
   = new DB2Connection("Database=DBHR;IsolationLevel=RepeatableRead");
```

You can also set the isolation level for a transaction if you are using transaction. Assuming you have defined a DB2Transaction named trans and a connection named conn, you could code:

```
trans = conn.BeginTransaction(IsolationLevel.ReadCommitted);
```

Finally, if you want to override an isolation level in a specific query, specify the override at the end of the query by using the WITH <isolation level abbreviation> clause. For example, to override the default isolation level of CS to use UR instead on a query, code the following and notice we've used WITH UR at the end of the query:

```
SELECT EMP_ID,
EMP_LAST_NAME,
EMP_FIRST_NAME
FROM EMPLOYEE
ORDER BY EMP_ID
WITH UR;
```

1 Note that RepeatableRead in ANSI means ReadStability in DB2. I mention it here because it could be a point of confusion. If you need what DB2 refers to as RepeatableRead, you would specify the ANSI value `TRANSACTION_SERIALIZABLE` instead of RepeatableRead.

COMMIT, ROLLBACK, and SAVEPOINTS

Central to understanding transaction management is the concept of a unit of work. A unit of work begins when a program is initiated. Multiple adds, changes and deletes may then take place during the same unit of work. The changes are not made permanent until a commit point is reached. A unit of work ends in one of three ways:

1. When a commit is issued.
2. When a rollback is issued.
3. When the program ends.

Let's look at each of these.

COMMIT

The COMMIT statement ends a transaction and makes the changes permanent and visible to other processes. Also, when a program ends, there is an implicit COMMIT. This is important to know; however an IBM recommended best practice is to do an explicit COMMIT at the end of the program.

ROLLBACK

A ROLLBACK statement ends a transaction without making changes permanent – the changes are simply discarded. This is done either intentionally by the application when it determines there is a reason to ROLLBACK the changes and it issues a ROLLBACK explicitly, or because the system traps an error that requires it to do a ROLLBACK of changes. In both cases, the rolled back changes are those that have been made since the last COMMIT point. If no COMMITs have been issued, then all changes made in the session are rolled back.

You can also issue a ROLLBACK TO <savepoint> if you are using SAVEPOINTS. We'll take a look at that shortly.

Here are some other points about ROLLBACK to know and remember:

- The abend of a process causes an implicit ROLLBACK.
- Global variable contents are not affected by ROLLBACK.

SAVEPOINT

The SAVEPOINT statement creates a point within a unit of recovery to which you can roll back changes. This is similar to using ROLLBACK to backout changes since the last COMMIT point, except a SAVEPOINT gives you even more control because it allows a partial

ROLLBACK **between** COMMIT points.

You might wonder what the point is of using a SAVEPOINT. Let's take an example. Suppose you have a program that does INSERT statements and you program logic to COMMIT every 500 inserts. If you issue a ROLLBACK, then all updates since the last COMMIT will be backed out. That's pretty straightforward.

But suppose you are updating information for vendors from a file of updates that is sorted by vendor, and if there is an error you want to rollback to where you started updating records for that vendor. And you want all other updates since the last COMMIT point to be applied to the database. This is different than rolling back to the last COMMIT point, and you can do it by setting a new savepoint each time the vendor changes. Issuing a SAVEPOINT enables you to execute several SQL statements as a single executable block between COMMIT statements. You can then undo changes back out to that savepoint by issuing a ROLLBACK TO SAVEPOINT statement.

Example

Let's do a simple example. First, create a new table and then add some records to the table. We'll create a copy of EMP_PAY.

```
CREATE TABLE HRSCHEMA.EMP_PAY_X
LIKE HRSCHEMA.EMP_PAY
IN DBHR.TSHR;
```

Now let's add some records. We'll add one record, then create a SAVEPOINT, add another record and then roll back to the SAVEPOINT. This should leave us with only the first record in the table.

```
INSERT INTO EMP_PAY_X
VALUES(1111,
45000.00,
1200.00);

SAVEPOINT A ON ROLLBACK RETAIN CURSORS;

INSERT INTO EMP_PAY_X
VALUES(2222,
55000.00,
1500.00);

ROLLBACK TO SAVEPOINT A;
```

We can verify that only the first record was added to the table:

```
SELECT * FROM HRSCHEMA.EMP_PAY_X;

EMP_ID  EMP_REGULAR_PAY  EMP_BONUS_PAY
------  ---------------  -------------
  1111         45000.00        1200.00
```

If you have multiple SAVEPOINTS and you ROLLBACK to one of them, then the ROLLBACK will include updates made after any later SAVEPOINTS. Let's illustrate this with an example. We'll set three SAVEPOINTs: A, B and C. We'll add a record, then issue savepoint and we'll do this three times. Then we'll ROLLBACK to the first SAVEPOINT which is A. What we're saying is that any updates made after A will be backed out, which includes the INSERTS made after SAVEPOINTS B and C. Let's try this:

```
INSERT INTO EMP_PAY_X
VALUES(2222,
55000.00,
1500.00);

SAVEPOINT A ON ROLLBACK RETAIN CURSORS;

INSERT INTO EMP_PAY_X
VALUES(3333,
65000.00,
2500.00);

SAVEPOINT B ON ROLLBACK RETAIN CURSORS;

INSERT INTO EMP_PAY_X
VALUES(4444,
75000.00,
2000.00);

SAVEPOINT C ON ROLLBACK RETAIN CURSORS;

ROLLBACK TO SAVEPOINT A;


SELECT * FROM HRSCHEMA.EMP_PAY_X;

EMP_ID  EMP_REGULAR_PAY  EMP_BONUS_PAY
------  ---------------  -------------
  1111         45000.00        1200.00
  2222         55000.00        1500.00
```

Now as you can see, only the first record (2222) was inserted because we specified ROLLBACK all the way to SAVEPOINT A. Note that the 1111 record was already in the table.

Here are a few more considerations for using SAVEPOINTS.

1. If you want to prevent a SAVEPOINT from being reused during the same unit of work, include the UNIQUE keyword when you define it.

2. When you issue a rollback, if you want to retain cursors, you must specify `ON ROLLBACK RETAIN CURSORS`. Similarly, if you want to retain any locks, you must specify `ON ROLLBACK RETAIN CURSORS`.

Units of Work

A unit of work is a set of database operations in an application that is ended by a commit, a rollback or the end of the application process. A commit or rollback operation applies only to the set of changes made within that unit of work. An application process can involve one or many units of work.

Once a commit action occurs, the database changes are permanent and visible to other application processes. Any locks obtained by the application process are held until the end of the unit of work. So if you update 10 records within one unit of work, the records are all locked until a commit point.

As explained elsewhere, in distributed environments where you update data stores on more than one system, a two-phase commit is performed. The two phase commit ensures that data is consistent between the two systems by either fully commiting or fully rolling back the unit of work. The two phase commit consists of a commit-request phase and an actual commit phase.

Autonomous Transactions

Autonomous Transactions Basics

Autonomous transactions are native SQL procedures which run with their own units of work, separate from the calling program. If a calling program issues a ROLLBACK to back out its changes, the committed changes of the autonomous procedure are not affected.

Autonomous procedures can be called by normal application programs, other stored procedures, user-defined functions or triggers. Autonomous procedures can also invoke triggers, perform SQL statements, and execute commit and rollback statements.

Restrictions

There are some restrictions on use of an autonomous procedure. The only kind of procedure that can function as an autonomous procedure is a native SQL procedure. An autonomous procedure cannot call another autonomous procedure. Also, since autonomous procedures commit work separately from the calling program or procedure, the autonomous procedure cannot see uncommitted changes in that calling program or procedure.

Here's one other limitation to be aware of. Since autonomous procedures do not share locks with the calling program or procedure, it is possible to have lock contention between these entities. The developer must consider carefully what type of work is involved and understand the possible results if both the calling and called entity are doing updates on the same data.

Applications

Autonomous procedures are useful for logging information about error conditions encountered by an application program. Similarly they can be used for creating an audit trail of activity for transactions.

Commit Processing

This section concerns the commit, rollback and recovery of an application or application program. We already covered the general use of COMMIT, ROLLBACK and SAVEPOINTs in prior subsections. Here we'll apply the COMMIT in a DB2 program.

Java Program with Commit

For the DB2 program, we should use the COMMIT statement at appropriate intervals. Let's use our update Java program. Let's set all of our employee records such that the `EMP_LAST_NAME` is in lower case.

```
UPDATE HRSCHEMA.EMPLOYEE SET EMP_LAST_NAME = LOWER(EMP_LAST_NAME);

SELECT EMP_LAST_NAME FROM HRSCHEMA.EMPLOYEE;

EMP_LAST_NAME
 -------------
 johnson
 stewart
 franklin
 schultz
 willard
 jenkins
 ford
 harris
 turnbull
 vereen
 yates
 duggan
 miller
```

Ordinarily we'd commit at intervals of hundreds or thousands of records. Since we only have 13 records, let's commit every 5 records. We will set up a commit counter called commit Counter and we'll increment it each time we update a record. Once the counter reaches 5 updates, we'll execute the commit method of the connection object. Here is our code:

```
package employee;
import java.sql.*;
public class JAVEMP3 {

        public static void main(String[] args) throws ClassNotFoundException
        {
            Class.forName("com.ibm.db2.jcc.DB2Driver");
            System.out.println("**** Loaded the JDBC driver");
             String    url
               ="jdbc:db2://localhost:50000/DBHR:retrieveMessagesFromServerOnGetMes-
               sage=true;";
            String user="rwinga01";
            String password="*********";

            Connection con=null;

                try {
                        con=DriverManager.getConnection(url, user, password);
                        System.out.println("**** Created the connection");
                        String query = "SELECT EMP_ID, EMP_LAST_NAME "
                                              + "FROM HRSCHEMA.EMPLOYEE "
                                              + "WHERE EMP_LAST_NAME <>
                                                   UPPER(EMP_LAST_NAME)" ;

                        Integer recordCount = 0;
                        Integer commitCount = 0;

                        Integer intEmpNo;
                        String  strLastName;
                        Statement stmt;
                        stmt = con.createStatement(ResultSet.TYPE_SCROLL_SENSITIVE,
                        ResultSet.CONCUR_UPDATABLE);
                        ResultSet rs=stmt.executeQuery(query);
                        while(rs.next())
                        {
                            intEmpNo     = rs.getInt(1);
                               strLastName   = rs.getString(2);
                            System.out.println("Employee " + intEmpNo
                               + " BEFORE Last name is: "
                               + strLastName);
                            strLastName  = strLastName.toUpperCase();
                            rs.updateString(2, strLastName);
                            rs.updateRow();
                            System.out.println("Employee "
                               + intEmpNo
                               + " AFTER  Last name is: "
                               + strLastName);
```

```
                    recordCount++;
                    commitCount++;
                    if(commitCount >= 5) {
                       con.commit();
                       commitCount = 0;
                    System.out.println("*** Commit taken at record " +
                    recordCount);
                    }

                    // Do final commit

                 }
                 con.commit();
                System.out.println("*** Final Commit taken at record "
                 + recordCount);

           } catch (SQLException e) {
                 System.out.println(e.getMessage());
                 System.out.println(e.getErrorCode());
                 e.printStackTrace();

                 }

        }

     }
```

And this is our output:

```
**** Loaded the JDBC driver
**** Created the connection
Employee 3217 BEFORE Last name is: johnson
Employee 3217 AFTER  Last name is: JOHNSON
Employee 7459 BEFORE Last name is: stewart
Employee 7459 AFTER  Last name is: STEWART
Employee 9134 BEFORE Last name is: franklin
Employee 9134 AFTER  Last name is: FRANKLIN
Employee 4720 BEFORE Last name is: schultz
Employee 4720 AFTER  Last name is: SCHULTZ
Employee 6288 BEFORE Last name is: willard
Employee 6288 AFTER  Last name is: WILLARD
*** Commit taken at record 5
Employee 1122 BEFORE Last name is: jenkins
Employee 1122 AFTER  Last name is: JENKINS
Employee 3333 BEFORE Last name is: ford
Employee 3333 AFTER  Last name is: FORD
Employee 7777 BEFORE Last name is: harris
Employee 7777 AFTER  Last name is: HARRIS
Employee 4175 BEFORE Last name is: turnbull
Employee 4175 AFTER  Last name is: TURNBULL
Employee 1111 BEFORE Last name is: vereen
```

```
Employee 1111 AFTER  Last name is: VEREEN
*** Commit taken at record 10
Employee 1112 BEFORE Last name is: yates
Employee 1112 AFTER  Last name is: YATES
Employee 1113 BEFORE Last name is: duggan
Employee 1113 AFTER  Last name is: DUGGAN
Employee 1114 BEFORE Last name is: miller
Employee 1114 AFTER  Last name is: MILLER
*** Final Commit taken at record 13
```

Sometimes you'll need to adjust the commit counter if it turns out you are not taking commits often enough (which can cause record locking for extended periods of time and potential for deadlock with other programs). Or you might even need to use another measure such as elapsed time instead of record count.

.NET Commit Example

.NET programs can commit but it requires creating a Transaction object which is tied to the Connection object. Here is an example of how to do this:

```
using System;
using System.Data;
using IBM.Data.DB2;

namespace NETEMP3
{
    class NETEMP3
    {
        static void Main(string[] args)
        {
            Console.WriteLine("Program NETEMP3 begins successfully");
            DB2Connection conn = new DB2Connection("Database=DBHR");

            String strLastName = "";
            String strSQL1
               = "SELECT EMP_ID, EMP_LAST_NAME FROM HRSCHEMA.EMPLOYEE "
               + "WHERE EMP_LAST_NAME <> UPPER(EMP_LAST_NAME)";

            String strSQL2 = "UPDATE HRSCHEMA.EMPLOYEE SET EMP_LAST_NAME = ? "
               + "WHERE EMP_ID = ?";
            int intEmpId = 0;

            conn.Open();

            DB2Transaction tran = conn.BeginTransaction();
            DB2Command cmd2 = new DB2Command(strSQL2, conn, tran);

            try
            {
```

```
                DB2Command cmd = new DB2Command(strSQL1, conn, tran);

                cmd2.Parameters.Add(new DB2Parameter("@lname", strLastName));
                cmd2.Parameters.Add(new DB2Parameter("@empId", intEmpId));

                DB2DataReader rs = cmd.ExecuteReader();

                if (rs.HasRows)
                {

                    while (rs.Read())

                    {
                        Console.WriteLine("Employee BEFORE surname is "
                             + rs["EMP_LAST_NAME"]);
                        intEmpId = rs.GetInt32(0);
                        strLastName = rs["EMP_LAST_NAME"].ToString();
                        strLastName = strLastName.ToUpper();
                        Console.WriteLine("Employee AFTER surname is "
                             + strLastName);
                        cmd2.Parameters["@empId"].Value = intEmpId;
                        cmd2.Parameters["@lname"].Value = strLastName;
                        cmd2.ExecuteNonQuery();

                    }

                    tran.Commit();
                    rs.Close();
                    Console.WriteLine("Successful commit ");

                }
            }

            catch (Exception e)
            {
                Console.WriteLine(e.Message);
                tran.Rollback();
            }

            finally
            {
              conn.Close();
                Console.WriteLine("Program NETEMP3 concluded successfully");

            }

        }
    }

}
```

Here's the output from this program.

```
Program NETEMP3 begins successfully
Employee BEFORE surname is johnson
Employee AFTER surname is JOHNSON
Employee BEFORE surname is stewart
Employee AFTER surname is STEWART
Employee BEFORE surname is franklin
Employee AFTER surname is FRANKLIN
Employee BEFORE surname is schultz
Employee AFTER surname is SCHULTZ
Employee BEFORE surname is willard
Employee AFTER surname is WILLARD
Employee BEFORE surname is jenkins
Employee AFTER surname is JENKINS
Employee BEFORE surname is ford
Employee AFTER surname is FORD
Employee BEFORE surname is harris
Employee AFTER surname is HARRIS
Employee BEFORE surname is turnbull
Employee AFTER surname is TURNBULL
Employee BEFORE surname is vereen
Employee AFTER surname is VEREEN
Employee BEFORE surname is yates
Employee AFTER surname is YATES
Employee BEFORE surname is miller
Employee AFTER surname is MILLER
Employee BEFORE surname is jenson
Employee AFTER surname is JENSON
Successful commit
Program NETEMP3 concluded successfully
```

Chapter Six Questions

1. To end a transaction without making the changes permanent, which DB2 statement should be issued?

    a. COMMIT
    b. BACKOUT
    c. ROLLBACK
    d. NO CHANGE

2. If you want to maximize data concurrency without seeing uncommitted data, which isolation level should you use?

    a. RR
    b. UR
    c. RS
    d. CS

3. Assume you have a long running process and you want to commit results after processing every 500 records, but still want the ability to undo any work that has taken place after the commit point. One mechanism that would allow you to do this is to issue a:

    a. SAVEPOINT
    b. COMMITPOINT
    c. BACKOUT
    d. None of the above

4. A procedure that commits transactions independent of the calling procedure is known as a/an:

    a. External procedure.
    b. SQL procedure.
    c. Autonomous procedure.
    d. Independent procedure.

5. To end a transaction and make the changes visible to other processes, which statement should be issued?

    a. ROLLBACK
    b. COMMIT
    c. APPLY
    d. CALL

6. Order the isolation levels, from greatest to least impact on performance.

    a. RR, RS, CS, UR
    b. UR, RR, RS, CS
    c. CS, UR, RR, RS
    d. RS, CS, UR, RR

7. Which isolation level is most appropriate when few or no updates are expected to a table?

    a. RR
    b. RS
    c. CS
    d. UR

Chapter Six Exercises

1. Write the Java or .NET concurrency code that ensures no records which have been retrieved during a unit of work can be changed by other processes until the unit of work completes. However, it is ok for other (new) records to be added to the table during the unit of work.

2. Write the Java or .NET code to end a unit of work and make the changes visible to other processes.

3. Write the Java or .NET code to end a unit of work and discard any updates that have been made since the last commit point.

CHAPTER SEVEN: TESTING & VALIDATING RESULTS

You obviously need to test and validate results for new and modified applications. This section is both a reminder to perform structured testing and some hints for how to go about it. If you have been an application developer for very long, most or all of this will not be new to you.

Test Structures

You normally have a test environment which is typically a separate instance of DB2 and is used primarily or exclusively for testing. You or your DBA will create test objects (tables, indexes, views) in the test environment. The DDL is usually saved and then modified as necessary to recreate the same object in another DB2 instance, or to drop and create a new version in the same instance.

If you do both production support and new development activities in the same test environment (not recommended but it's the case in many shops), it is important to have a strategy for dealing with these work flows so they don't impact each other. For example if you add a column to a table in your test system, that may be fine for the development work. But if someone else is changing the production version of a program (that happens to use the table) to resolve a production problem, it may not work in the test environment because of the new column. Or if it does work in test, when you migrate the fix to production it might not work there. Ideally, use separate test environments for new development and production support.

Test Data

It is vital to know the business rules of the application so that you select a robust set of test data for your application. To test successfully, all branches of your program or package must be tested, and then all components must be tested together (integration testing). It is especially important that in addition to testing new code and SQL, you also perform regression testing on existing code. Take the time to develop a good structured, comprehensive test plan that can be reused many times.

You can create brand new test data, or you can extract data from production and load it to test, or you can do both. Here are a few basic methods for loading test data.

1. Create the data in a flat file and write an application program to load it using INSERT statements.

2. Use the DB2 UNLOAD and LOAD utilities to extract data from production and load to test. We'll go over details for this later in this section.

Testing SQL Statements

Before coding SQL statements in an application program, you should test them using Data Studio. If you have been using DB2 for LUW for any length of time you are probably familiar with Data Studio. If you are not familiar, here is a sample query executed from the Data Studio tool.

```
SELECT * FROM HRSCHEMA.EMPLOYEE
WHERE EMP_ID IN (3217,9134);
```

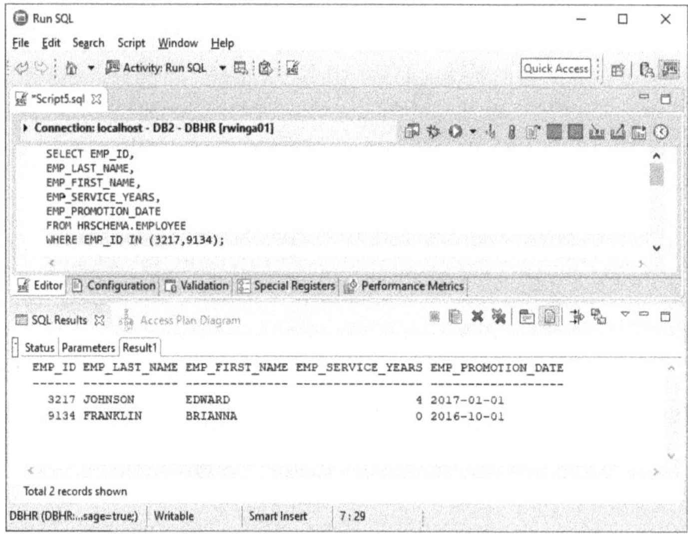

Debugging Programs

When you are getting unexpected results from your program, make sure to review these items from your compiler listing:

- Output from the language compiler or interpreter. If you receive any Java or .NET warnings or errors, make sure to resolve them. Often something that appears unimportant may be causing a run time error.

- Output from the bind process. If you are using SQLJ, did you have any error messages? How about warning messages? Resolve these.

Testing Stored Procedures

How to Test a Stored Procedure

There are several ways you can test a stored procedure. You can of course write an application program to do this. The easiest and most common way is to use Data Studio.

Suppose we have a stored procedure called GETEMP that takes an employee id as input and returns the employee's last and first names. To run it, click Stored Procedures in the object tree, right click on the name of the stored procedure, and then click on **RUN**.

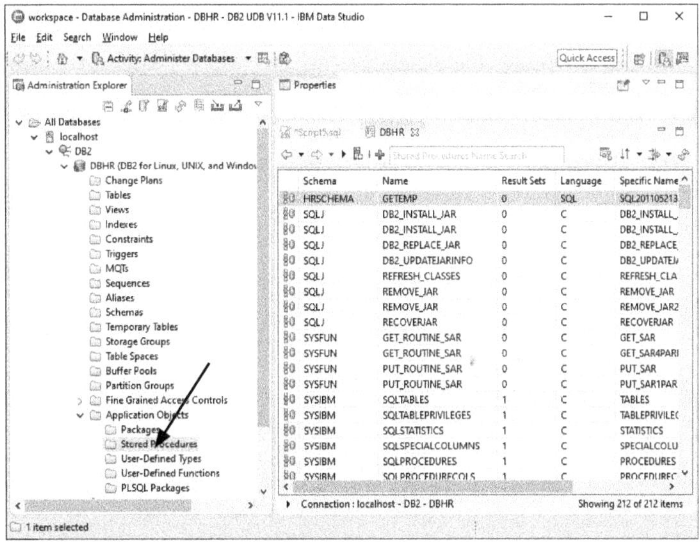

Next fill in the employee number you want data for. Let's use 3217. Then click **RUN**.

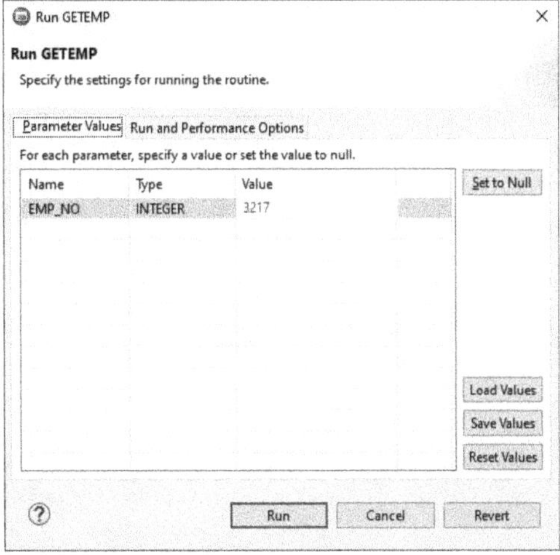

You'll see the results by clicking the **Parameters** tab.

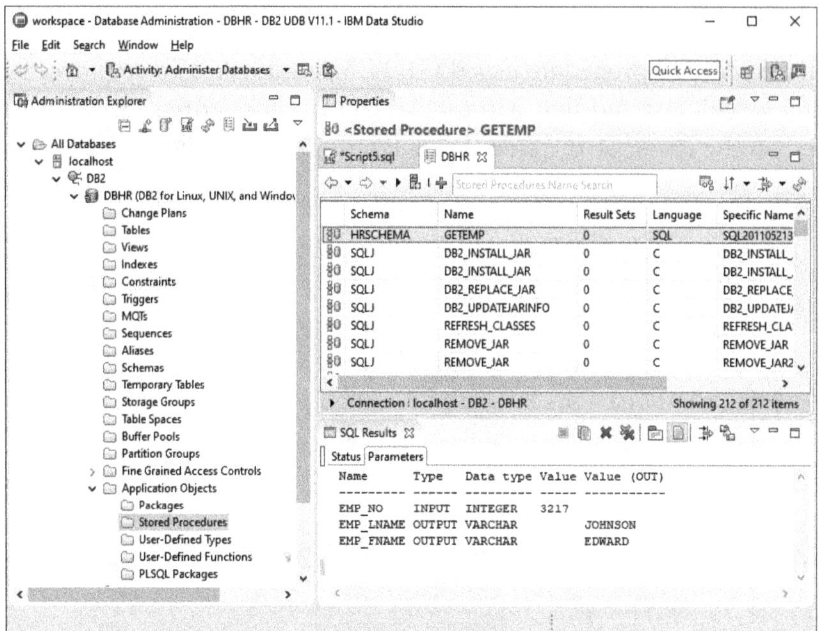

Once you've validated that the stored procedure works correctly, you can use it in programs. We'll cover stored procedures in detail in Chapter 8.

DB2 EXPORT and IMPORT utilities

EXPORT UTILITY

Use the EXPORT utility to copy the contents of a table into a flat file. You do this by right clicking the table in the database tree, selecting **UNLOAD**, then choosing **With EXPORT Utility**, and then choosing whether you want a delimited or IXF (DB2 interchange) type file. Let's choose a delimited file and we'll route it to our My Documents folder using file name hrschema.emppay.unload.txt.

At this point you can click RUN and your data will be unloaded.

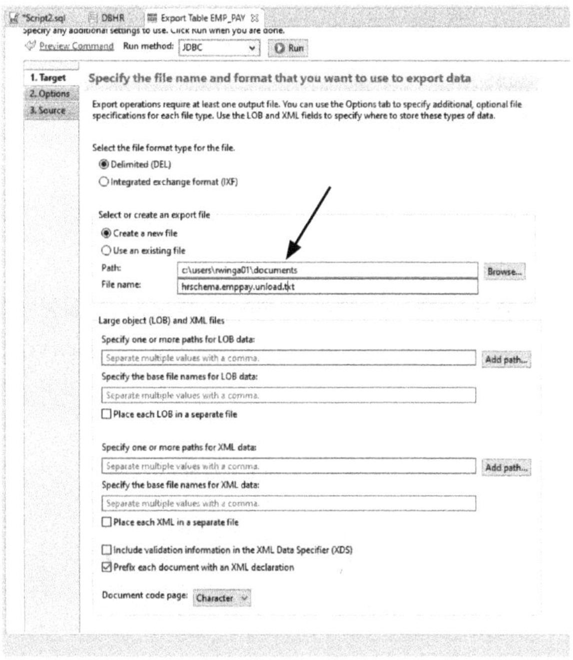

You can scroll to the bottom and if there are no errors, you will see a message indicating the export is a success.

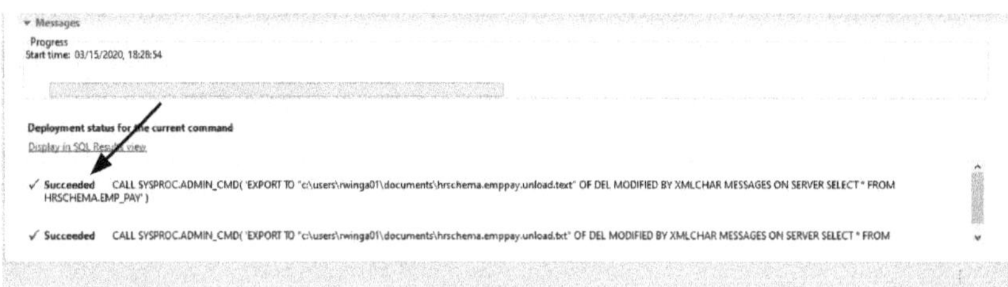

Go ahead and take a look at your data in Notepad or WordPad:

```
3217,+065000.00,+005500.00
7459,+085000.00,+004500.00
9134,+075000.00,+002500.00
6288,+070000.00,+002000.00
4720,+080000.00,+002500.00
```

If you want a subset of the data in the table, when building the EXPORT click on the Source tab and you can adjust the SQL to limit the data in whatever way you wish. Let's say we only want to unload three rows.

Click on the **Source** tab and enter or modify your SQL.

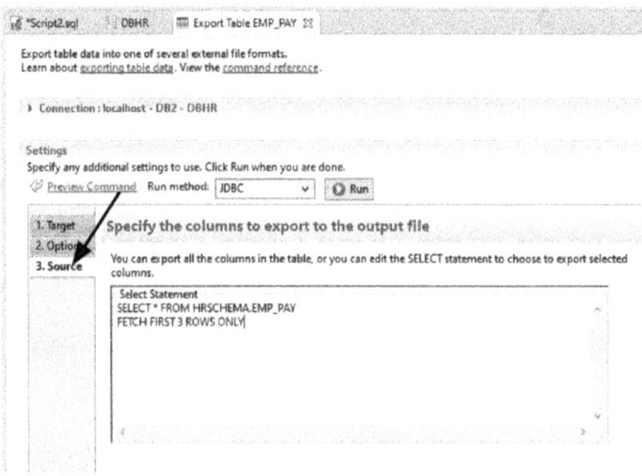

Now when you run it, your file will only include three rows, which you can verify by browsing the file with Notepad or WordPad or your favorite text editor:

```
3217,+065000.00,+005500.00
7459,+085000.00,+004500.00
9134,+075000.00,+002500.00
```

IMPORT UTILITY

Use the IMPORT utility when you want to load a DB2 table from an input file. This is useful when you are working in a test system and need to restore a baseline set of data. The input file can be in DB2 format or one of the other supported formats.

To illustrate, let's make a copy of the structure of the `EMP_PAY` table and we'll call it `EMP_PAY2.`

```
CREATE TABLE HRSCHEMA.EMP_PAY2
LIKE HRSCHEMA.EMP_PAY;
```

Now let's find and right click the new `EMP_PAY2` table in the database tree (you may need to refresh the tree by right clicking and specifying Refresh). Select **LOAD** and then select **With Import Utility**. You will see this panel.

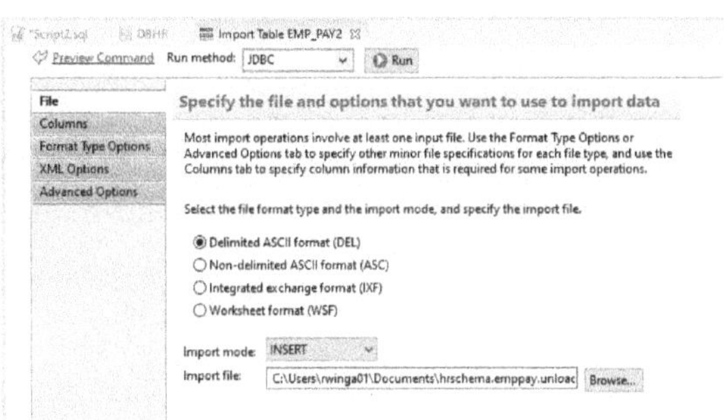

Here you must specify the import file type (in this case delimited) as well as the import mode.

If you choose INSERT mode, then records will only be inserted – if a record with the same key already exists on the table, the new record will be discarded.

If you choose Insert/Replace, then the record will be inserted if it does not exist and replaced if it does exist – this is essentially a Merge action.

Finally you can choose REPLACE in which case the record must already exist in the table and if it does, it will be updated (if it does not already exist, the new record will be discarded).

Finally, specify the input file name, and then you can click on Run.

Check for the succeeded message, and then you can query the `EMP_PAY2` table to verify that the records were loaded. In this case we will right click on the table and select **Data --> Browse Data.** Here is the result:

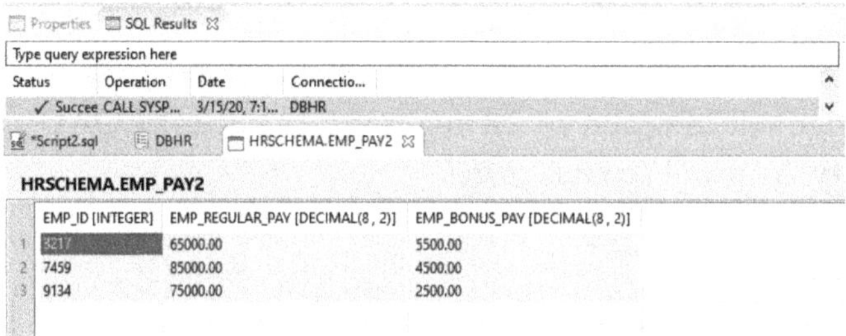

Again the import utility is most useful for running testing where you need to reload the same data. I recommend you use it often in both development and maintenance environments.

Application Performance

Here are some factors that lead to efficient design. You may or may not be expected to help tune your team's applications (in some shops the DBAs to this), but you should absolutely be aware of efficient programming practices. Here are some suggestions.

1. Make sure data concurrency is set optimally.

    - Commit your work regularly to reduce contention due to locking. In some cases this means committing after a certain amount of time as opposed to a number of transactions.

    - Bind most applications with the `ISOLATION(CS)` which is considered optimal to prevent unnecessary locking and to release locks as soon as possible.

    - Include logic in your application program to retry after a deadlock or timeout to attempt recovery from the contention situation without assistance.

2. Use stored procedures to improve performance– they are compiled and run on the server for optimum performance.

3. Check with the database administrator to see if the table needs to be reorganized.

4. Evaluate your SQL, checking to see if you are using stage 1 predicates versus stage 2 predicates.

We'll consider application performance in more detail in the next chapter of this book.

Chapter Seven Questions

1. Suppose you have created a test version of a production table, and you want to to use the UNLOAD utility to extract the first 1,000 rows from the production table to load to the test version. Which keyword would you use in the UNLOAD statement?

    a. WHEN
    b. SELECT
    c. SAMPLE
    d. SUBSET

2. Which of the following is an SQL error code indicating a DB2 package is not found within the DB2 plan?

    a. -803
    b. -805
    c. -904
    d. -922

3. Which DB2 utility updates the statistics used by the DB2 Optimizer to choose a data access path?

    a. REORG
    b. RUNSTATS
    c. REBIND
    d. OPTIMIZE

4. Which of the following is NOT a way you could test a DB2 SQL statement?

    a. Running the statement from the DB2 command line processor.
    b. Running the statement from a Java program.
    c. Running the statement from IBM Data Studio.
    d. All of the above are valid ways to test an SQL statement.

Chapter Seven Exercises:

1. Write an EXPORT control statement to unload a sample of 500 records from the HRSCHEMA.EMP_PAY table for use in loading a test table.

2. Suppose that the HRSCHEMA.EMPLOYEE table exists on a system whose location name is DENVER. Write a query that uses three part naming to implicitly connect to DENVER and then retrieve all information for EMP_ID 3217.

CHAPTER EIGHT: ADVANCED TOPICS

STORED PROCEDURES

A stored procedure is a set of compiled statements that is stored on the DB2 server. The stored procedures typically include SQL statements to access data in a DB2 table. Stored procedures are similar to sub-programs in that they can be called by other programs. Specifically, stored procedures are invoked by the CALL statement as in:

```
CALL <stored procedure name><(parameters)>
```

Stored procedures can be called from an application program such as COBOL, from a Rexx exec, from QMF or from Data Studio. Stored procedures are created using the `CREATE PROCEDURE` statement. The details of the stored procedure depend on whether it is external or native. We'll look at examples of each.

Types of stored procedures

There are three types of stored procedures:

- Native SQL procedure
- External procedure
- Sourced procedure

Native SQL procedures

A native SQL procedure is a procedure that consists exclusively of SQL statements, and is created entirely within the CREATE PROCEDURE statement. Native SQL procedures are not associated with an external program.

External stored procedures

An external stored procedure is one written in a programming language such as COBOL or Java. When you want to create an external stored procedure, the following programming languages can be used:

- C/C++
- Java
- COBOL
- .NET CLR (common language runtime)
- OLE DB (table functions only)

Sourced procedures

A sourced procedure is based upon another procedure. In a federated system, a *federated procedure* is a sourced procedure whose source procedure is at a supported data source.

Examples of Stored Procedures

Native SQL Stored Procedure

Let's start with a procedure that will return the first and last names of an employee, given an employee number. We will pass employee number as an IN parameter and receive the employee's first and last names as OUT parameters. Since we are only using SQL statements, we will specify the SQL language in the definition, and specify our intent to read data.

```
CREATE PROCEDURE HRSCHEMA.GETEMP (IN EMP_NO INT,
  OUT EMP_LNAME VARCHAR(30),
  OUT EMP_FNAME VARCHAR(20))

LANGUAGE SQL
READS SQL DATA

 BEGIN
    SELECT EMP_LAST_NAME,
           EMP_FIRST_NAME
    INTO EMP_LNAME,
         EMP_FNAME
    FROM HRSCHEMA.EMPLOYEE
    WHERE EMP_ID = EMP_NO;
END
```

When you create a stored procedure, you must authorize its use with a grant. We'll be generous and grant to PUBLIC but in a real work environment you would most likely grant it to members of a certain group.

```
GRANT EXECUTE ON PROCEDURE HRSCHEMA.GETEMP TO PUBLIC;
```

Now we need a program to call the stored procedure. We will develop a Java program to do that. If you are not familiar with Java coding for DB2, I recommend obtaining the book **DB2 11.1 for LUW: Basic Training for Application Developers**.

Meanwhile, here is our code listing for calling the `GETEMP` stored procedure. In this case we've hard-coded employee number 3217 to pass to the `GETEMP` stored procedure.

```
package employee;
import java.sql.*;
public class JAVEMP7 {
```

```
public static void main(String[] args) throws ClassNotFoundException
{
        Class.forName("com.ibm.db2.jcc.DB2Driver");
        System.out.println("**** Loaded the JDBC driver");
    String url
        ="jdbc:db2://localhost:50000/DBHR:retrieveMessagesFromServerOn-
GetMessage=true;";
    String user= "rwinga01";
    String password="*********";

    Connection con=null;
    CallableStatement cs=null;

        try {
                con=DriverManager.getConnection(url, user, password);
                System.out.println("**** Created the connection");
                int empId = 3217;
                String lName;
                String fName;

            System.out.println("Executing JAVEMP7 to call stored
                procedure GETEMP");

                cs = con.prepareCall("{call HRSCHEMA.GETEMP(?,?,?)}");
                cs.registerOutParameter(2, Types.VARCHAR);
                cs.registerOutParameter(3, Types.VARCHAR);
                cs.setInt(1, 3217);

                cs.execute();

                lName = cs.getString(2);
                fName = cs.getString(3);

                System.out.println("Employee ID : " + empId);
                System.out.println("Last Name   : " + lName);
                System.out.println("First Name  : " + fName);
                System.out.println("Completed execution of JAVEMP7");

        } catch (SQLException e) {
                System.out.println(e.getMessage());
                System.out.println(e.getErrorCode());
                e.printStackTrace();

        }
    }
}
```

Now when we run the procedure we get the following results:

```
**** Loaded the JDBC driver
**** Created the connection
```

```
Executing JAVEMP7 to call stored procedure GETEMP
Employee ID : 3217
Last Name   : JOHNSON
First Name  : EDWARD
Completed execution of JAVEMP7
```

The .NET version of the program is as follows:

```
using System;
using System.Collections.Generic;
using System.Data;
using System.Linq;
using System.Text;
using System.Threading.Tasks;
using IBM.Data.DB2;
using IBM.Data.DB2Types;

/* Program to connect to DB2 for calling stored procedure  */
namespace NETEMP7
{
    class NETEMP7
    {
        static void Main(string[] args)
        {
            DB2Connection conn = null;
            DB2Command cmd = null;

            try
            {
                conn = new DB2Connection("Database=DBHR");
                conn.Open();
                Console.WriteLine("Successful connection!");
                cmd = conn.CreateCommand();
                String procName = "HRSCHEMA.GETEMP";
                cmd.CommandType = System.Data.CommandType.StoredProcedure;
                cmd.CommandText = procName;

                // Define parameters and establish direction for output
                // parameters

                cmd.Parameters.Add(new DB2Parameter("@empid", 3217));
                cmd.Parameters.Add(new DB2Parameter("@lname", ""));
                cmd.Parameters.Add(new DB2Parameter("@fname", ""));
                cmd.Parameters[1].Direction = ParameterDirection.Output;
                cmd.Parameters[2].Direction = ParameterDirection.Output;


                /* Execute the stored procedure */

                Console.WriteLine("Call stored procedure named " + procName);
                cmd.ExecuteNonQuery();
```

```
                /*  Now capture and display the results */

                String empID = cmd.Parameters[0].Value.ToString();
                String lName = cmd.Parameters[1].Value.ToString();
                String fName = cmd.Parameters[2].Value.ToString();

                Console.WriteLine("Employee id : " + empID);
                Console.WriteLine("Last Name   : " + lName);
                Console.WriteLine("First Name  : " + fName);

            }

            catch (Exception e)
            {
                Console.WriteLine(e.Message);
            }

            finally
            {
                conn.Close();
            }

        }
    }
}
```

And here is the result of the run:

```
Successful connection!
Call stored procedure named HRSCHEMA.GETEMP
Employee id : 3217
Last Name   : JOHNSON
First Name  : EDWARD
The program '[9516] NETEMP7.exe' has exited with code 0 (0x0).
```

External Stored Procedure

Now let's do the same procedure but we'll make it an external procedure implemented with a Java program. First, let's define the procedure. We'll call it **get_emp_info_java,** and the main method will be **get_info**.

```
CREATE OR REPLACE PROCEDURE HRSCHEMA.get_emp_info_java
(IN empID Integer,
OUT EMP_LNAME VARCHAR(30),
OUT EMP_FNAME VARCHAR(20),
OUT sqlCode Integer,
OUT sqlMsg    VARCHAR (255))
FENCED
READS SQL DATA
LANGUAGE JAVA
EXTERNAL NAME  'employee.get_emp_info_java.get_info'
```

```
DYNAMIC RESULT SETS 0
PROGRAM TYPE SUB
PARAMETER STYLE JAVA
```

We'll need to grant security on the new stored procedure:

```
GRANT EXECUTE ON PROCEDURE HRSCHEMA.get_emp_info_java TO PUBLIC;
```

Next, let's write the program to implement this procedure. Our class name must be **get_emp_info_java** and we must create a **get_info** method to query the DB2 data.

```
package employee;
import java.sql.*;

public class get_emp_info_java {

        public static void get_info (int empID,
        String[] lName,
        String[] fName,
        int[] sqlCode,
        String[] sqlMsg) throws ClassNotFoundException, SQLException

        {
                Class.forName("com.ibm.db2.jcc.DB2Driver");
                DriverManager.registerDriver(new com.ibm.db2.jcc.DB2Driver());

            Connection con=null;

                try {
                        System.out.println("**** Loaded the JDBC driver");
                        con =
                        DriverManager.getConnection("jdbc:default:connection");
                        System.out.println("**** Created the connection");


                        String query = "SELECT EMP_LAST_NAME, "
                                     + "EMP_FIRST_NAME "
                                     + "FROM HRSCHEMA.EMPLOYEE "
                                     + "WHERE EMP_ID = " + empID;

                        System.out.println(query);

                        Statement stmt;
                        stmt = con.createStatement();
                        ResultSet rs=stmt.executeQuery(query);
                        while(rs.next())
                        {
                            lName[0] = rs.getString(1);
                            fName[0] = rs.getString(2);
                        }
```

```
            } catch (SQLException e) {
                sqlMsg[0] = e.getMessage();
                sqlCode[0] = e.getErrorCode();
                    System.out.println(e.getMessage());
                    System.out.println(e.getErrorCode());
                    e.printStackTrace();

            }
        }
}
```

Finally, we must either move the class file to the default location where DB2 stores function type routines, or else we must fully qualify the path name. We'll choose to move the class file to the function area. The DB2 location for functions is the **sqllib/function** directory in the home directory of the database instance owner. We're also specified in the Java code that our program is part of package **employee**. For this reason we must create subdirectory employee under the sqllib/function directory – DB2 will look for the get_emp_info_java class file there.

Important note: Once you've implemented or changed the procedure you must do one of two things to ensure that DB2 loads the newest version of your procedure. Either

1. Issue the following command:

   ```
   CALL SQLJ.REFRESH_CLASSES()
   ```

   OR

2. Bounce your DB2 instance by issuing `DB2STOP` and then `DB2START` commands.

Now we need a program to call the external stored procedure. We can clone the one we wrote to call the native stored procedure. That was `JAVEMP7` and all we need to do is change the name of the procedure we are calling. The new program name is `JAVEMP8`.

```
package employee;
import java.sql.*;

public class JAVEMP8 {

        public static void main(String[] args) throws ClassNotFoundException
        {
                Class.forName("com.ibm.db2.jcc.DB2Driver");
                System.out.println("**** Loaded the JDBC driver");
                //String jdbcClassName="com.ibm.db2.jcc.DB2Driver";
            String url
                ="jdbc:db2://localhost:50000/DBHR:retrieveMessagesFromServerOn-
        GetMessage=true;";
            String user="rwinga01";
```

```
        String password="*********";

        Connection con=null;
        CallableStatement cs=null;

          try {
                //Class.forName(jdbcClassName);
                con=DriverManager.getConnection(url, user, password);

                System.out.println("**** Created the connection");
                int empId = 3217;
                String lName;
                String fName;
                int sqlCode;
                String sqlMsg;

             System.out.println("Executing JAVEMP8 to call stored proce-
             dure get_emp_info_java");

                cs = con.prepareCall("{call HRSCHEMA.get_emp_info_
                java(?,?,?,?,?)}");

                cs.setInt(1, empId);
                cs.registerOutParameter(2, Types.VARCHAR);
                cs.registerOutParameter(3, Types.VARCHAR);
                cs.registerOutParameter(4, Types.INTEGER);
                cs.registerOutParameter(5, Types.VARCHAR);
                cs.execute();

                lName   = cs.getString(2);
                fName   = cs.getString(3);
                sqlCode = cs.getInt(4);
                sqlMsg  = cs.getString(5);


                System.out.println("Employee ID : " + empId);
               System.out.println("Last Name   : " + lName);
               System.out.println("First Name  : " + fName);
               System.out.println("SQLCODE     : " + sqlCode);
               System.out.println("Message     : " + sqlMsg);
               System.out.println("Completed execution of JAVEMP8");

          } catch (SQLException e) {
                System.out.println(e.getMessage());
                System.out.println(e.getErrorCode());
                e.printStackTrace();

          }
     }
}
```

Now when we run this program, it will call the stored procedure and display these results:

```
**** Loaded the JDBC driver
**** Created the connection
Executing JAVEMP8 to call stored procedure get_emp_info_java
Employee ID : 3217
Last Name   : JOHNSON
First Name  : EDWARD
SQLCODE     : 0
Message     : null
Completed execution of JAVEMP8
```

Sourced Stored Procedure

Suppose you want to call a federated procedure named EMPNROLL on server ORCLFED1. You know that the schema name for the procedure is RECRUIT and the package name is P1. On your local system you want to name the package EMP_ENROLL under schema HRSCHEMA. Here is the DDL to do this:

```
CREATE PROCEDURE HRSCHEMA.EMP_ENROLL
SOURCE RECRUIT.P1.EMPNROLL
FOR SERVER ORCLFED1
WITH RETURN TO CLIENT ALL
```

You can then call the procedure HRSCHEMA.EMP_ENROLL which will invoke procedure RECRUIT.P1.EMPNROLL on the federated server.

Note: Before a federated procedure can be registered for a data source, the federated server must be configured to access that data source. This includes: registering the wrapper for the data source, creating the server definition for the data source, and creating the user mappings between the federated server and the data source server for the data sources that require user mapping. For details, see the IBM Knowledge Center for DB2 11.1.

More Stored Procedure Examples

Let's do a few more examples of stored procedures, and in this case we'll create some data access routines. Specifically we'll create stored procedures to retrieve information for an employee, to add or update an employee, and to delete an employee.

For retrieving employee data, we'll simply expand our GETEMP procedure to include all of the original fields we created the table with. But in addition, let's first add a new column to the EMPLOYEE table which is the employee social security number.

```
ALTER TABLE HRSCHEMA.EMPLOYEE
ADD COLUMN EMP_SSN CHAR(09);

CREATE INDEX HRSCHEMA.NDX_SSN ON HRSCHEMA.EMPLOYEE (EMP_SSN);
```

Now, let's call the new procedure GET_EMP_INFO.

```
CREATE OR REPLACE PROCEDURE HRSCHEMA.GET_EMP_INFO
(IN EMP_NO INT,
OUT EMP_LNAME VARCHAR(30),
OUT EMP_FNAME VARCHAR(20),
OUT EMP_SRVC_YRS INT,
OUT EMP_PROM_DATE DATE,
OUT EMP_PROF XML,
OUT EMP_SSN  CHAR(09))

LANGUAGE SQL
READS SQL DATA

BEGIN

   SELECT EMP_LAST_NAME,
          EMP_FIRST_NAME,
          EMP_SERVICE_YEARS,
          EMP_PROMOTION_DATE,
          EMP_PROFILE,
          EMP_SSN
    INTO EMP_LNAME,
         EMP_FNAME,
         EMP_SRVC_YRS,
         EMP_PROM_DATE,
         EMP_PROF,
         EMP_SSN
    FROM HRSCHEMA.EMPLOYEE
    WHERE EMP_ID = EMP_NO;
 END
```

Next, we'll create a procedure that merges the input data into the table, either adding it if it is a new record, or updating it if an existing record.

```
CREATE PROCEDURE HRSCHEMA.MRG_EMP_INFO
(IN EMP_NO INT,
 IN EMP_LNAME VARCHAR(30),
 IN EMP_FNAME VARCHAR(20),
 IN EMP_SRVC_YRS INT,
 IN EMP_PROM_DATE DATE,
 IN EMP_PROF XML,
 IN EMP_SOC_SEC  CHAR(09))
 LANGUAGE SQL
 MODIFIES SQL DATA

 BEGIN
    MERGE INTO HRSCHEMA.EMPLOYEE AS T
    USING
     (VALUES (EMP_NO,
     EMP_LNAME,
     EMP_FNAME,
```

```
      EMP_SRVC_YRS,
      EMP_PROM_DATE,
      EMP_PROF,
      EMP_SOC_SEC))
      AS S
      (EMP_ID,
       EMP_LAST_NAME,
       EMP_FIRST_NAME,
       EMP_SERVICE_YEARS,
       EMP_PROMOTION_DATE,
       EMP_PROFILE,
       EMP_SSN)
      ON S.EMP_ID = T.EMP_ID

      WHEN MATCHED
         THEN UPDATE
            SET EMP_ID             = S.EMP_ID,
                EMP_LAST_NAME      = S.EMP_LAST_NAME,
                EMP_FIRST_NAME     = S.EMP_FIRST_NAME,
                EMP_SERVICE_YEARS  = S.EMP_SERVICE_YEARS,
                EMP_PROMOTION_DATE = S.EMP_PROMOTION_DATE,
                EMP_PROFILE        = S.EMP_PROFILE,
                EMP_SSN            = S.EMP_SSN

      WHEN NOT MATCHED
         THEN INSERT
            VALUES (S.EMP_ID,
            S.EMP_LAST_NAME,
            S.EMP_FIRST_NAME,
            S.EMP_SERVICE_YEARS,
            S.EMP_PROMOTION_DATE,
            S.EMP_PROFILE,
            S.EMP_SSN) ;
END
```

Finally, let's take care of the delete function.

```
CREATE PROCEDURE HRSCHEMA.DLT_EMP_INFO
(IN EMP_NO INT)

 LANGUAGE SQL
 MODIFIES SQL DATA

 BEGIN
    DELETE FROM HRSCHEMA.EMPLOYEE
    WHERE EMP_ID = EMP_NO;

 END
```

Before we can use these procedures we must grant access to them. In our case we will grant to public, but normally you will grant access only to your developer and user groups.

```
GRANT EXECUTE ON PROCEDURE HRSCHEMA.GET_EMP_INFO TO PUBLIC;
GRANT EXECUTE ON PROCEDURE HRSCHEMA.MRG_EMP_INFO TO PUBLIC;
GRANT EXECUTE ON PROCEDURE HRSCHEMA.DLT_EMP_INFO TO PUBLIC;
```

Next we need a Java program to test each of these stored procedures. Here is one that works:

```
package employee;
import java.sql.*;
public class JAVEMPA {

        public static void main(String[] args) throws ClassNotFoundException
        {
            System.out.println("Executing JAVEMPA to call several stored
            procedures");

                Class.forName("com.ibm.db2.jcc.DB2Driver");
                System.out.println("**** Loaded the JDBC driver");
            String url
                ="jdbc:db2://localhost:50000/DBHR:retrieveMessagesFromServerOn-
        GetMessage=true;";
            String user="rwinga01";
            String password="*********";

            Connection con=null;
            CallableStatement cs=null;

                try {
                con=DriverManager.getConnection(url, user, password);
                System.out.println("**** Created the connection");

                int empId           = 7938;
                String lName        = "WINFIELD";
                String fName        = "STANLEY";
                int srvcYears       = 3;
                Date promDate       = null;
                SQLXML empProfile   = null;
                String empSSN       = "382734509";
                System.out.println("Calling stored procedure
                  HRSCHEMA.MRG_EMP_INFO to add or update a record");

                cs = con.prepareCall("{call
                  HRSCHEMA.MRG_EMP_INFO(?,?,?,?,?,?,?)}");
             cs.setInt(1, empId);
             cs.setString(2, lName);
             cs.setString(3, fName);
             cs.setInt(4, srvcYears);
             cs.setNull(5, Types.DATE);
             cs.setNull(6, Types.SQLXML);
             cs.setString(7, empSSN);

             cs.execute();
```

```
System.out.println("Employee ID : " + empId);
System.out.println("Last Name   : " + lName);
System.out.println("First Name  : " + fName);
System.out.println("Service Years  : " + srvcYears);
System.out.println("Promotion Date  : " + promDate);
System.out.println("Profile  : " + empProfile);
System.out.println("SSN  : " + empSSN);

System.out.println("Completed successful execution of procedure
   HRSCHEMA.MRG_EMP_INFO");
System.out.println("********************");

System.out.println("Calling stored procedure
   HRSCHEMA.GET_EMP_INFO to display a record");

cs = con.prepareCall("{call
   HRSCHEMA.GET_EMP_INFO(?,?,?,?,?,?,?)}");
cs.setInt(1, empId);
cs.registerOutParameter(2, Types.VARCHAR);
         cs.registerOutParameter(3, Types.VARCHAR);
cs.registerOutParameter(4, Types.VARCHAR);
cs.registerOutParameter(5, Types.DATE);
cs.registerOutParameter(6, Types.SQLXML);
cs.registerOutParameter(7, Types. VARCHAR);

cs.execute();

lName      = cs.getString(2);
fName      = cs.getString(3);
srvcYears  = cs.getInt(4);
promDate   = cs.getDate(5);
empProfile = cs.getSQLXML(6);
empSSN     = cs.getString(7);

System.out.println("Employee ID  : " + empId);
System.out.println("Last Name    : " + lName);
System.out.println("First Name   : " + fName);
System.out.println("Serv Years   : " + srvcYears);
System.out.println("Prom Date    : " + promDate);
System.out.println("Profile     : " + empProfile);
System.out.println("SSN          : " + empSSN);
System.out.println("Completed successful execution of procedure
HRSCHEMA.GET_EMP_INFO");

System.out.println("********************");
System.out.println("Calling stored procedure HRSCHEMA.DLT_EMP_INFO
to delete a record");

empId = 7938;

cs = con.prepareCall("{call HRSCHEMA.DLT_EMP_INFO(?)}");
cs.setInt(1, empId);
```

```
                cs.execute();

                System.out.println("Employee ID " + empId + " was successfully
                deleted");

                System.out.println("Completed successful execution of procedure
                HRSCHEMA.DLT_EMP_INFO");

                System.out.println("*******************");
                System.out.println("Completed execution of JAVEMPA");

            } catch (SQLException e) {
                  System.out.println(e.getMessage());
                  System.out.println(e.getErrorCode());
                  e.printStackTrace();

            }
        }
}
```

Here is the output from the program run:

```
Executing JAVEMPA to call several stored procedures
**** Loaded the JDBC driver
**** Created the connection
Calling stored procedure HRSCHEMA.MRG_EMP_INFO to add or update a record
Employee ID : 7938
Last Name   : WINFIELD
First Name  : STANLEY
Service Years  : 3
Promotion Date  : null
Profile  : null
SSN  : 382734509
Completed successful execution of procedure HRSCHEMA.MRG_EMP_INFO
********************
Calling stored procedure HRSCHEMA.GET_EMP_INFO to display a record
Employee ID : 7938
Last Name   : WINFIELD
First Name  : STANLEY
Service Years  : 3
Promotion Date  : null
Profile  : null
SSN  : 382734509
Completed successful execution of procedure HRSCHEMA.GET_EMP_INFO
********************
Calling stored procedure HRSCHEMA.DLT_EMP_INFO to delete a record
Employee ID 7938 was successfully deleted
Completed successful execution of procedure HRSCHEMA.DLT_EMP_INFO
********************
Completed successful execution of procedure HRSCHEMA.MRG_EMP_INFO
********************
Completed execution of JAVEMPA
```

Here is our .NET version of the program to call the data access stored procedures.

```
using System;
using System.Collections.Generic;
using System.Data;
using System.Linq;
using System.Text;
using System.Threading.Tasks;
using IBM.Data.DB2;
using IBM.Data.DB2Types;

/* Program to connect to DB2 for calling stored procedure  */
namespace NETEMPA
{
    class NETEMPA
    {
        static void Main(string[] args)
        {
            DB2Connection conn = null;
            DB2Command cmd = null;
            DB2Command cmd2 = null;
            DB2Command cmd3 = null;

            try
            {
                conn = new DB2Connection("Database=DBHR");
                conn.Open();
                Console.WriteLine("Executing NETEMPA to call several stored
                procedures");
                Console.WriteLine("Successful connection!");

                Console.WriteLine("Calling stored procedure
                HRSCHEMA.MRG_EMP_INFO to add or update a record");

                cmd = conn.CreateCommand();
                String procName = "HRSCHEMA.MRG_EMP_INFO";
                cmd.CommandType = System.Data.CommandType.StoredProcedure;
                cmd.CommandText = procName;

                // Load the work variables

                int empId = 7938;
                String lName = "WINFIELD";
                String fName = "STANLEY";
                int srvcYears = 3;
                DateTime? promDate = null;  // if no date, we will set the input
                                            // parameter value to NULL of date

                String empProfile = null;  // if no employee profile, we will
                                            // set the input parameter value to
                                            // NULL of XML
```

```
String empSSN = "382734509";
String strPromDate = "";
String strEmpProfile = "";

// Define parameters, establish direction for output parameters

cmd.Parameters.Add(new DB2Parameter("@empid", empId));
cmd.Parameters.Add(new DB2Parameter("@lname", lName));
cmd.Parameters.Add(new DB2Parameter("@fname", fName));
cmd.Parameters.Add(new DB2Parameter("@srvcYears", srvcYears));
cmd.Parameters.Add(new DB2Parameter("@promDate", promDate));
cmd.Parameters.Add(new DB2Parameter("@empProfile", empProfile));
cmd.Parameters.Add(new DB2Parameter("@empSSN", empSSN));

/* Execute the stored procedure */

cmd.ExecuteNonQuery();

Console.WriteLine("Successfully called stored procedure "
+ procName);

Console.WriteLine("****************************************");

Console.WriteLine("Calling stored procedure
HRSCHEMA.GET_EMP_INFO to retrieve a record");

cmd2 = conn.CreateCommand();
procName = "HRSCHEMA.GET_EMP_INFO";
cmd2.CommandType = System.Data.CommandType.StoredProcedure;
cmd2.CommandText = procName;

cmd2.Parameters.Add(new DB2Parameter("@empid",DB2Type.Integer));
cmd2.Parameters.Add(new DB2Parameter("@lname", DB2Type.VarChar));
cmd2.Parameters.Add(new DB2Parameter("@fname", DB2Type.VarChar));
cmd2.Parameters.Add(new DB2Parameter("@srvcYears", DB2Type.Integer));
cmd2.Parameters.Add(new DB2Parameter("@promDate", DB2Type.DateTime));
cmd2.Parameters.Add(new DB2Parameter("@empProfile", DB2Type.Xml));
cmd2.Parameters.Add(new DB2Parameter("@empSSN", DB2Type.VarChar));

cmd2.Parameters[1].Direction = ParameterDirection.Output;
cmd2.Parameters[2].Direction = ParameterDirection.Output;
cmd2.Parameters[3].Direction = ParameterDirection.Output;
cmd2.Parameters[4].Direction = ParameterDirection.Output;
cmd2.Parameters[5].Direction = ParameterDirection.Output;
cmd2.Parameters[6].Direction = ParameterDirection.Output;

cmd2.Parameters[0].Value = 7938;

cmd2.ExecuteNonQuery();

/*  Now capture and display the results */

String strEmpId        = cmd2.Parameters[0].Value.ToString();
lName                  = cmd2.Parameters[1].Value.ToString();
```

```
    fName                 = cmd2.Parameters[2].Value.ToString();
    String strSrvcYears   = cmd2.Parameters[3].Value.ToString();

    if (cmd2.Parameters[4].Value != DBNull.Value)
       strPromDate    = cmd2.Parameters[4].Value.ToString();

    if (cmd2.Parameters[5].Value != DBNull.Value)
        strEmpProfile = cmd2.Parameters[5].Value.ToString();

    if (cmd2.Parameters[6].Value != DBNull.Value)
       empSSN = cmd2.Parameters[6].Value.ToString();

    Console.WriteLine("Employee id : " + strEmpId);
    Console.WriteLine("Last Name   : " + lName);
    Console.WriteLine("First Name  : " + fName);
    Console.WriteLine("Srvc Years  : " + strSrvcYears);
    Console.WriteLine("Prom Date   : " + strPromDate);
    Console.WriteLine("Profile     : " + strEmpProfile);
    Console.WriteLine("Soc Sec     : " + empSSN);

    Console.WriteLine("Successfully called stored procedure "
    + procName);

    // NOW DELETE THE RECORD

    Console.WriteLine("Calling stored procedure
    HRSCHEMA.DLT_EMP_INFO to delete a record");

    cmd3 = conn.CreateCommand();
    procName = "HRSCHEMA.DLT_EMP_INFO";
    cmd3.CommandType = System.Data.CommandType.StoredProcedure;
    cmd3.CommandText = procName;

    cmd3.Parameters.Add(new DB2Parameter("@empid",DB2Type.Integer));
    cmd3.Parameters[0].Value = 7938;
    cmd3.ExecuteNonQuery();

    Console.WriteLine("Employee " + empId + " was successfully
    deleted");
    Console.WriteLine("Successfully called stored procedure "
    + procName);

}

catch (Exception e)
{
    Console.WriteLine(e.Message);
}

finally
{
    conn.Close();
}
```

```
            }
        }
    }
```

And here is the output:

```
Executing NETEMPA to call several stored procedures
Successful connection!
Calling stored procedure HRSCHEMA.MRG_EMP_INFO to add or update a record
Successfully called stored procedure HRSCHEMA.MRG_EMP_INFO
*************************************************
Calling stored procedure HRSCHEMA.GET_EMP_INFO to retrieve a record
Employee id : 7938
Last Name   : WINFIELD
First Name  : STANLEY
Srvc Years  : 3
Prom Date   :
Profile     :
Soc Sec     : 382734509
Successfully called stored procedure HRSCHEMA.GET_EMP_INFO
Calling stored procedure HRSCHEMA.DLT_EMP_INFO to delete a record
Employee 7938 was successfully deleted
Successfully called stored procedure HRSCHEMA.DLT_EMP_INFO
The program '[8080] NETEMPA.exe' has exited with code 0 (0x0).
```

Error handling with Stored Procedures

One thing we have not done yet is encounter an error in a native SQL stored procedure. There are a few ways to handle this, but I find the quickest and easiest is to use the `GET DIAGNOSTICS` feature of DB2 to return an error message.

Let's change our `GET_EMP_INFO` stored procedure to include an output parameter to pass the error message. Then we'll need to add the `GET DIAGNOSTICS` code right after the BEGIN statement. Finally we'll assign the trapped message to the `SQL_MSG` parameter. I've bolded the new code. Go ahead and drop the `GET_EMP_INFO` procedure, and then recreate it with the code below.

```
CREATE PROCEDURE HRSCHEMA.GET_EMP_INFO
(IN EMP_NO INT,
OUT EMP_LNAME VARCHAR(30),
OUT EMP_FNAME VARCHAR(20),
OUT EMP_SRVC_YRS INT,
OUT EMP_PROM_DATE DATE,
OUT EMP_PROF XML,
OUT EMP_SSN  CHAR(09),
OUT SQL_MSG VARCHAR(250))
```

```
LANGUAGE SQL
READS SQL DATA

BEGIN
   DECLARE SQLMSG VARCHAR(250);
   DECLARE CONTINUE HANDLER FOR SQLEXCEPTION, SQLWARNING, NOT FOUND
   GET DIAGNOSTICS EXCEPTION 1 SQLMSG = MESSAGE_TEXT;

   SELECT EMP_LAST_NAME,
          EMP_FIRST_NAME,
          EMP_SERVICE_YEARS,
          EMP_PROMOTION_DATE,
          EMP_PROFILE,
          EMP_SSN
    INTO EMP_LNAME,
         EMP_FNAME,
         EMP_SRVC_YRS,
         EMP_PROM_DATE,
         EMP_PROF,
         EMP_SSN
    FROM HRSCHEMA.EMPLOYEE
    WHERE EMP_ID = EMP_NO;

    SET SQL_MSG = SQLMSG;

 END
```

Now let's run the revised procedure with a nonexistent employee number. We can run it in Data Studio by right clicking on the stored procedure and selecting **Run.** You'll be prompted to enter an employee id. Enter 9999. Click **Run.**

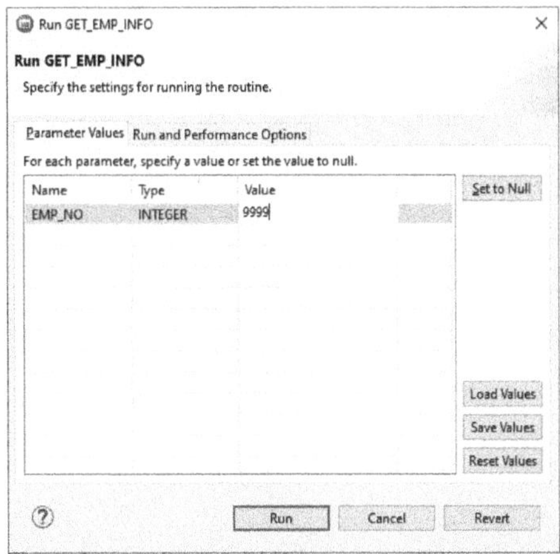

As you can see, the result is that we have a message indicating the record was not found.

SQL Results

Status | Parameters

Name	Type	Data type	Value	Value (OUT)
EMP_NO	INPUT	INTEGER	9999	
EMP_LNAME	OUTPUT	VARCHAR		*NULL*
EMP_FNAME	OUTPUT	VARCHAR		*NULL*
EMP_SRVC_YRS	OUTPUT	INTEGER		*NULL*
EMP_PROM_DATE	OUTPUT	DATE		*NULL*
EMP_PROF	OUTPUT	XML		*NULL*
EMP_SSN	OUTPUT	CHAR		*NULL*
SQL_MSG	OUTPUT	VARCHAR		SQL0100W No row was found for FETCH, UPDATE or DELETE; or the result of a query is an empty table. SQLSTATE=02000

Now let's try it again with a record that does exist, such as 3217. In this case the data information is returned and the `SQL_MSG` variable is null. That tells us the operation succeeded.

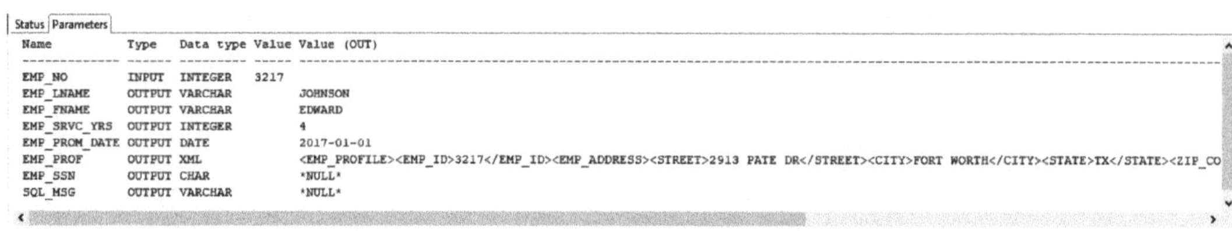

Status | Parameters

Name	Type	Data type	Value	Value (OUT)
EMP_NO	INPUT	INTEGER	3217	
EMP_LNAME	OUTPUT	VARCHAR		JOHNSON
EMP_FNAME	OUTPUT	VARCHAR		EDWARD
EMP_SRVC_YRS	OUTPUT	INTEGER		4
EMP_PROM_DATE	OUTPUT	DATE		2017-01-01
EMP_PROF	OUTPUT	XML		<EMP_PROFILE><EMP_ID>3217</EMP_ID><EMP_ADDRESS><STREET>2913 PATE DR</STREET><CITY>FORT WORTH</CITY><STATE>TX</STATE><ZIP_CO
EMP_SSN	OUTPUT	CHAR		*NULL*
SQL_MSG	OUTPUT	VARCHAR		*NULL*

Now let's go ahead and change the `MRG_EMP_INFO` and `DLT_EMP_INFO` procedures. We can add exactly the same new OUT parameter and `GET DIAGNOSTICS` logic.

```
CREATE PROCEDURE HRSCHEMA.MRG_EMP_INFO
(IN EMP_NO INT,
 IN EMP_LNAME VARCHAR(30),
 IN EMP_FNAME VARCHAR(20),
 IN EMP_SRVC_YRS INT,
 IN EMP_PROM_DATE DATE,
 IN EMP_PROF XML,
 IN EMP_SOC_SEC  CHAR(09),
 OUT SQL_MSG     VARCHAR(250))

 LANGUAGE SQL
 MODIFIES SQL DATA

 BEGIN
    DECLARE SQLMSG VARCHAR(250);
    DECLARE CONTINUE HANDLER FOR SQLEXCEPTION, SQLWARNING, NOT FOUND
    GET DIAGNOSTICS EXCEPTION 1 SQLMSG = MESSAGE_TEXT;

    MERGE INTO HRSCHEMA.EMPLOYEE AS T
    USING
     (VALUES (EMP_NO,
     EMP_LNAME,
     EMP_FNAME,
     EMP_SRVC_YRS,
     EMP_PROM_DATE,
```

```
    EMP_PROF,
    EMP_SOC_SEC))
    AS S
    (EMP_ID,
     EMP_LAST_NAME,
     EMP_FIRST_NAME,
     EMP_SERVICE_YEARS,
     EMP_PROMOTION_DATE,
     EMP_PROFILE,
     EMP_SSN)
    ON S.EMP_ID = T.EMP_ID

    WHEN MATCHED
       THEN UPDATE
          SET EMP_ID             = S.EMP_ID,
              EMP_LAST_NAME      = S.EMP_LAST_NAME,
              EMP_FIRST_NAME     = S.EMP_FIRST_NAME,
              EMP_SERVICE_YEARS  = S.EMP_SERVICE_YEARS,
              EMP_PROMOTION_DATE = S.EMP_PROMOTION_DATE,
              EMP_PROFILE        = S.EMP_PROFILE,
              EMP_SSN            = S.EMP_SSN

    WHEN NOT MATCHED
       THEN INSERT
          VALUES (S.EMP_ID,
          S.EMP_LAST_NAME,
          S.EMP_FIRST_NAME,
          S.EMP_SERVICE_YEARS,
          S.EMP_PROMOTION_DATE,
          S.EMP_PROFILE,
          S.EMP_SSN) ;

    SET SQL_MSG = SQLMSG;

  END

CREATE PROCEDURE HRSCHEMA.DLT_EMP_INFO
(IN EMP_NO INT,
 OUT SQL_MSG     VARCHAR(250))

 LANGUAGE SQL
 MODIFIES SQL DATA

 BEGIN
    DECLARE SQLMSG VARCHAR(250);
    DECLARE CONTINUE HANDLER FOR SQLEXCEPTION, SQLWARNING, NOT FOUND
    GET DIAGNOSTICS EXCEPTION 1 SQLMSG = MESSAGE_TEXT;

    DELETE FROM HRSCHEMA.EMPLOYEE
    WHERE EMP_ID = EMP_NO;

    SET SQL_MSG = SQLMSG;
 END
```

Finally, we must modify the programs that use these procedures to account for the new OUT parameter. We will also change those programs to report an error if the `SQL_MSG` value is not `NULL`.

Here is the revised `JAVEMPA` listing.

```
package employee;
import java.sql.*;
public class JAVEMPA {

        public static void main(String[] args) throws ClassNotFoundException
        {
            System.out.println("Executing JAVEMPA to call several stored procedures");

            Class.forName("com.ibm.db2.jcc.DB2Driver");
            System.out.println("**** Loaded the JDBC driver");
            String url
               ="jdbc:db2://localhost:50000/DBHR:retrieveMessagesFromServerOnGetMes-
               sage=true;";
            String user="rwinga01";
            String password="*********";

            Connection con=null;
            CallableStatement cs=null;

               try {
                       con=DriverManager.getConnection(url, user, password);
                       System.out.println("**** Created the connection");

                       int empId          = 7938;
                       String lName       = "WINFIELD";
                       String fName       = "STANLEY";
                       int srvcYears      = 3;
                       Date promDate      = null;    // if no date, we will set the input
                                                     // parameter value to NULL of date

                       SQLXML empProfile = null;    // if no employee profile, we will
                                                     // set the input parameter value to
                                                    // NULL of XML

                       String empSSN        = "382734509";
                       String sqlMsg       = null;
                       System.out.println("Calling stored procedure
                        HRSCHEMA.MRG_EMP_INFO to add or update a record");

                       cs = con.prepareCall("{call
                          HRSCHEMA.MRG_EMP_INFO(?,?,?,?,?,?,?,?)}");

                       cs.registerOutParameter(8, Types.VARCHAR);

                        cs.setInt(1, empId);
                        cs.setString(2, lName);
```

```
        cs.setString(3, fName);
        cs.setInt(4, srvcYears);
        cs.setNull(5, Types.DATE);
        cs.setNull(6, Types.SQLXML);
        cs.setString(7, empSSN);
        cs.setString(8, null);

        cs.execute();

        sqlMsg     = cs.getString(8);

        System.out.println("Employee ID      : " + empId);
        System.out.println("Last Name        : " + lName);
        System.out.println("First Name       : " + fName);
        System.out.println("Service Years    : " + srvcYears);
        System.out.println("Promotion Date   : " + promDate);
        System.out.println("Profile          : " + empProfile);
        System.out.println("SSN              : " + empSSN);

 if (sqlMsg != null) {
        System.out.println("Unsuccessful execution of procedure
        _HRSCHEMA.MRG_EMP_INFO");
        System.out.println(sqlMsg);
}
else
        System.out.println("Completed successful execution of procedure
        HRSCHEMA.MRG_EMP_INFO");

        System.out.println("********************");

        System.out.println("Calling stored procedure
         HRSCHEMA.GET_EMP_INFO to display a record");

        cs = con.prepareCall("{call
         HRSCHEMA.GET_EMP_INFO(?,?,?,?,?,?,?,?)}");

        cs.setInt(1, empId);
        cs.registerOutParameter(2, Types.VARCHAR);
        cs.registerOutParameter(3, Types.VARCHAR);
        cs.registerOutParameter(4, Types.VARCHAR);
        cs.registerOutParameter(5, Types.DATE);
        cs.registerOutParameter(6, Types.SQLXML);
        cs.registerOutParameter(7, Types.VARCHAR);
        cs.registerOutParameter(8, Types.VARCHAR);

        cs.execute();

        lName       = cs.getString(2);
        fName       = cs.getString(3);
        srvcYears   = cs.getInt(4);
        promDate    = cs.getDate(5);
        empProfile  = cs.getSQLXML(6);
        empSSN      = cs.getString(7);
```

```
                sqlMsg     = cs.getString(8);

                System.out.println("Employee ID  : " + empId);
                System.out.println("Last Name    : " + lName);
                System.out.println("First Name   : " + fName);
                System.out.println("Serv Years   : " + srvcYears);
                System.out.println("Prom Date    : " + promDate);
                System.out.println("Profile     : " + empProfile);
                System.out.println("SSN          : " + empSSN);

        if (sqlMsg != null) {
          System.out.println("Unsuccessful execution of procedure
           HRSCHEMA.GET_EMP_INFO");
          System.out.println(sqlMsg);
        }

    else
           System.out.println("Completed successful execution of procedure
            HRSCHEMA.GET_EMP_INFO");

           System.out.println("********************");
           System.out.println("Calling stored procedure HRSCHEMA.DLT_EMP_INFO to
            delete a record");

            cs = con.prepareCall("{call HRSCHEMA.DLT_EMP_INFO(?,?)}");
            cs.setInt(1, empId);
            cs.registerOutParameter(2, Types.VARCHAR);

            cs.execute();

            if (sqlMsg != null) {
                  System.out.println("Unsuccessful execution of procedure
                   HRSCHEMA.DLT_EMP_INFO");
                  System.out.println(sqlMsg);
        }
        else
                  System.out.println("Completed successful execution of procedure
                  HRSCHEMA.DLT_EMP_INFO");

                 System.out.println("********************");
                 System.out.println("Completed execution of JAVEMPA");

          } catch (SQLException e) {
                  System.out.println(e.getMessage());
                  System.out.println(e.getErrorCode());
                  e.printStackTrace();

          }
    }
}
```

And here is the revised .NET listing:

```
using System;
using System.Collections.Generic;
using System.Data;
using System.Linq;
using System.Text;
using System.Threading.Tasks;
using IBM.Data.DB2;
using IBM.Data.DB2Types;

/* Program to connect to DB2 for calling stored procedure  */
namespace NETEMPA
{
    class NETEMPA
    {
        static void Main(string[] args)
        {
            DB2Connection conn = null;
            DB2Command cmd = null;
            DB2Command cmd2 = null;
            DB2Command cmd3 = null;

            try
            {
                conn = new DB2Connection("Database=DBHR");
                conn.Open();
                Console.WriteLine("Executing NETEMPA to call several stored
                 procedures");
                Console.WriteLine("Successful connection!");
                Console.WriteLine("Calling stored procedure
               HRSCHEMA.MRG_EMP_INFO to add or update a record");

                cmd = conn.CreateCommand();
                String procName = "HRSCHEMA.MRG_EMP_INFO";
                cmd.CommandType = System.Data.CommandType.StoredProcedure;
                cmd.CommandText = procName;

                // Load the work variables

                int empId = 7938;
                String lName = "WINFIELD";
                String fName = "STANLEY";
                int srvcYears = 3;
                DateTime? promDate = null;  // if no date, we will set the input
                                            // parameter value to NULL of date

                String empProfile = null;   // if no employee profile, we will
                                            // set the input parameter value to
                                            // NULL of XML

                String empSSN = "382734509";
                String strPromDate = "";
                String strEmpProfile = "";
                String sqlMsg = null;
```

```
// Define parameters, establish direction for output parameters

cmd.Parameters.Add(new DB2Parameter("@empid", empId));
cmd.Parameters.Add(new DB2Parameter("@lname", lName));
cmd.Parameters.Add(new DB2Parameter("@fname", fName));
cmd.Parameters.Add(new DB2Parameter("@srvcYears", srvcYears));
cmd.Parameters.Add(new DB2Parameter("@promDate", promDate));
cmd.Parameters.Add(new DB2Parameter("@empProfile", empProfile));
cmd.Parameters.Add(new DB2Parameter("@empSSN", empSSN));
cmd.Parameters.Add(new DB2Parameter("@sqlMsg", sqlMsg));

cmd.Parameters[7].Direction = ParameterDirection.Output;

/* Execute the stored procedure */

cmd.ExecuteNonQuery();

sqlMsg = cmd.Parameters[7].Value.ToString();
if (sqlMsg != "")
{
    Console.WriteLine("Unsuccessfully called stored procedure "
      + procName);
    Console.WriteLine(sqlMsg);
}
else
{
    Console.WriteLine("Successfully called stored procedure "
      + procName);
    Console.WriteLine("The following information was merged ");
    Console.WriteLine("Employee id : " + empId);
    Console.WriteLine("Last Name   : " + lName);
    Console.WriteLine("First Name  : " + fName);
    Console.WriteLine("Srvc Years  : " + srvcYears);
    Console.WriteLine("Prom Date   : " + strPromDate);
    Console.WriteLine("Profile    : " + strEmpProfile);
    Console.WriteLine("Soc Sec     : " + empSSN);
}

Console.WriteLine("****************************************** ");

Console.WriteLine("Calling stored procedure
HRSCHEMA.GET_EMP_INFO to retrieve a record");

cmd2 = conn.CreateCommand();
procName = "HRSCHEMA.GET_EMP_INFO";
cmd2.CommandType = System.Data.CommandType.StoredProcedure;
cmd2.CommandText = procName;

cmd2.Parameters.Add(new DB2Parameter("@empid",DB2Type.Integer));
cmd2.Parameters.Add(new DB2Parameter("@lname", DB2Type.VarChar));
cmd2.Parameters.Add(new DB2Parameter("@fname", DB2Type.VarChar));
cmd2.Parameters.Add(new DB2Parameter("@srvcYears", DB2Type.Integer));
cmd2.Parameters.Add(new DB2Parameter("@promDate", DB2Type.DateTime));
```

```
cmd2.Parameters.Add(new DB2Parameter("@empProfile", DB2Type.Xml));
cmd2.Parameters.Add(new DB2Parameter("@empSSN", DB2Type.VarChar));
cmd2.Parameters.Add(new DB2Parameter("@sqlMsg", sqlMsg));

cmd2.Parameters[1].Direction = ParameterDirection.Output;
cmd2.Parameters[2].Direction = ParameterDirection.Output;
cmd2.Parameters[3].Direction = ParameterDirection.Output;
cmd2.Parameters[4].Direction = ParameterDirection.Output;
cmd2.Parameters[5].Direction = ParameterDirection.Output;
cmd2.Parameters[6].Direction = ParameterDirection.Output;
cmd2.Parameters[7].Direction = ParameterDirection.Output;
cmd2.Parameters[0].Value = 7938;

cmd2.ExecuteNonQuery();

/*  Now capture and display the results */

String strEmpId      = cmd2.Parameters[0].Value.ToString();
lName                = cmd2.Parameters[1].Value.ToString();
fName                = cmd2.Parameters[2].Value.ToString();
String strSrvcYears  = cmd2.Parameters[3].Value.ToString();
if (cmd2.Parameters[4].Value != DBNull.Value)
   strPromDate   = cmd2.Parameters[4].Value.ToString();
if (cmd2.Parameters[5].Value != DBNull.Value)
    strEmpProfile = cmd2.Parameters[5].Value.ToString();
if (cmd2.Parameters[6].Value != DBNull.Value)
   empSSN = cmd2.Parameters[6].Value.ToString();

sqlMsg = cmd2.Parameters[7].Value.ToString();
if (sqlMsg != "")
{
    Console.WriteLine("Unsuccessfully called stored procedure "
       + procName);
    Console.WriteLine(sqlMsg);
}
else
{
    Console.WriteLine("Successfully called stored procedure "
      + procName);
    Console.WriteLine("Employee id : " + strEmpId);
    Console.WriteLine("Last Name   : " + lName);
    Console.WriteLine("First Name  : " + fName);
    Console.WriteLine("Srvc Years  : " + strSrvcYears);
    Console.WriteLine("Prom Date   : " + strPromDate);
    Console.WriteLine("Profile     : " + strEmpProfile);
    Console.WriteLine("Soc Sec     : " + empSSN);
}


// NOW DELETE THE RECORD

Console.WriteLine("Calling stored procedure
HRSCHEMA.DLT_EMP_INFO to delete a record");

cmd3 = conn.CreateCommand();
```

```
                    procName = "HRSCHEMA.DLT_EMP_INFO";
                    cmd3.CommandType = System.Data.CommandType.StoredProcedure;
                    cmd3.CommandText = procName;

                    cmd3.Parameters.Add(new DB2Parameter("@empid", DB2Type.Integer));
                    cmd3.Parameters.Add(new DB2Parameter("@sqlMsg", sqlMsg));
                    cmd2.Parameters[1].Direction = ParameterDirection.Output;
                    cmd3.Parameters[0].Value = 7938;
                    cmd3.ExecuteNonQuery();

                    sqlMsg = cmd3.Parameters[1].Value.ToString();
                    if (sqlMsg != "")
                    {
                        Console.WriteLine("Unsuccessfully called stored procedure "
                           + procName);
                        Console.WriteLine(sqlMsg);
                    }
                    else
                    {
                        Console.WriteLine("Successfully called stored procedure "
                          + procName);
                        Console.WriteLine("Employee " + empId + " was successfully
                        deleted");
                    }

                }

                catch (Exception e)
                {
                    Console.WriteLine(e.Message);
                }

                finally
                {
                    conn.Close();
                }

            }
        }
    }
```

This concludes our discussion of stored procedures. As you can tell this is a very powerful technology that promotes reusability and can help minimize custom coding. I suggest that you make good use of it.

USER DEFINED FUNCTIONS

A user defined function (UDF) is a function written by an application programmer or DBA, as opposed to those functions provided out of the box by DB2. UDFs extend DB2 functionality by allowing new functions to be created. As a function, a UDF always returns a value, and is invoked by referencing the function name and passing any parameters.

Types of UDF

There are five varieties of UDFs as follows:

- SQL Scalar , Table or Row
- External Scalar
- External Table
- OLE DB External Table
- Sourced

Examples of UDFs

SQL Scalar Function

An SQL scalar function will return a single value using only SQL statements. There is no external program. You may recall earlier we established a business rule that an employee's "level" was based on their years of service. We used an SQL with a CASE statement to return a value of JUNIOR, ADVANCED or SENIOR. Here's the SQL we used earlier:

```
SELECT EMP_ID,
EMP_LAST_NAME,
EMP_FIRST_NAME,
CASE
   WHEN EMP_SERVICE_YEARS  < 1 THEN 'ENTRY'
   WHEN EMP_SERVICE_YEARS  < 5 THEN 'ADVANCED'
   ELSE 'SENIOR'
END CASE
FROM HRSCHEMA.EMPLOYEE;
```

```
EMP_ID EMP_LAST_NAME EMP_FIRST_NAME CASE
------ ------------- -------------- --------
  3217 JOHNSON       EDWARD         ADVANCED
  7459 STEWART       BETTY          SENIOR
  9134 FRANKLIN      BRIANNA        ENTRY
  4720 SCHULTZ       TIM            SENIOR
  6288 WILLARD       JOE            SENIOR
  1122 JENKINS       DEBORAH        SENIOR
  3333 FORD          JAMES          SENIOR
  7777 HARRIS        ELISA          ADVANCED
  4175 TURNBULL      FRED           ADVANCED
  1111 VEREEN        CHARLES        SENIOR
  1112 YATES         JANENE         SENIOR
  1114 MILLER        PHYLLIS        SENIOR
  1115 JENSON        PAUL           SENIOR
```

Now let's say we have several programs that need to generate these values. We could copy the same SQL to each program, but what if the logic changes in the future? Either the cutoff years or the named literals could change. In that case it would be convenient to only have to make the change in one place. A UDF can accomplish that objective.

We'll create a UDF that accepts an integer which is the employee's years of service, and then we'll return the literal value that represents the employees level of service in the company. First, we must define the UDF to DB2. We need to specify at least:

- The name of the function
- Input parameters
- Return parameter type

Now let's code the UDF:

```
CREATE FUNCTION HRSCHEMA.EMP_LEVEL (YRS_SRVC INT)
    RETURNS VARCHAR(10)
    READS SQL DATA
    RETURN
   (SELECT
    CASE
       WHEN YRS_SRVC  < 1 THEN 'ENTRY     '
       WHEN YRS_SRVC  < 5 THEN 'ADVANCED  '
       ELSE 'SENIOR    '
    END CASE
    FROM SYSIBM.SYSDUMMY1)
```

Note that we specify `SYSIBM.SYSDUMMY1` as the data source to satisfy the SELECT syntax (since our information is not coming from a table, we reference dummy table IBM set up for just this purpose).

Once you've run the DDL successfully, let's run a query against the new UDF:

```
SELECT HRSCHEMA.EMP_LEVEL(7)
AS EMP_LVL
FROM SYSIBM.SYSDUMMY1;

EMP_LVL
----------
SENIOR
```

The above is a very simple example, and the SQL in this case does not actually access a table. Let's do one more that will access a table. How about a UDF that will return the full name of an employee given the employee's id number? So we need to pass an integer employee id, and we'll return a VARCHAR value.

```
CREATE FUNCTION HRSCHEMA.EMP_FULLNAME (EMP_NO INT)
RETURNS VARCHAR(40)
READS SQL DATA
RETURN
SELECT
EMP_FIRST_NAME || ' ' || EMP_LAST_NAME AS FULL_NAME
FROM HRSCHEMA.EMPLOYEE
WHERE EMP_ID  = EMP_NO;
```

Now let's run the query to use this UDF:

```
SELECT HRSCHEMA.EMP_FULLNAME(3217) AS FULLNAME
FROM SYSIBM.SYSDUMMY1;

FULLNAME
--------------
EDWARD JOHNSON
```

SQL Table Function

An SQL table function returns a table of values. Let's look at the common table expression we used earlier. Remember it goes like this:

```
WITH EMP_PAY_SUM (EMP_ID, EMP_PAY_TOTAL) AS
(SELECT EMP_ID,
SUM(EMP_PAY_AMT)
AS EMP_PAY_TOTAL
FROM HRSCHEMA.EMP_PAY_HIST
GROUP BY EMP_ID)

SELECT EMP_ID,
EMP_PAY_TOTAL
FROM EMP_PAY_SUM;

EMP_ID EMP_PAY_TOTAL
------ -------------
  3217      10833.32
  4720      13333.32
  6288      11666.64
  7459      14166.64
  9134      12500.00
```

Now let's define a table UDF that will return the same results. We don't pass any IN parameters, but we will return a TABLE of values.

```
CREATE FUNCTION HRSCHEMA.EMP_PAY_SUM ()
   RETURNS TABLE (EMP_ID  INTEGER, EMP_PAY_TOTAL DECIMAL (9,2))
   READS SQL DATA
```

```
RETURN
   SELECT EMP_ID, SUM(EMP_PAY_AMT)
   AS EMP_PAY_TOTAL
   FROM HRSCHEMA.EMP_PAY_HIST
   GROUP BY EMP_ID;
```

Now we can reference this UDF in a query. Notice that we invoke the built-in **TABLE** function to return the values generated by the EMP_PAY_SUM user defined function.

```
SELECT * FROM TABLE(HRSCHEMA.EMP_PAY_SUM()) AS EPS

EMP_ID EMP_PAY_TOTAL
------ -------------
  3217      10833.32
  4720      13333.32
  6288      11666.64
  7459      14166.64
  9134      12500.00
```

External Scalar Function

An external scalar function is one that returns a single scalar value, usually based on some parameter value that is passed in. The function is implemented using a program, hence the designation as an "external" function.

You may recall earlier we created an SQL scalar UDF that returned a string value for an employee "level" based on the years of service. We could create a similar external UDF and implement it in Java.

```
CREATE FUNCTION HRSCHEMA.get_emp_level (INT)
RETURNS VARCHAR(10)
EXTERNAL NAME 'employee.get_emp_level!get_level'
LANGUAGE JAVA
NO SQL
FENCED
PARAMETER STYLE JAVA
```

Here's our Java program:

```
package employee;
import java.sql.*;

public class get_emp_level {

       public static String get_level (int years) throws
              ClassNotFoundException, SQLException

       {
              String level = "";
```

```
                try {
                        switch (years) {
                                case 0:  level = "ENTRY";
                        break;
                                case 1: case 2: case 3: case 4:
                                        level = "ADVANCED";
                        break;

                        default: level = "SENIOR";
                        break;

                        }
                }

                catch (Exception e) {
                        System.out.println(e.getMessage());
                        e.printStackTrace();

                }
                return level;
        }
}
```

Now you must move the class file to the SQLLIB/Function/employee subdirectory under your DB2 installation directory. This is where DB2 will look for it unless you have specified a fully qualified path in the stored procedure definition.

Also remember: Once you've implemented (and each time you change) the Java program, you must do one of two things to ensure that DB2 loads the newest version of your procedure before you test it:

1. Issue the following command:

   ```
   CALL SQLJ.REFRESH_CLASSES()
   ```

2. Bounce your DB2 instance by issuing `DB2STOP` and then `DB2START`.

Now we can call this function, either from another program or from Data Studio:

```
SELECT HRSCHEMA.get_emp_level(0) AS EMP_LVL
FROM SYSIBM.SYSDUMMY1;

EMP_LVL
-------
ENTRY


SELECT HRSCHEMA.get_emp_level(2) AS EMP_LVL
FROM SYSIBM.SYSDUMMY1;
```

```
EMP_LVL
--------
ADVANCED


SELECT HRSCHEMA.get_emp_level(7) AS EMP_LVL
FROM SYSIBM.SYSDUMMY1;

EMP_LVL
-------
SENIOR
```

External Table Function

An external table function returns a table of values. Here we could use such a function as a replacement for the common table expression we used earlier in this study guide. Let's again return to that.

```
WITH EMP_PAY_SUM (EMP_ID, EMP_PAY_TOTAL) AS
(SELECT EMP_ID,
SUM(EMP_PAY_AMT)
AS EMP_PAY_TOTAL
FROM HRSCHEMA.EMP_PAY_HIST
GROUP BY EMP_ID)

SELECT EMP_ID,
EMP_PAY_TOTAL
FROM EMP_PAY_SUM

EMP_ID EMP_PAY_TOTAL
------ -------------
  3217      10833.32
  4720      13333.32
  6288      11666.64
  7459      14166.64
  9134      12500.00
```

Normally common table expressions are used with complex SQL to simplify things. Ours is not very complex, but we could simplify even further by using a UDF instead of the common table expression. Let's register our new function as follows:

```
CREATE FUNCTION HRSCHEMA.get_emp_pay ()
RETURNS TABLE  (EMP_ID INT, EMP_PAY_TOTAL DECIMAL(8,2))
EXTERNAL NAME 'employee.get_emp_pay!get_pay'
LANGUAGE JAVA
PARAMETER STYLE DB2GENERAL
NOT DETERMINISTIC
READS SQL DATA
SCRATCHPAD 10
NO FINAL CALL
```

```
DISALLOW PARALLEL
NO DBINFO
```

Now we need to implement this UDF with a Java program. For a table function in Java, you must implement the Java UDF interface. Otherwise you will get some very odd error messages when you try to call your function. Here is what you must do:

1. Import COM.ibm.db2.app.UDF

2. Implement the UDF interface by including these methods:

    a. SQLUDF_TF_FIRST (perform any initialization)

    b. SQLUDF_TF_OPEN (open the data ResultSet)

    c. SQLUDF_TF_FETCH (cycle through the ResultSet, assigning data to the parameter variables)

    d. SQLUDF_TF_CLOSE (perform any needed close logic)

    e. SQLUDF_TF_FINAL (perform any finalization)

Note: these methods are called behind the scenes by DB2. You do not need to call them, they just have to be present in your program. Ok here is our program:

```
package employee;
import java.sql.*;
import java.math.*;
import COM.ibm.db2.app.UDF;

public class get_emp_pay extends UDF

{
        Connection con=null;
        Statement stmt = null;
        ResultSet rs = null;

        int empNo = 0;
        BigDecimal empPay = new BigDecimal(0.0);

        get_emp_pay() {

           }

        public void get_pay (int empId, BigDecimal emp_pay_tot)
                throws ClassNotFoundException, SQLException, Exception
        {
                Class.forName("com.ibm.db2.jcc.DB2Driver");
                DriverManager.registerDriver(new com.ibm.db2.jcc.DB2Driver());
```

```
        int callType = getCallType();
            switch(callType) {

                case SQLUDF_TF_FIRST:
                    // initialize anything that needs it

                    break;

                case SQLUDF_TF_OPEN:

                    // get the result set
                    con
                    = DriverManager.getConnection("jdbc:default:connection");
                    String query = "SELECT EMP_ID, "
                                 + "SUM(EMP_PAY_AMT) AS EMP_PAY_TOTAL "
                                 + "FROM HRSCHEMA.EMP_PAY_HIST "
                                 + "GROUP BY EMP_ID";

                    stmt = con.createStatement();
                    rs=stmt.executeQuery(query);

                    break;

                case SQLUDF_TF_FETCH:

                    // cycle through the result set

                    if (rs.next())
                    {
                        empNo  = rs.getInt(1);
                        empPay = rs.getBigDecimal(2);
                        set(1, empNo);
                        set(2, empPay);
                    }
                    else setSQLstate("02000");

                    break;

                case SQLUDF_TF_CLOSE:
                    break;

                case SQLUDF_TF_FINAL:
                    break;

                default:
              throw new Exception("UNEXPECTED call type of "+callType);

            } // close switch

    } // close get_pay

}
```

Move you class file to the SQLLIB/Function/employee directory and run the command to refresh the class files which is: `CALL SQLJ.REFRESH_CLASSES()`. Now let's run it:

```
SELECT * FROM TABLE(HRSCHEMA.get_emp_pay()) AS EMP_PAY

EMP_ID EMP_PAY_TOTAL
------ -------------
  3217      10833.32
  4720      13333.32
  6288      11666.64
  7459      14166.64
  9134      12500.00
```

OLE DB External Table

You can register an OLE DB external table function which will access data from an OLE DB provider. However, if you are running 64 bit Windows, you must also be running 64-bit versions of Microsoft Office and have the 64-bit version of OLE DB 12 installed. If not, when your function tries to connect to the OLE DB provider, you will get an error saying the OLE DB provider is not installed.

Let's do an example. Suppose we have a new employee table stored in Microsoft Access. Let's assume an MS Access database called Recruiting which has a table called NewEmp (new employees). The table has only three columns: empid, lastName and firstName. There are three rows in the table.

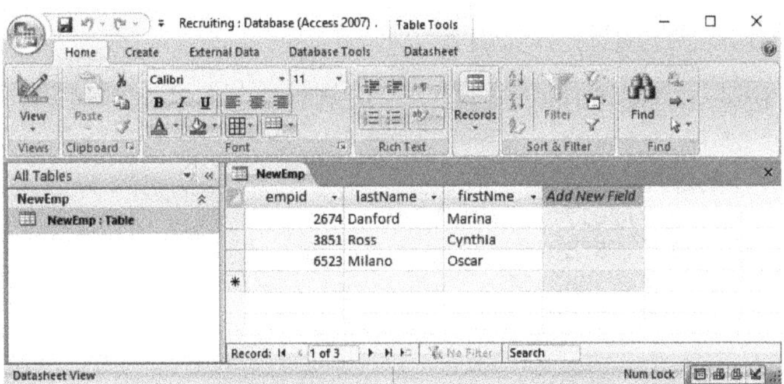

Now, let's register a new DB2 external function as follows:

```
CREATE FUNCTION HRSCHEMA.get_new_emp_info ()
RETURNS TABLE (empid INTEGER,
               lastName VARCHAR(20),
               firstName VARCHAR(15))
LANGUAGE OLEDB
EXTERNAL NAME '!NewEmp!Provider=Microsoft.ACE.OLEDB.12.0;
   Data Source=c:\AccessDB\Recruiting.accdb!COLLATING_SEQUENCE=Y'
```

Next we must grant access to the function:

```
GRANT EXECUTE ON FUNCTION HRSCHEMA.get_new_emp_info to public
```

Finally, let's use the function to retrieve our data:

```
SELECT * FROM TABLE(get_new_emp_info());

Empid  lastName      firstName
------ ------------- -------------
  2674 DANFORD       MARINA
  3851 ROSS          CYNTHIA
  6523 MILANO        OSCAR
```

Sourced Function

A sourced function redefines or extends an existing DB2 function. It is typically written to enable the processing of user defined data types in a function. For example, suppose you define a Canadian dollar type as follows:

```
CREATE DISTINCT TYPE HRSCHEMA.CANADIAN_DOLLAR AS DECIMAL (9,2);
```

Now create a table using this type:

```
CREATE TABLE HRSCHEMA.CAN_PAY_TBL
(EMP_ID INT,
PAY_DATE DATE,
PAY_AMT HRSCHEMA.CANADIAN_DOLLAR)
IN TSHR;
```

Now let's add four rows to the table.

```
INSERT INTO HRSCHEMA.CAN_PAY_TBL
VALUES(3217, '01/01/2017', 5500.50),
(3217, '02/01/2017', 5500.50),
(3217, '03/01/2017', 5500.50),
(3217, '04/01/2017', 5500.50);
```

Now let's say we want to query a sum of the `PAY_AMT` rows:

```
SELECT SUM(PAY_AMT) FROM HRSCHEMA.CAN_PAY_TBL;

No authorized routine named "SUM" of type "FUNCTION" having
compatible arguments was found.. SQLCODE=-440, SQLSTATE=42884,
DRIVER=4.18.60
```

Unfortunately we received an error because the SUM function in DB2 does not know about a `CANADIAN_DOLLAR` data type, so the value we passed is an "incompatible argument".

To fix this error we must extend the SUM function to work with CANADIAN_DOLLAR data type by creating a user defined function based on the SUM function but accepting a CANADIAN_DOLLAR argument. Try this one:

```
CREATE FUNCTION SUM(HRSCHEMA.CANADIAN_DOLLAR)
RETURNS DECIMAL (9,2)
SOURCE SYSIBM.SUM(DECIMAL);
```

Now you have a SUM function for which CANADIAN_DOLLAR is an input parameter. When DB2 processes the query it will use the new user defined version of the SUM function because that's the one that matches your query arguments. Your SUM query will work now.

```
SELECT SUM(PAY_AMT) FROM HRSCHEMA.CAN_PAY_TBL;

1
--------
22002.00
```

TRIGGERS

A trigger performs a set of actions when an INSERT, UPDATE or DELETE takes place. Triggers are stored in the database which is a significant advantage of using them instead of application logic.

The CREATE TRIGGER statement defines a trigger and builds a trigger package at the current server. Advantages of using a trigger include:

- Ability to write to other tables for audit trail.
- Ability to read other tables for validation.
- Ability to compare data before and after update operations.

Types of triggers

There are three types of triggers:

1. INSERT
2. UPDATE
3. DELETE

A MERGE action also fires INSERT and UPDATE triggers (if they exist) depending on whether the MERGE causes an INSERT or UPDATE.

Timings of triggers

There are three timings of triggers as well:

1. BEFORE
2. AFTER
3. INSTEAD OF

A `BEFORE` trigger performs its action before the SQL operation (INSERT, UPDATE or DELETE) that fired the trigger. An `AFTER` trigger performs its action after the SQL operation (INSERT, UPDATE or DELETE) that fired the trigger. An `INSTEAD OF` trigger is completely different – it enables ADD, UPDATE or DELETE operation through what would normally be a read-only view. We'll explain that more when we get to the INSTEAD of example below.

The basic syntax of the `CREATE TRIGGER` statement is:

```
CREATE TRIGGER <trigger name> <AFTER / BEFORE / INSTEAD OF>
ON <table name> REFERENCING <see examples>
FOR EACH ROW
<action to take>
```

Examples of Triggers

Sample After Trigger

One common use of triggers is to automatically add records to a history table when there is a change to the records in a base table. In this case we will create a history table to store previous versions of pay rates in the `EMP_PAY` table. We can create the history table like this:

```
CREATE TABLE HRSCHEMA.EMP_PAY_HST LIKE HRSCHEMA.EMP_PAY;
```

And then we'll add an additional column to the history table to keep track of when the record was added:

```
ALTER TABLE HRSCHEMA.EMP_PAY_HST
ADD AUDIT_DATE TIMESTAMP DEFAULT CURRENT TIMESTAMP;
```

Now we will create a trigger so that when a change is made to an `EMP_PAY` record, we will write the old version of the record to the history table. The trigger knows about the old and new versions of the record we are modifying so we specify the OLD version of the record and the fields to be added to the history table.

```
CREATE TRIGGER HRSCHEMA.TRG_EMP_PAY
AFTER UPDATE ON HRSCHEMA.EMP_PAY
REFERENCING OLD AS oldcol NEW AS newcol
```

```
FOR EACH ROW MODE DB2SQL
INSERT INTO HRSCHEMA.EMP_PAY_HST(
EMP_ID,
EMP_REGULAR_PAY,
EMP_BONUS_PAY,
AUDIT_DATE)
VALUES
(oldcol.EMP_ID,
oldcol.EMP_REGULAR_PAY,
oldcol.EMP_BONUS_PAY,
CURRENT TIMESTAMP);
```

Now let's look at an EMP_PAY record, modify it, and then see if the old version get's added to the history table:

```
SELECT * FROM HRSCHEMA.EMP_PAY
WHERE EMP_ID = 3217;

EMP_ID  EMP_REGULAR_PAY  EMP_BONUS_PAY
------  ---------------  -------------
  3217         65000.00        5500.00
```

Let's change the EMP_REGULAR_PAY to 57000.

```
UPDATE HRSCHEMA.EMP_PAY
SET EMP_REGULAR_PAY = 57000
WHERE EMP_ID = 3217;
```

Now if we select from the history table, we see the previous version of the record. In this case we made the change on February 24, 2017.

```
SELECT * FROM HRSCHEMA.EMP_PAY_HST
WHERE EMP_ID = 3217;

EMP_ID  EMP_REGULAR_PAY  EMP_BONUS_PAY  AUDIT_DATE
------  ---------------  -------------  -------------------
  3217         65000.00        5500.00  2017-02-24-07.08.39.
```

Note: the temporal tables introduced in DB2 10 provide excellent functionality for storing record history for system time enabled tables. Keep this in mind when designing your tables. We'll look at temporal tables later in this text. However, the trigger technique described above is still a very reliable way of automating the capture of record history.

Sample BEFORE Trigger

For this example, we are going to enforce data integrity between two tables without defining referential constraints. Instead of using foreign keys we will use a trigger.

Assume two tables:

- DEPTMENT which has department codes and descriptions.
- EMP_DATA_X which has an employee id, first and last names, and a department code.

Here's the DDL to create those tables, and to add a few records.

```
CREATE TABLE HRSCHEMA.DEPTMENT
(DEPT_CODE   CHAR(04) NOT NULL,
 DEPT_NAME   VARCHAR (20) NOT NULL)
IN TSHR;

INSERT INTO HRSCHEMA.DEPTMENT
VALUES ('DPTA','DEPARTMENT A');
INSERT INTO HRSCHEMA.DEPTMENT
VALUES ('DPTB','DEPARTMENT B');

CREATE TABLE HRSCHEMA.EMP_DATA_X
(EMP_ID   INT,
EMP_LNAME  VARCHAR(30),
EMP_FNAME  VARCHAR(20),
EMP_DEPT   CHAR (04))
IN TSHR;

INSERT INTO
HRSCHEMA.EMP_DATA_X
VALUES(8888,
'JONES',
'WILLIAM',
'DPTA');
```

Let's say we have a business rule that the department column in EMP_DAT_X can only have values that exist in the DEPTMENT table. Of course we could create a referential constraint with a foreign key, but let's say we prefer to implement this rule as a trigger instead. The trigger should prevent invalid updates and return an error message if a user tries to update a EMP_DATA_X record using an EMP_DEPT value that is not in the DEPTMENT table.

This trigger would accomplish this job:

```
CREATE TRIGGER HRSCHEMA.BLOCK_DEPT_UPDATE
NO CASCADE BEFORE UPDATE OF
EMP_DEPT ON HRSCHEMA.EMP_DATA_X
REFERENCING NEW AS N
FOR EACH ROW MODE DB2SQL
WHEN (N.EMP_DEPT NOT IN (SELECT DEPT_CODE FROM HRSCHEMA.DEPTMENT))
   BEGIN ATOMIC
      SIGNAL SQLSTATE '85101' ('Invalid department code');
   END
```

Currently the data in these tables looks like this:

```
SELECT * FROM HRSCHEMA.DEPTMENT;

DEPT_CODE  DEPT_NAME
---------  ------------
DPTA       DEPARTMENT A
DPTB       DEPARTMENT B


SELECT * FROM HRSCHEMA.EMP_DATA_X

EMP_ID  EMP_LNAME             EMP_FNAME            EMP_DEPT
------  ---------             ---------            --------
  8888  JONES                 WILLIAM              DPTA
```

If we try the following SQL it will fail because department code "DPTC" does not exist in the DEPTMENT table.

```
UPDATE HRSCHEMA.EMP_DATA_X
SET EMP_DEPT = 'DPTC'
WHERE EMP_ID = 8888;

Application raised error or warning with diagnostic text:
"Invalid department code".. SQLCODE=-438, SQLSTATE=85101,
DRIVER=4.18.60
```

And the result is as we expected, plus the error text is what we defined in the trigger. Triggers give you flexibility in terms of the error message – you can customize it as you see fit. That's one advantage of using triggers rather than foreign keys.

Sample INSTEAD OF Trigger

An INSTEAD OF trigger is different than all other types of triggers. The purpose of an INSTEAD OF trigger is to allow updates to take place from what is normally a read-only view. You may know that a view that includes more than one table is read only. Let's look at an example of creating ansd updating data using a view with an INSTEAD OF trigger.

We'll start with a query that joins certain columns in the EMPLOYEE table with the EMP_PAY table.

```
SELECT
A.EMP_ID,
A.EMP_LAST_NAME,
B.EMP_REGULAR_PAY
FROM HRSCHEMA.EMPLOYEE A, HRSCHEMA.EMP_PAY B
WHERE A.EMP_ID = B.EMP_ID;
```

```
EMP_ID  EMP_LAST_NAME          EMP_REGULAR_PAY
------  -------------          ---------------
  3217  JOHNSON                       55000.00
  7459  STEWART                       80000.00
  9134  FRANKLIN                      80000.00
  4720  SCHULTZ                       80000.00
  6288  WILLARD                       70000.00
```

Now let's create a view based on this query:

```
CREATE VIEW HRSCHEMA.EMP_PROFILE_PAY AS
SELECT A.EMP_ID, A.EMP_LAST_NAME, B.EMP_REGULAR_PAY
FROM HRSCHEMA.EMPLOYEE A, HRSCHEMA.EMP_PAY B
WHERE A.EMP_ID = B.EMP_ID;
```

And now we can query the data using the view:

```
SELECT * FROM HRSCHEMA.EMP_PROFILE_PAY;

EMP_ID  EMP_LAST_NAME          EMP_REGULAR_PAY
------  -------------          ---------------
  3217  JOHNSON                       55000.00
  7459  STEWART                       80000.00
  9134  FRANKLIN                      80000.00
  4720  SCHULTZ                       80000.00
  6288  WILLARD                       70000.00
```

Now suppose we want to use this view to **update** the EMP_REGULAR_PAY column. Let's try and see what happens:

```
UPDATE HRSCHEMA.EMP_PROFILE_PAY
SET EMP_REGULAR_PAY = 65000
WHERE EMP_ID = 3217;
```

The target fullselect, view, typed table, materialized query table, range-clustered table, or staging table in the INSERT, DELETE, UPDATE, MERGE, or TRUNCATE statement is a target for which the requested operation is not permitted.. SQLCODE=-150, SQLSTATE=42807, DRIVER=4.18.60

As you can see, we are not allowed to perform updates using this view. The fact that the view involves more than one table makes it a read-only view. However, we can perform updates through this view if we create an INSTEAD OF trigger on the view. The DDL looks like this:

```
CREATE TRIGGER HRSCHEMA.EMP_PROF_PAY_UPDATE
INSTEAD OF UPDATE ON HRSCHEMA.EMP_PROFILE_PAY
REFERENCING NEW AS NEWEMP OLD AS OLDEMP
FOR EACH ROW
MODE DB2SQL
BEGIN ATOMIC
```

```
    UPDATE HRSCHEMA.EMP_PAY AS E
    SET (EMP_REGULAR_PAY)
        = (NEWEMP.EMP_REGULAR_PAY)
    WHERE NEWEMP.EMP_ID = E.EMP_ID ;
END
```

This trigger will intercept the UPDATE request from the `EMP_PROFILE_PAY` view and perform a direct update to the `EMP_PAY` table. Now let's try our update:

```
UPDATE HRSCHEMA.EMP_PROFILE_PAY
SET EMP_REGULAR_PAY = 65000
WHERE EMP_ID = 3217;

Updated 1 rows.
```

Finally let's select the row we just changed using the view:

```
SELECT * FROM HRSCHEMA.EMP_PROFILE_PAY
WHERE EMP_ID = 3217;

EMP_ID       EMP_LAST_NAME  EMP_REGULAR_PAY
------       -------------  ---------------
  3217       JOHNSON               65000.00
```

And as we can see, the `EMP_REGULAR_PAY` did get changed. Of course when all is said and done, you could simply have updated the base table to begin with. However, views give you more control over what users and/or programmers are allowed to see and change in a table. The point of the `INSTEAD OF` triggers is to allow you to use a view as the interface for adds, changes and deletes.

Important to Remember Trigger Information

- A trigger is fired by an INSERT, UPDATE or DELETE of a record in a table.

- By default, the LOAD operation does not fire triggers. However, if the `SHRLEVEL CHANGE` option is included on the LOAD statement, triggers will be fired.

- You cannot use `FOR EACH STATEMENT` with `BEFORE` or `INSTEAD OF` timing. `FOR EACH STATEMENT` means your trigger logic is to be applied only once after the triggering statement finishes processing the affected rows.

- If you do not specify a list of column names in the trigger, an update operation on any column of the subject table will fire the trigger.

- A trigger can call a local stored procedure.

- A trigger cascade occurs when the SQL statements executed by one trigger fires one or more other triggers (for example, a trigger action on one table might write a row to another table which in turn has an INSERT trigger on it that performs some other action).

- If a column is included on a table for which a trigger is defined, the column cannot be dropped from the table unless the trigger is first dropped.

- If you alter a column definition for a table in which a trigger is defined on that column, the trigger packages are invalidated.

- If you drop a table for which a trigger has been defined, the trigger is also dropped automatically.

SPECIAL TABLES

Temporal and Archive Tables

Temporal Tables

Temporal tables were introduced to DB2 in version 10. Briefly, a temporal table is one that keeps track of "versions" of data over time and allows you to query data according to the time frame. It is important to understand what problems you can solve with the technologies, such as automatically preventing overlapping rows for business time. We'll get to that in the examples.

Some benefits of DB2's built in support for managing temporal data include:

- Reduces application logic
- Can automatically maintain a history of table changes
- Ensures consistent handling of time related events

Now let's look at the two varieties of time travel in DB2, which are business time (sometimes referred to as application time) and system time.

Business Time

An employee's pay typically changes over time. Besides an employee's current salary, there may be scenarios under which an HR department or supervisor might need to know what pay rate was in effect for an employee at some time in the past. We might also need to allow for cases where the employee terminated for some period of time and then returned. Or maybe they took a non-paid leave of absence. This is the concept of business time and it can be fairly complex depending on the business rules. It basically means a period of time in which a data value is accurate. You could think of it as having an effective date and discontinue date.

A table can only have one business time period. When a BUSINESS_TIME period is defined for a table, DB2 generates a check constraint in which the end column value must be greater than the begin column value. Once a table is version enabled, the following clauses allow you to pull data for a particular business time period:

```
FOR BUSINESS_TIME FROM ... TO ...
FOR BUSINESS_TIME BETWEEN... AND...
```

For example:

```
SELECT * FROM HRSCHEMA.EMP_PAY
FOR BUSINESS_TIME BEWTWEEN '2017-01-01' AND '2017-02-01'
ORDER BY EMP_ID;
```

System Time

System time simply means the time during which a piece of data is in the database, i.e., when the data was added, changed or deleted. Sometimes it is important to know this. For example a user might enter an employee's salary change on a certain date but the effective date of the salary change might be earlier or later than the date it was actually entered into the system. An audit trail table often has a timestamp that can be considered system time at which a transaction occurred.

Like with business time, once a table is version-enabled for system time, the following clauses allow you to pull data for a particular system period:

```
FOR SYSTEM_TIME FROM ... TO ...
FOR SYSTEM_TIME BETWEEN... AND...
```

For example, maybe we want to know see several series of EMPLOYEE table records that were changed over a period of a month. Assuming a system version enabled table, this would work:

```
SELECT * FROM HRSCHEMA.EMPLOYEE
FOR SYSTEM_TIME BEWTWEEN '2017-01-01' AND '2017-02-01'
ORDER BY EMP_ID;
```

Bitemporal Support

In some cases you may need to support both business and system time in the same table. DB2 supports this and it is called bitemporal support. Now let's move on to some examples of all three types of temporal tables.

Business Time Example

You create a temporal table by adding columns for the start and ending period for which the data is valid. Let's do an example. We could modify our existing EMP_PAY table and we'll do that, but first let's look at how we would have defined it if we originally made it a temporal table.

Our original DDL for creating EMP_PAY looks like this:

```
CREATE TABLE HRSCHEMA.EMP_PAY(
EMP_ID INT NOT NULL,
EMP_REGULAR_PAY DECIMAL (8,2) NOT NULL,
EMP_BONUS_PAY DECIMAL   (8,2))
PRIMARY KEY (EMP_ID)) IN TSHR;
```

Let's create another version of the EMP_PAY table and this one will be temporal. Our new table name will be EMP_PAYX. Besides the pay related columns, we must define two date variables to represent starting and ending business time. Below we use BUS_START and BUS_END for this. Next you must reference these columns in the PERIOD BUSINESS_TIME clause. Finally you will make BUSINESS_TIME part of the primary key and (I strongly recommend) specify that overlapping time periods are not allowed.

Here is our DDL:

```
CREATE TABLE HRSCHEMA.EMP_PAYX(
EMP_ID INT NOT NULL,
EMP_REGULAR_PAY DECIMAL (8,2) NOT NULL,
EMP_BONUS_PAY DECIMAL   (8,2),
BUS_START   DATE  NOT NULL,
BUS_END     DATE  NOT NULL,

PERIOD BUSINESS_TIME(BUS_START, BUS_END),
PRIMARY KEY (EMP_ID, BUSINESS_TIME WITHOUT OVERLAPS)) IN TSHR;
```

Now let's insert a few rows into the table. Keep in mind that we now have a start and end date for which the information is valid. That could pose a problem if our end date is really "until further notice". Some applications solve that problem by establishing a date in the distant future as the standard end date for current data. We'll use 12/31/2099 for this example. For convenience we can use the existing EMP_PAY table to load EMP_PAYX using a query:

```
INSERT INTO HRSCHEMA.EMP_PAYX
SELECT EMP_ID,
EMP_REGULAR_PAY,
EMP_BONUS_PAY,
```

```
'2017-01-01',
'2099-12-31'
FROM HRSCHEMA.EMP_PAY;
```

Here's our resulting data:

```
EMP_ID EMP_REGULAR_PAY EMP_BONUS_PAY BUS_START  BUS_END
------ --------------- ------------- ---------- ----------
  3217        65000.00       5500.00 2017-01-01 2099-12-31
  7459        85000.00       4500.00 2017-01-01 2099-12-31
  9134        75000.00       2500.00 2017-01-01 2099-12-31
  6288        70000.00       2000.00 2017-01-01 2099-12-31
  4720        80000.00       2500.00 2017-01-01 2099-12-31
```

Now let's suppose employee 3217 has been given a raise to 70K per year effective 2/1/2017. First we need to set the end business date on the existing record.

```
UPDATE HRSCHEMA.EMP_PAYX
SET BUS_END     = '2017-02-01'
WHERE EMP_ID = 3217;
```

Notice that we used 2017-02-01 as the `BUS_END` date on this record. This is true even though we will use 2017-02-01 as the beginning time for the next record. This could be a bit confusing, and please read the next couple of paragraphs carefully.

IMPORTANT: both system and business time are inclusive of start date and exclusive of end date. For example, to set an employee salary effective date January 1, 2017 and ending at midnight on January 31, 2017 you would use start date 2017-01-01. But you would use end date 2017-02-01. If you instead set the `BUS_END` for 2017-01-31, then a query for `BUSINESS_TIME` as of January 31 would not include that record because the end time is exclusive. Normally you want your query to be inclusive of the date you specify, so you'd need to specify 2017-02-01 as the end date in this case.

If the above is a little confusing, it is because generally date related evaluations in DB2 do not work this way (if you say BETWEEN two dates, it means inclusive at both ends), but temporal tables do work this way. So be sure that you get this! For setting business and system time, the start date is **inclusive** but the end date is **exclusive**.

Now let's add the new row which we'll make effective on 2017-02-01 and ending on 2099-12-31:

```
INSERT INTO HRSCHEMA.EMP_PAYX
VALUES (3217,
70000.00,
5500.00,
```

```
'2017-02-01',
'2099-12-31');
```

Here's the result:

```
SELECT * FROM HRSCHEMA.EMP_PAYX ORDER BY EMP_ID;

EMP_ID EMP_REGULAR_PAY EMP_BONUS_PAY BUS_START  BUS_END
------ --------------- ------------- ---------- ----------
  3217        65000.00       5500.00 2017-01-01 2017-02-01
  3217        70000.00       5500.00 2017-02-01 2099-12-31
  4720        80000.00       2500.00 2017-01-01 2099-12-31
  6288        70000.00       2000.00 2017-01-01 2099-12-31
  7459        85000.00       4500.00 2017-01-01 2099-12-31
  9134        75000.00       2500.00 2017-01-01 2099-12-31
```

Note that there are now two records for employee 3217. If you were to query this data as of 02/01/2017, you would get a different result than if you queried it for business time 01/15/2017. Recall that querying data in temporal tables is supported by specific temporal clauses, including:

- AS OF
- FROM
- BETWEEN

Here are two queries, and notice the different results in EMP_REGULAR_PAY for employee 3217 depending on the business time you specified.

```
SELECT * FROM HRSCHEMA.EMP_PAYX
FOR BUSINESS_TIME AS OF '2017-02-01'
ORDER BY EMP_ID;

EMP_ID EMP_REGULAR_PAY EMP_BONUS_PAY BUS_START  BUS_END
------ --------------- ------------- ---------- ----------
  3217        70000.00       5500.00 2017-02-01 2099-12-31
  4720        80000.00       2500.00 2017-01-01 2099-12-31
  6288        70000.00       2000.00 2017-01-01 2099-12-31
  7459        85000.00       4500.00 2017-01-01 2099-12-31
  9134        75000.00       2500.00 2017-01-01 2099-12-31
```

```
SELECT * FROM HRSCHEMA.EMP_PAYX
FOR BUSINESS_TIME AS OF '2017-01-15'
ORDER BY EMP_ID;

EMP_ID EMP_REGULAR_PAY EMP_BONUS_PAY BUS_START  BUS_END
------ --------------- ------------- ---------- ----------
  3217        65000.00       5500.00 2017-01-01 2017-02-01
  4720        80000.00       2500.00 2017-01-01 2099-12-31
  6288        70000.00       2000.00 2017-01-01 2099-12-31
```

```
7459         85000.00       4500.00 2017-01-01 2099-12-31
9134         75000.00       2500.00 2017-01-01 2099-12-31
```

Here's another benefit of this design. Since you defined the primary key with non-overlapping business times, DB2 will not allow you to enter any overlapping start and end dates (which can result in the old -811 multiple rows for a fullselect). That feature prevents one of the most pervasive and time-consuming application design errors I've observed over the years.

System Time Example

When you want to capture actions taken on a table at a particular time, use system time. Suppose you want to keep a snapshot of every record BEFORE it is changed. DB2's temporal table functionality also includes automated copying of a "before" image of each record to a history table if you specify one. This feature can be used in lieu of using triggers which are also often used to store a history of each version of a record.

Let's take the example of our `EMPLOYEE` table. For audit purposes, let's say we want to capture all changes made to a record by saving the previous version of each record in a history table. DB2's temporal table functionality makes this pretty easy. Follow these steps:

- Add system time fields to the base table
- Add a transaction id field to the base table
- Alter you base table to use system time
- Create a history table
- Version-enable the base table pointing it to the history table

Adding system time to the table is as simple as adding the time and transaction id columns needed to track system time. Let's do this for the `EMPLOYEE` table.

```
ALTER TABLE HRSCHEMA.EMPLOYEE
ADD COLUMN SYS_START TIMESTAMP(12)
GENERATED ALWAYS AS ROW BEGIN NOT NULL;

ALTER TABLE HRSCHEMA.EMPLOYEE
ADD COLUMN SYS_END TIMESTAMP(12)
GENERATED ALWAYS AS ROW END NOT NULL;

ALTER TABLE HRSCHEMA.EMPLOYEE
ADD COLUMN TRANS_ID TIMESTAMP(12) NOT NULL GENERATED
ALWAYS AS TRANSACTION START ID;

ALTER TABLE HRSCHEMA.EMPLOYEE
ADD PERIOD SYSTEM_TIME (SYS_START, SYS_END);
```

Now let's explore the history table feature. There may be cases in which you want to maintain a record of all changes made to table. You can do this automatically by defining a history table and enabling your base table for versioning. Let's create a history table EMPLOYEE_HISTORY and we'll make it identical to EMPLOYEE.

```
CREATE TABLE HRSCHEMA.EMPLOYEE_HISTORY LIKE HRSCHEMA.EMPLOYEE;
```

Now we can enable versioning in the EMPLOYEE table like this:

```
ALTER TABLE HRSCHEMA.EMPLOYEE
ADD VERSIONING
USE HISTORY TABLE
HRSCHEMA.EMPLOYEE_HISTORY;
```

At this point we can make a change to one of the EMPLOYEE records and subsequently we expect to see the old version of the record in the history table.

```
UPDATE HRSCHEMA.EMPLOYEE
SET EMP_FIRST_NAME = 'FREDERICK'
WHERE EMP_ID = 4175;
```

Assume that today is January 30, 2017 so that's when we changed our data. When you query with a specified system time, DB2 implicitly joins the base table and the history table. For example, let's pull data for employee 4175 as of 1/15/2017:

```
SELECT EMP_ID, EMP_FIRST_NAME, SYS_START, SYS_END
FROM HRSCHEMA.EMPLOYEE FOR SYSTEM_TIME AS OF '2017-01-15' WHERE EMP_ID = 4175;

EMP_ID  EMP_FIRST_NAME  SYS_START              SYS_END
------  --------------  ---------------------  --------------------------
 4175   FRED            0001-01-01 00:00:00.0  2017-01-30 17:29:38.608073
```

Notice that the previous version of the record is pulled up (FRED instead of FREDERICK) even though that version is no longer in the EMPLOYEE table. Since we specified system time as of 1/15/2017, DB2 found the relevant record in the history table and included it in the result of the query.

Now let's perform the same query for system time as of February 1, 2017.

```
SELECT EMP_ID, EMP_FIRST_NAME, SYS_START, SYS_END
FROM HRSCHEMA.EMPLOYEE
FOR SYSTEM_TIME AS OF '2017-02-01' WHERE EMP_ID = 4175;
 EMP_ID    EMP_FIRST_NAME  SYS_START                     SYS_END
 ------    --------------  --------------------------    ---------------------
   4175    FREDERICK       2017-01-30 17:29:38.608073    9999-12-30 00:00:00.0
```

Now you've got the most current record with the modified name FREDERICK. That is pretty cool feature and most if it happens automatically once you set it up. It can really help save time when researching particular values that were in the table sometime in the past.

NOTE: You can only use a history table with a system time enabled table.

Bi-Temporal Example

Finally, let's do an example where you need both business time and system time enabled for the same table. Let's go back go our EMP_PAY table and create yet another version called EMP_PAYY:

```
CREATE TABLE HRSCHEMA.EMP_PAYY(
EMP_ID INT NOT NULL,
EMP_REGULAR_PAY DECIMAL (8,2) NOT NULL,
EMP_BONUS_PAY DECIMAL   (8,2),
BUS_START   DATE  NOT NULL,
BUS_END     DATE  NOT NULL,
SYS_START   TIMESTAMP(12)
   GENERATED ALWAYS AS ROW BEGIN NOT NULL,
SYS_END     TIMESTAMP(12)
   GENERATED ALWAYS AS ROW END NOT NULL,
TRANS_ID TIMESTAMP(12) NOT NULL
   GENERATED ALWAYS AS TRANSACTION START ID,

PERIOD BUSINESS_TIME(BUS_START, BUS_END),
PERIOD SYSTEM_TIME (SYS_START, SYS_END),
PRIMARY KEY (EMP_ID, BUSINESS_TIME WITHOUT OVERLAPS)) IN TSHR;
```

You'll still need to create the history table and then version enable `EMP_PAYY`.

```
CREATE TABLE HRSCHEMA.EMP_PAYY_HISTORY LIKE HRSCHEMA.EMP_PAYY;
```

Now we can enable versioning in the `EMP_PAYY` table like this:

```
ALTER TABLE HRSCHEMA.EMP_PAYY
ADD VERSIONING
USE HISTORY TABLE HRSCHEMA.EMP_PAYY_HISTORY;
```

This concludes our discussion of DB2's support for temporal tables and time travel queries. This is a very powerful technology and I encourage you to learn it to take advantage of its features that improve your client's access to actionable business information. It can also ease the application development and production support efforts!

Materialized Query Tables

A materialized query table (MQT) basically stores the result set of a query. It is used to store aggregate results from one or more other tables. MQTs are often used to improve performance for certain aggregation queries by providing pre-computed results. Consequently,

MQTs are most often used in analytic or data warehousing environments.

MQTs are either system-maintained or user maintained. For a system maintained table, the data can be updated using the `REFRESH TABLE` statement. A user-maintained MQT can be updated using the LOAD utility, and also the UPDATE, INSERT, and DELETE SQL statements.

Let's do an example of an MQT that summarizes monthly payroll. Earlier we created a table named `EMP_PAY_HIST` which is a history of each employee's salary for each paycheck. The table is defined as follows:

Column Name	**Definition**
EMP_ID	Numeric
EMP_PAY_DATE	Date
EMP_PAY_AMT	Decimal(8,2)

Assume that the data is as follows:

```
SELECT * FROM HRSCHEMA.EMP_PAY_HIST ORDER BY EMP_PAY_DATE, EMP_ID;

EMP_ID                        EMP_PAY_DATE    EMP_PAY_AMT
------                        ------------    -----------
  3217                        2017-01-15          2708.33
  4720                        2017-01-15          3333.33
  6288                        2017-01-15          2916.66
  7459                        2017-01-15          3541.66
  9134                        2017-01-15          3125.00
  3217                        2017-01-31          2708.33
  4720                        2017-01-31          3333.33
  6288                        2017-01-31          2916.66
  7459                        2017-01-31          3541.66
  9134                        2017-01-31          3125.00
  3217                        2017-02-15          2708.33
  4720                        2017-02-15          3333.33
  6288                        2017-02-15          2916.66
  7459                        2017-02-15          3541.66
  9134                        2017-02-15          3125.00
  3217                        2017-02-28          2708.33
  4720                        2017-02-28          3333.33
  6288                        2017-02-28          2916.66
  7459                        2017-02-28          3541.66
  9134                        2017-02-28          3125.00
```

Finally, let's assume we regularly need an aggregated total of each employee's year to date pay.

We could do this with a materialized query table. Let's build the query that will summarize the employee pay from the beginning of the year to current date:

```
SELECT EMP_ID, SUM(EMP_PAY_AMT) AS EMP_PAY_YTD
FROM HRSCHEMA.EMP_PAY_HIST
GROUP BY EMP_ID
ORDER BY EMP_ID;

EMP_ID           EMP_PAY_YTD
------           -----------
  3217               9166.64
  7459              13333.32
  9134              13333.32
```

Now let's create the MQT using this query and we'll make it a system managed MQT:

```
CREATE TABLE HRSCHEMA.EMP_PAY_TOT (EMP_ID, EMP_PAY_YTD) AS
(SELECT EMP_ID, SUM(EMP_PAY_AMT) AS EMP_PAY_YTD
FROM HRSCHEMA.EMP_PAY_HIST
GROUP BY EMP_ID)
DATA INITIALLY DEFERRED
REFRESH DEFERRED
MAINTAINED BY SYSTEM
ENABLE QUERY OPTIMIZATION;
```

We can now populate the table by issuing the `REFRESH TABLE` statement as follows:

```
REFRESH TABLE HRSCHEMA.EMP_PAY_TOT;
```

Finally we can query the MQT as follows:

```
SELECT * FROM HRSCHEMA.EMP_PAY_TOT;

EMP_ID                       EMP_PAY_YTD
 ------                      -----------
   3217                         10833.32
   4720                         13333.32
   6288                         11666.64
   7459                         14166.64
   9134                         12500.00
```

Again, a main benefit of MQTs is that once you generate the results, many people can retrieve them without the overhead of re-generating the aggregate result set. This is especially helpful and efficient when summarizing large quantities of data.

Temporary Tables

Sometimes you may need to create a DB2 table for the duration of a session but no longer than that. For example you may have a programming situation where it is convenient to have a temporary table which you can load for these operations:

- To join the data in the temporary table with another table
- To store intermediate results that you can query later in the program
- To load data from a flat file into a relational format

Let's assume that you only need the temporary table for the duration of a session or iteration of a program since temporary tables are dropped automatically as soon as the session ends.

Temporary tables are created using either the CREATE statement or the DECLARE statement. The differences will be explored in the Application Design section of this book. For now we will just look at an example of creating a table called `EMP_INFO` using both methods:

```
CREATE GLOBAL TEMPORARY TABLE EMP_INFO(
EMP_ID   INT,
EMP_LNAME  VARCHAR(30),
EMP_FNAME  VARCHAR(30));

DECLARE GLOBAL TEMPORARY TABLE EMP_INFO(
EMP_ID   INT,
EMP_LNAME  VARCHAR(30),
EMP_FNAME  VARCHAR(30));
```

When using the LIKE clause to create a temporary table, the implicit table definition includes only the column name, data type and nullability characteristic of each of the columns of the source table, and any column defaults. The temporary table does NOT have any unique constraints, foreign key constraints, triggers, indexes, table partitioning keys, or distribution keys.

CREATED Temporary Tables

Created temporary tables:

- Have an entry in the system catalog (`SYSIBM.SYSTABLES`)
- Cannot have indexes
- Their columns cannot use default values (except NULL)
- Cannot have constraints
- Cannot be used with DB2 utilities
- Cannot be used with the UPDATE statement
- If DELETE is used at all, it will delete all rows from the table
- Do not provide for locking or logging

DECLARED Temporary Tables

A declared temporary table offers some advantages over created temporary tables.

- Can have indexes and check constraints
- Can use the UPDATE statement
- Can do positioned deletes

So declared temporary tables offer more flexibility than created temporary tables. However, when a session ends, DB2 will automatically delete both the rows in the table and the table definition. So if you want a table definition that persists in the DB2 catalog for future use, you would need to use a created temporary table.

Temporary Table Example

Let's do a simple but more practical example. Suppose you have employee pay data in a flat file that you want to summarize. You want to know the maximum, minimum and average salary for a group of records. There are many ways to accomplish this. One way is with a temporary DB2 table that we will load the data. Then we'll use a query to summarize the data using DB2 functions.

Here's a table definition, followed by the insertion of three records. Next we'll select the records just to ensure they are there. Finally, we'll run a query to generate the maximum, minimum and average salary. If you run this in Data Studio make sure to select all the operations before clicking on the Run button.

```
DECLARE GLOBAL TEMPORARY TABLE SESSION.EMP_PAY(
EMP_ID   INT,
EMP_ANN_SALARY DECIMAL (8,2) NOT NULL)ON COMMIT PRESERVE ROWS;

INSERT INTO SESSION.EMP_PAY
VALUES (3217, 85000.00),
       (7459, 80000.00),
       (9134, 70000.00);  COMMIT;

SELECT * FROM SESSION.EMP_PAY;



EMP_ID   EMP_ANN_SALARY
------   --------------
  3217         85000.00
  7459         80000.00
  9134         70000.00

SELECT MAX(EMP_ANN_SALARY) AS MAX_SALARY,
MIN(EMP_ANN_SALARY) AS MIN_SALARY,
AVG(EMP_ANN_SALARY) AS AVG_SALARY  FROM SESSION.EMP_PAY;
```

```
MAX_SALARY     MIN_SALARY     AVG_SALARY
----------     ----------     ------------------------------
  85000.00       70000.00     78333.333333333333333333333333
```

DB2 temporary tables can be useful in many circumstances, especially where you've got aggregated totals that you need to apply to particular records based on another column. For example, suppose we have determined the average employee salary by department, and we want to apply that average value to a new `AVG_SAL` column on each employee's pay record. Let's say we've already created the new column to the employee record. How to get the average salary data there?

One way to accomplish the above would be to load the average salary by department to a temporary table, and then cycle through each employee record using the employee's department to do a lookup in the temporary table. Once you have the average salary for that department from the temporary table, you can then apply it to the employee's record. Of course this is a trivial example, but you can use temporary tables to help solve more complex problems and without the overhead of keeping a permanent table.

Things to remember about temporary tables:

- Use temporary tables when you need the data only for the duration of the session.
- Created temporary tables can provide excellent performance because they do not use locking or logging.
- Declared temporary tables can also be very efficient because you can choose not to log, and they only allow limited locking.
- The schema for a temporary table is always SESSION.
- If you create a temporary table and you wish to replace any existing temporary table that has the same name, use the WITH REPLACE clause.
- If you create a temporary table from another table using the LIKE clause, the temporary table will NOT have any unique constraints, foreign key constraints, triggers, indexes, table partitioning keys, or distribution keys from the original table.

XML SCHEMA VALIDATION

DB2 does a basic check to ensure that the data you use to populate an XML column is well formed. We saw an example of this check back in section 1. As long as the XML is well formed, DB2 will apply it to the XML column.

In addition to ensuring well formed XML, you can also validate the XML against an XML schema. XML schema help you ensure that the XML content meets the data and business rules you define. Again, if you are not very familiar with these concepts, I encourage you to read the recommended XML books. Now let's create an example to go with the `EMP_PROFILE` column of the `EMPLOYEE` table.

First, let's review the XML column and some sample content.

```
<?xml version="1.0" encoding="UTF-8"?>
<EMP_PROFILE>
      <EMP_ID>4175</EMP_ID>
      <EMP_ADDRESS>
            <STREET>6161 MARGARET LANE</STREET>
            <CITY>ERINDALE</CITY>
            <STATE>AR</STATE>
            <ZIP_CODE>72653</ZIP_CODE>
      </EMP_ADDRESS>
      <BIRTH_DATE>1991-07-14</BIRTH_DATE>
</EMP_PROFILE>
```

Now let's establish some rules for the content of the `EMP_PROFILE` structure:

- EMP_ID is required and must be an integer greater than zero.
- STREET, CITY, STATE are required and must be string values.
- ZIP_CODE must be an integer greater than zero.
- BIRTH_DATE is optional, but if entered must be a valid date.

An XML schema for this structure might look like this:

```
<?xml version="1.0" encoding="UTF-8" ?>
<xs:schema xmlns:xs="http://www.w3.org/2001/XMLSchema">
<xs:element name="EMP_PROFILE">
   <xs:complexType>
     <xs:sequence>
        <xs:element name="EMP_ID" type="xs:positiveInteger" />
        <xs:element name="EMP_ADDRESS">
        <xs:complexType>
        <xs:sequence>
           <xs:element name="STREET" type="xs:string" />
           <xs:element name="CITY" type="xs:string" />
           <xs:element name="STATE" type="xs:string" />
```

```
                <xs:element name="ZIP_CODE" type="xs:positiveInteger"/>
            </xs:sequence>
            </xs:complexType>
            </xs:element>
             <xs:element name="BIRTH_DATE" minOccurs="0"
               type="xs:date" />
         </xs:sequence>
      </xs:complexType>
   </xs:element>
   </xs:schema>
```

Ok, we have our schema, what's next?

Registering the XML Schema

Now we need to know how to define the XML schema to DB2. The schema must be added to the XML Schema Repository (XSR) before it can be used to validate XML docs. The XSR is a repository for all XML schemas that are required to validate and process the XML documents that are stored in XML columns. DB2 creates the XSR tables during installation or migration.

The `REGISTER XMLSCHEMA` command performs the first step of the XML schema registration process, by registering the primary XML schema document. The final step of the XML schema registration process requires that the `COMPLETE XMLSCHEMA` command run successfully for the XML schema. Alternatively, if there are no other XML schema documents to be included, issue the `REGISTER XMLSCHEMA` command with the `COMPLETE` keyword to complete registration in one step.

You can enter the `REGISTER XMLSCHEMA` command using the command line processor, or you can use some built-in stored procedures. Let's call our validation schema `HRSCHEMA.EMP_PROFILE`. Let's initially locate the schema definition in a local file named `EMP_PROFILE`.xsd in the C:\TEMP directory.

Now's let's execute the registration from the DB2 Command Line Processor (CLP).

```
REGISTER XMLSCHEMA 'http://employeeschema/emp_profile.xsd'
FROM 'file:///c:/TEMP/emp_profile.xsd' AS HRSCHEMA.EMP_PROFILE
```

You can add additional documents using the `ADD XMLSCHEMA DOCUMENT` command. In our case we only used a single document, so we can proceed to the completion step. Use this command to complete the registration process.

```
COMPLETE XMLSCHEMA HRSCHEMA.EMP_PROFILE
```

Validating XML Against a Schema

Once registered and completed, we can validate an XML value against the specified schema. Let try inserting an XML column with an invalid birth date. You'll use the XMLVALDATE function with the ACCORDING TO XMLSCHEMA clause.

```
UPDATE HRSCHEMA.EMPLOYEE SET EMP_PROFILE
 = XMLVALIDATE (XMLPARSE ( DOCUMENT ('<EMP_PROFILE>
  <EMP_ID>3217</EMP_ID>
  <EMP_ADDRESS>
      <STREET>2913 PATE DR</STREET>
      <CITY>FORT WORTH</CITY>
      <STATE>TX</STATE>
      <ZIP_CODE>76105</ZIP_CODE>
    </EMP_ADDRESS>
    <BIRTH_DATE>1952-02-30</BIRTH_DATE>
    </EMP_PROFILE>')) ACCORDING TO XMLSCHEMA ID HRSCHEMA.EMP_PROFILE)
        WHERE EMP_ID = 3217;

Incorrect XML data. Expected data of type "dateTime" and found value "1952-02-
30" which is not a valid value for that type.. SQLCODE=-16105, SQLSTATE=2200M,
DRIVER=4.18.60
```

So when we use validation, the UPDATE command fails. Now let's correct the birth date. We'll change it from February 30 to February 28, and we'll see that it now executes correctly:

```
UPDATE HRSCHEMA.EMPLOYEE SET EMP_PROFILE
 = XMLVALIDATE (XMLPARSE (DOCUMENT ('<EMP_PROFILE>
  <EMP_ID>3217</EMP_ID>
  <EMP_ADDRESS>
      <STREET>2913 PATE DR</STREET>
      <CITY>FORT WORTH</CITY>
      <STATE>TX</STATE>
      <ZIP_CODE>76105</ZIP_CODE>
    </EMP_ADDRESS>
    <BIRTH_DATE>1952-02-28</BIRTH_DATE>
    </EMP_PROFILE>')) ACCORDING TO XMLSCHEMA ID HRSCHEMA.EMP_PROFILE)
WHERE EMP_ID = 3217;

Updated 1 rows.
```

Automating Schema Validation with Triggers

Finally, let's look at how we could configure to automatically do the XML validation instead of us having to code the validation in the SQL. You can do this by creating a `BEFORE` trigger on the `EMP_PROFILE XML` column. Consider this DDL:

```
CREATE TRIGGER UPD_EMP_PROFILE
NO CASCADE BEFORE UPDATE ON HRSCHEMA.EMPLOYEE
REFERENCING NEW AS N
FOR EACH ROW MODE DB2SQL
BEGIN ATOMIC
   SET (N.EMP_PROFILE) = XMLVALIDATE(N.EMP_PROFILE
   ACCORDING TO XMLSCHEMA ID HRSCHEMA.EMP_PROFILE);
END
```

Now let's try modifying the `EMP_PROFILE` column using a bad birth date as we did earlier. This time we will not specify the validation in the SQL.

```
UPDATE HRSCHEMA.EMPLOYEE
SET EMP_PROFILE
 = '<EMP_PROFILE>
  <EMP_ID>3217</EMP_ID>
  <EMP_ADDRESS>
  <STREET>2913 PATE DR</STREET>
  <CITY>FORT WORTH</CITY>
  <STATE>TX</STATE>
  <ZIP_CODE>76105</ZIP_CODE>
 </EMP_ADDRESS>
 <BIRTH_DATE>1952-02-30</BIRTH_DATE>
 </EMP_PROFILE>'
WHERE EMP_ID = 3217

Incorrect XML data. Expected data of type "dateTime" and found value "1952-02-
30" which is not a valid value for that type.. SQLCODE=-16105, SQLSTATE=2200M,
DRIVER=4.18.60
```

Again, we get an error because the birth date value is not a valid date and the XML schema requires it to be. Our trigger forced the validation. From now on, any record we try to update in the `EMPLOYEE` table will fail if it includes an `EMP_PROFILE` value that does not conform to the schema we set up. Of course you would also want to create an INSERT trigger to force validation of records that are being added. Use the same logic for the trigger.

Note: Instead of using the command line processor to register your schema, you can also use a set of built-in stored procedures. These are:

> `SYSPROC.XSR_REGISTER` - Begins registration of an XML schema. You call this stored procedure when you add the first XML schema document to an XML schema.
>
> `SYSPROC.XSR_ADDSCHEMADOC` - Adds additional XML schema documents to an XML schema that you are in the process of registering. You can call
>
> `SYSPROC.XSR_ADDSCHEMADOC` only for an existing XML schema that is not yet com-

plete.

SYSPROC.XSR_COMPLETE - Completes the registration of an XML schema.

PERFORMANCE

It's vital to design your DB2 applications with optimal performance in mind. Resolving performance issues can also be part of your job as a maintenance developer when enhanced functionality or larger data volumes begin to cause performance issues. This chapter looks at several areas you should be familiar with to develop and tune your applications.

Creating and Using Explain Data

The EXPLAIN statement helps you to gather information about the access paths DB2 uses when retrieving or updating data. This in turn can help you to diagnose performance related issues, such as excessively long running queries.

Creating Plan Tables

Explain tables must be created before the EXPLAIN data cab be captured. This can be accomplished in these ways:

1. Call the SYSINSTALLOBJECTS stored procedure with the EXPLAIN parameter:

```
CALL SYSPROC.SYSINSTALLOBJECTS('EXPLAIN', 'C',
CAST (NULL AS VARCHAR(128)), CAST (NULL AS VARCHAR(128)))
```

The last two parameters are the tablespace and schema name. If you accept the default, the EXPLAIN tables will be created in the SYSTOOLSPACE tablespace using the SYSTOOLS schema. You can also specify another schema and tablespace.

```
CALL SYSPROC.SYSINSTALLOBJECTS('EXPLAIN', 'C', 'TSHR', 'ROBERT');
```

In my case I could use my logon id rwinga01 because that will typically be the default later if I don't specify a schema on EXPLAIN commands.

2. Run the EXPLAIN.DDL command file as follows:

```
Connect to <database>

-tf EXPLAIN.DDL
```

Choose either of the above methods to create EXPLAIN tables (using either the SYSTOOLS schema or your own id).

The EXPLAIN Statement

The EXPLAIN statement generates explain data that can be analyzed to determine access paths the DB2 Optimizer has chosen for a query or plan. There are a couple of ways to use the EXPLAIN function. We'll look at one utility called **db2exfmt** that runs from the DB2 command line processor (CLP). Then we'll look at the Visual Explain utility that runs in the Data Studio IDE.

Creating Explain Data

You can generate data in the EXPLAIN tables in a couple of ways. We'll first generate it manually using SQL, and then later we'll use Visual Explain to add the data automatically.

Here is the query we want to explain.

```
SELECT EMP_ID, EMP_LAST_NAME
FROM HRSCHEMA.EMPLOYEE WHERE EMP_ID IN (3217, 9134)
ORDER BY EMP_ID;
```

To generate EXPLAIN data for it, use the following:

```
EXPLAIN PLAN SET QUERYNO = 1
FOR SELECT EMP_ID, EMP_LAST_NAME FROM HRSCHEMA.EMPLOYEE
WHERE EMP_ID IN (3217, 9134)
ORDER BY EMP_ID;
```

The above generates explain data into the various EXPLAIN tables. If you were to query the `EXPLAIN_INSTANCE` table, you would see a new row added with various values pointed to other EXPLAIN tables. This could get interesting, looking at all the tables. We'll detail the contents of each of the EXPLAIN tables later in the chapter. However, we need to use our DB2 tools to get some meaningfull interpetation of the data!

EXPLAIN Using db2exfmt

The db2exfmt utility formats EXPLAIN data for interpetation. Let's use the data for the query we just added. We'll need to get the EXPLAIN timestamp value of the EXPLAIN statement we just executed. There are two ways to do this. One way is to issue the following SQL against the `EXPLAIN_INSTANCE` table.

```
SELECT EXPLAIN_TIME FROM SYSTOOLS.EXPLAIN_INSTANCE;

EXPLAIN_TIME
-----------------------
2017-11-28 22:04:24.102
```

There is a second, easier way. If we know know that the last EXPLAIN statement is the one we want data for, we do not need the timestamp per se, we can just specify -1 for the timestamp parameter in the db2exfmt and DB2 will use the latest timestamp.

Ok, let's open a DB2 Command Line Processor window. We will execute the db2exfmt utility and direct the output to a file in the C:\MISC directory called **db2exfmt.out.txt**.
Here is the command to enter:

```
db2exfmt -d DBHR -g  -w -1 -n % -s %  -o C:\MISC\db2exfmt.out.txt
```

The meaning of the parameter values is a follows:

-d the database name
-g a graph will be generated.
-w the timestamp if specified or -1 for the latest timestamp on the table
-n package name
-s schema, defaults to userid if none specified
-o the file to which the output is to be directed

Here is the listing from the Command window:

```
C:\windows\system32>db2exfmt -d DBHR -g -w -1 -n % -s % -o C:\MISC\db2exfmt.out.txt
DB2 Universal Database Version 11.1, 5622-044 (c) Copyright IBM Corp. 1991, 2015

Licensed Material - Program Property of IBM
IBM DATABASE 2 Explain Table Format Tool

Connecting to the Database.
Connect to Database Successful.
Enter section number (0 for all, Default 0) ==>
Output is in C:\MISC\db2exfmt.out.txt.
Executing Connect Reset -- Connect Reset was Successful.
```

And now we can open file **db2exfmt.out.txt** and here is our formated EXPLAIN data.

```
DB2 Universal Database Version 11.1, 5622-044 (c) Copyright IBM Corp. 1991, 2015
Licensed Material - Program Property of IBM
IBM DATABASE 2 Explain Table Format Tool

******************** EXPLAIN INSTANCE ********************

DB2_VERSION:       11.01.1
FORMATTED ON DB:   DBHR
SOURCE_NAME:       SYSSH200
SOURCE_SCHEMA:     NULLID
SOURCE_VERSION:
EXPLAIN_TIME:      2017-11-28-22.04.24.102000
EXPLAIN_REQUESTER: RWINGA01
```

```
Database Context:
----------------
        Parallelism:          None
        CPU Speed:            1.968101e-007
        Comm Speed:           0
        Buffer Pool size:     77082
        Sort Heap size:       2752
        Database Heap size:   3336
        Lock List size:       19205
        Maximum Lock List:    98
        Average Applications: 1
        Locks Available:      602268

Package Context:
---------------
        SQL Type:           Dynamic
        Optimization Level: 5
        Blocking:           Block All Cursors
        Isolation Level:    Cursor Stability

---------------- STATEMENT 1  SECTION 65 ----------------
        QUERYNO:       1
        QUERYTAG:
        Statement Type:   Select
        Updatable:        No
        Deletable:        No
        Query Degree:     1


Original Statement:
------------------
SELECT EMP_ID, EMP_LAST_NAME
      FROM HRSCHEMA.EMPLOYEE
         WHERE EMP_ID IN (3217, 9134)
         ORDER BY EMP_ID

Optimized Statement:
-------------------
SELECT
  Q3.EMP_ID AS "EMP_ID",
  Q3.EMP_LAST_NAME AS "EMP_LAST_NAME"
FROM
  HRSCHEMA.EMPLOYEE AS Q3
WHERE
  Q3.EMP_ID IN (3217, 9134)
ORDER BY
  Q3.EMP_ID;
```

```
Access Plan:
-----------
        Total Cost:             6.77946
        Query Degree:           1


                Rows
               RETURN
               (   1)
                Cost
                 I/O
                 |
                 2
               FETCH
               (   2)
               6.77946
                 1
           /----+----\
          2            9
       IXSCAN     TABLE: HRSCHEMA
       (   3)        EMPLOYEE
      0.0113598         Q3
          0
         |
          9
    INDEX: SYSIBM
  SQL171127154214400
         Q3
```

The most important part of this output is the access plan. What the access plan above tells us is that the dynamic query we explained will use an index -- specifically the SQL171127154214400 index (primary key index on EMPLOYEE) to access the data using an index scan (IXSCAN) to FETCH data from table HRSCHEMA.EMPLOYEE.

This fact is good because our EXPLAIN data tells us an index exists and is being used to locate the data rows needed by the query. That typically means we can expect pretty good performance.

Now let's take a different scenario where we are trying to solve a performance issue. Let's assume we are pulling data from the EMP_PAY_CHECK table and the query is running very slow (I know there aren't enough records in our table to cause a problem, but please humor me for educational purposes). Here is the query:

```
SELECT EMP_ID, EMP_REGULAR_PAY
FROM HRSCHEMA.EMP_PAY_CHECK WHERE (EMP_ID = 3217 OR EMP_ID = 9134);
```

Let's create an EXPLAIN statement for this as follows:

```
EXPLAIN PLAN SET QUERYNO = 2
FOR SELECT EMP_ID, EMP_REGULAR_PAY
FROM HRSCHEMA.EMP_PAY_CHECK
WHERE (EMP_ID = 3217 OR EMP_ID = 9134);
```

Now we can run our same db2exfmt command again to get the formatted EXPLAIN data. When we open the db2exfmt.out.txt file, let's just focus on the access plan which is as follows:

```
Access Plan:
-----------
      Total Cost:             6.77763
      Query Degree:           1


          Rows
         RETURN
         (   1)
          Cost
           I/O
           |
            2
         TBSCAN
         (   2)
         6.77763
            1
           |
            6
    TABLE: HRSCHEMA
     EMP_PAY_CHECK
          Q3
```

A brief look at the result tells us what our problem might be. DB2 is using access type TB-SCAN which means a table scan. We are walking through the entire table record-by-record to find the qualifying rows. For a very small table this may be ok, but for tables that have any significant amount of data, this is the slowest way to access the data.

One way to make the query more efficient is to create an index on the `EMP_ID` column. Let's try to solve this problem by creating an index on `EMP_ID` and then see the result.

```
CREATE UNIQUE INDEX HRSCHEMA.EMP_PC_NDX
ON HRSCHEMA.EMP_PAY_CHECK (EMP_ID);
```

Now we can rerun the explain statement:

```
EXPLAIN PLAN SET QUERYNO = 2
FOR SELECT EMP_ID, EMP_REGULAR_PAY
FROM HRSCHEMA.EMP_PAY_CHECK
WHERE (EMP_ID = 3217 OR EMP_ID = 9134);
```

Now let's rerun our db2exfmt command and see the results.

```
Access Plan:
-----------
	Total Cost:		6.77829
	Query Degree:		1


              Rows
             RETURN
             (   1)
              Cost
               I/O
               |
                2
             FETCH
             (   2)
             6.77829
                1
          /----+----\
         2              6
      IXSCAN     TABLE: HRSCHEMA
      (   3)       EMP_PAY_CHECK
    0.0101946           Q3
         0
        |
         6
  INDEX: HRSCHEMA
    EMP_PC_NDX
        Q3
```

Ok, we're in business now. Our access type is IXSCAN which means an index scan. Specifically DB2 will use the index `EMP_PC_NDX` that we just created. We are matching on one column, the `EMP_ID`. This is good. Again, using an index should mean fairly good performance.

Now let's do a third example. Suppose we have a new requirement to order the results of our previous query by regular pay in descending order. And this time we want all the employees (not a specific one). What effect will that have? Let's run the EXPLAIN statement again, changing the SQL and also the QUERYNO so that it is unique.

```
EXPLAIN PLAN SET QUERYNO = 3
FOR SELECT EMP_ID, EMP_REGULAR_PAY
FROM HRSCHEMA.EMP_PAY_CHECK
ORDER BY EMP_REGULAR_PAY DESC;
```

Now let's run our db2exfmt command again, and review the access plan from the output:

```
Access Plan:
-----------
        Total Cost:             6.7796
        Query Degree:           1

              Rows
             RETURN
             (   1)
              Cost
               I/O
               |
                6
             TBSCAN
             (   2)
             6.7796
                1
               |
                6
             SORT
             (   3)
             6.77935
                1
               |
                6
             FETCH
             (   4)
             6.77871
                1
          /----+----\
        6              6
     IXSCAN     TABLE: HRSCHEMA
     (   5)       EMP_PAY_CHECK
   0.0098964           Q1
        0
       |
        6
 INDEX: HRSCHEMA
   EMP_PC_NDX
       Q1
```

Now we see there are multiple steps in the access plan. We are doing an index scan of the rows, then sorting the retrieved rows, then a table scan of the resulting rows (since we are reading all the records we don't actually need the `EMP_ID` index). The sort is not necessarily so good. Once again, if the table contents is fairly small then the SORT step is not likely to be a problem. But if there are many rows in the table we may take a performance hit for having to do the sort to satisfy the ORDER BY clause.

One way to resolve such an issue could be to create another index that includes both the employee id and the regular pay column. First I want to pause to state that adding new indexes is not always the best approach to resolving a performance issue. There are tradeoffs between the overhead cost of additional indexes versus the cost of a long running query. Moreover there are other things to look into besides indexing, such as how the SQL statement is structured to begin with and whether it is using index or data sargable predicates (and avoiding residual predicates if possible).

Having issued the above caveat, let's assume that we decide we need to create the new index and then run the EXPLAIN statement again. Here is the DDL for the new index which will be organized by `EMP_REGULAR_PAY`.

```
CREATE INDEX HRSCHEMA.EMP_PC_RP_NDX
ON HRSCHEMA.EMP_PAY_CHECK
(EMP_REGULAR_PAY DESC, EMP_ID ASC);
```

Here's the EXPLAIN statement for our query:

```
EXPLAIN PLAN SET QUERYNO = 4
FOR SELECT EMP_REGULAR_PAY, EMP_ID
FROM HRSCHEMA.EMP_PAY_CHECK
ORDER BY EMP_REGULAR_PAY DESC;
```

Finally, let's rerun our db2exfmt and look at our results.

```
Access Plan:
-----------
       Total Cost:          0.0098964
       Query Degree:        1

      Rows
     RETURN
     (   1)
      Cost
       I/O
        |
        6
     IXSCAN
     (   2)
    0.0098964
        0
        |
        6
  INDEX: HRSCHEMA
   EMP_PC_RP_NDX
       Q1
```

This access path looks a lot better. We've eliminated the sort step because we have an index that is available to order the results. Notice also that we are doing an **index-only** scan with the new index EMP_PC_RP_NDX. Since all the columns we are requesting are available in the index itself, we don't need to touch the base table. This should be a very efficient scan!

Obviously not all performance issues are this easy to solve as the ones above. But you now have experience with the EXPLAIN tool with the db2exfmt utility. This should give you the information you need to help solve such problems.

Visual Explain Using Data Studio

You can also use Data Studio to evaluate the EXPLAIN statement results and it will give you a more visually appealing result. Let's run through a couple of queries just to see how we would do it. First, let's drop the indexes we just created for the EMP_PAY_CHECK table.

```
DROP INDEX HRSCHEMA.EMP_PC_RP_NDX ;
DROP INDEX HRSCHEMA.EMP_PC_NDX ;
```

Now let's get back to running SQL against the EMP_PAY_CHECK table and that we discover it is a very long running query.

```
SELECT * FROM HRSCHEMA.EMP_PAY_CHECK WHERE EMP_ID = 3217;
```

To check out the access path, we can use Visual Explain with Data Studio to check the access path chosen by the DB2 Optimizer. Simply open Data Studio, select the **RUN SQL** perspective, type the query in, select the query text, right click and select **Open Visual Explain**. You can accept the defaults on the configuration panels and then click **Finish**.

You will see this screen:

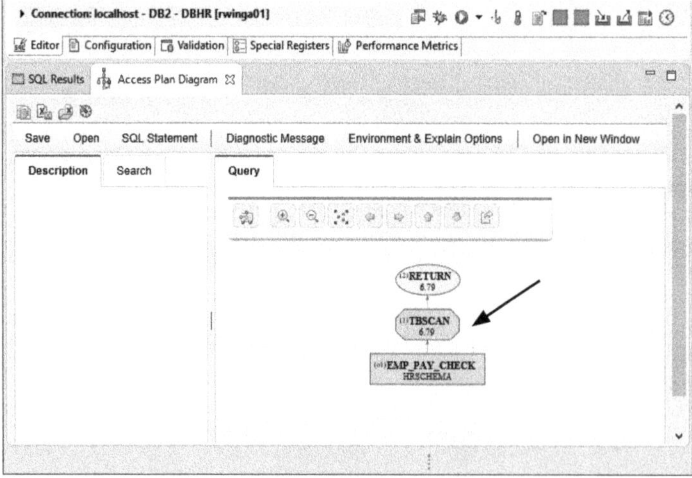

Note the `TBSCAN` in the middle of the diagram. This indicates you are doing a table scan, which means you are not using an index. Unless your table is very small, doing a table scan can definitely create a performance problem. Now let's go back and try to solve our performance problem again by recreating the index.

```
CREATE UNIQUE INDEX HRSCHEMA.EMP_PC_NDX ON HRSCHEMA.EMP_PAY_CHECK (EMP_ID);
```

And now retry the explain table with our query and check the result.

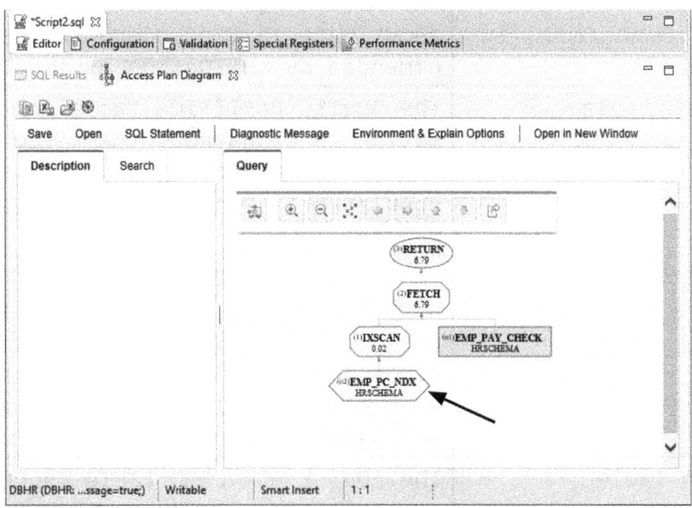

And now we can see the optimizer chose our new index – note the access path is to use an index scan (IXSCAN) using index `EMP_PC_NDX`. This should increase the performance considerably.

You could use Visual Explain for all the queries we looked at earlier. Some people prefer the db2exfmt output. Others like the visualization of the results you get with Visual Explain.

As we wrap up this performance troublshooting topic, the following page shows a list of all the EXPLAIN table types and what they contain. This should give you an idea of how the information in them can help you.

All Explain Tables and Their Contents

The following is a list of all the EXPLAIN tables and what they are for, as provided in the IBM product documentation. [1]

ADVISE_INDEX table	The ADVISE_INDEX table represents the recommended indexes.
ADVISE_INSTANCE table	The ADVISE_INSTANCE table contains information about db2advis execution, including start time.
ADVISE_MQT table	The ADVISE_MQT table contains information about materialized query tables (MQT) recommended by the Design Advisor.
ADVISE_PARTITION table	The ADVISE_PARTITION table contains information about database partitions recommended by the Design Advisor, and can only be populated in a partitioned database environment.
ADVISE_TABLE table	The ADVISE_TABLE table stores the data definition language (DDL) for table creation, using the final Design Advisor recommendations for materialized query tables (MQTs), multidimensional clustered tables (MDCs), and database partitioning.
ADVISE_WORKLOAD table	The ADVISE_WORKLOAD table represents the statement that makes up the workload.
EXPLAIN_ACTUALS table	The EXPLAIN_ACTUALS table contains Explain section actuals information.
EXPLAIN_ARGUMENT table	The EXPLAIN_ARGUMENT table represents the unique characteristics for each individual operator, if there are any.
EXPLAIN_DIAGNOSTIC table	The EXPLAIN_DIAGNOSTIC table contains an entry for each diagnostic message produced for a particular instance of an explained statement in the EXPLAIN_STATEMENT table.
EXPLAIN_DIAGNOSTIC_DATA table	The EXPLAIN_DIAGNOSTIC_DATA table contains message tokens for specific diagnostic messages that are recorded in the EXPLAIN_DIAGNOSTIC table. The message tokens provide additional information that is specific to the execution of the SQL statement that generated the message.

1 https://www.ibm.com/support/knowledgecenter/SSEPGG_11.1.0/com.ibm.db2.luw.sql.ref.doc/doc/r0008441.html

EXPLAIN_FORMAT_OUTPUT table	When db2exfmt utility formats explain tables, its output sometimes contains symbols around operators.
EXPLAIN_INSTANCE table	The EXPLAIN_INSTANCE table is the main control table for all Explain information. Each row of data in the Explain tables is explicitly linked to one unique row in this table.
EXPLAIN_OBJECT table	The EXPLAIN_OBJECT table identifies those data objects required by the access plan generated to satisfy the SQL statement.
EXPLAIN_OPERATOR table	The EXPLAIN_OPERATOR table contains all the operators needed to satisfy the query statement by the query compiler.
EXPLAIN_PREDICATE table	The EXPLAIN_PREDICATE table identifies which predicates are applied by a specific operator.
EXPLAIN_STATEMENT table	The EXPLAIN_STATEMENT table contains the text of the SQL statement as it exists for the different levels of Explain information.
EXPLAIN_STREAM table	The EXPLAIN_STREAM table represents the input and output data streams between individual operators and data objects. The data objects themselves are represented in the EXPLAIN_OBJECT table. The operators involved in a data stream are to be found in the EXPLAIN_OPERATOR table.
OBJECT_METRICS table	The OBJECT_METRICS table contains runtime statistics collected for each object referenced in a specific execution of a section (identified by executable ID) at a specific time (identified by execution time).

SQL Performance Considerations

For best performance it is necessary to consider how SQL predicates are processed. In DB2 LUW there are four categories of predicates that are grouped according to when they are evaluated in the process, and which provide best performance.

Use of Predicates

Range Delimiting Predicates

Range delimiting predicates provide a start and/or stop range of values for an index search. For example:

```
EMP_ID >= 5000 AND EMP_ID <= 9999.
```

Index Sargable Predicates

These predicates do not bracket the search but they still use an index because the columns are part of an index key. Examples based on our EMPLOYEE table would be:

```
EMP_ID = 3217;
EMP_SSN = 439459834;
```

Data Sargable Predicates

These predicates search on non-indexed columns. For example:

```
EMP_LAST_NAME = 'JONES';
```

Residual Predicates

Residual predicates are the least optimal in terms of performance because they require additional processing beyond access to a base table. Examples of these include use of ANY, ALL, SOME or IN. Reading LOB data is also residual if the data is stored in a LOB table space (which is typical).

The point of the above is for you to make sure to make use the best predicates available. Here are some best practices you can use:

1. If an index can be used, write your query to use it. If possible, code to bracket the index search. If not, it is still very efficient to use an index. If a column is not indexed but is often used for searching, consider indexing it.

2. For both index sargable and data sargable predicates, write your queries such that the most restrictive predicates are evaluated first. For example if you have a nation-wide company and you run a query to return all male employees who live

in the state of Texas, the query is more efficient if written as:

```
WHERE EMP_ADDRESS_STATE = 'TX' AND EMP_SEX = 'M'
```

instead of:

```
WHERE EMP_SEX = 'M' AND EMP_ADDRESS_STATE = 'TX'
```

It likely that there are many more employees who are male than the number of employees who live in Texas. By specifying the state of residence as the first predicate you automatically eliminate all the employees living anywhere else, hence you restrict the initial selection to a smaller number of rows than if you specified all male employees.

3. Make sure that your table statistics are current. This is required to support your queries' access paths. Perform **RUNSTATS** periodically to refresh the statistics, and then rebind packages to take advantage of changes.

4. If you have designed your queries efficiently and still encounter issues, you can also perform the following actions to resolve specific problems.

    - Enable queries to be re-optimized at run time (use the REOPT bind option).

    - Specify optimization parameters at the statement level.

    - Specify an access path at the statement level.

Parallel Processing

If a table is partitioned, you can often improve performance by enabling parallel processing. The `DEGREE` option determines whether to attempt to run a query using parallel processing to maximize performance. Setting the `CURRENT DEGREE` special register to `ANY` enables parallel processing. Specifying `DEGREE(ANY)` when binding a package also enables parallel processing.

Whether or not you enable parallel processing depends on your overall application design and objectives.

- If you have regular sequential processing of the entire table (such as to unload a table or search the entire population for multiple conditions), parallel processing can help.

- If you are searching a table based on a key or bracketed key values, and the key is also the parition key, DB2 will limit the seach to only those partitions that would include the key you are looking for.

You need to know your processing need and your data model to determine whether parallel processing will be helpful. Your DBA can also help you determine whether parallel processing would be helpful for your application.

DB2 Trace

When you are trying to solve a problem it helps a lot to have accurate information about what DB2 is doing. One way of obtaining useful information is to use the **db2trc** utility. This is not something an application developer is usually responsible for or expected to know about. Per IBM, it is usually something for DBA or IBM support personnel to evaluate. So for this text book I will leave it alone.

Federation Support Overview

Data federation provides the ability to access data on different systems with a single view. This means you can view and join data not only from different DB2 servers, but also from non-DB2 data sources such as Oracle, MS Excel or data located on cloud. Normally this is a DBA topic, yet you may find this information valuable.

Components

DB2 federation is accomplished using:

1. A DB2 federated server.
2. A federated DB2 database on the DB2 federated server.
3. Data sources such as other DB2 servers or non-DB2.
4. DB2 federation wrappers for data sources.
5. Server definitions.
6. User mappings.
7. Nicknames for objects (tables, views).

Obviously you also will have users who access the federated server and database to access data in the data sources.

Federation with DB2 Data Source

Setting federation between two or more DB2 servers is not difficult. Often the two servers are different DB2 instances on different machines. I want you to be able to follow along on your local machine, so we'll use the same instance with two different databases. The procedure is exactly the same.

Assume our instance name is DB2 and we have an HR database named DBHR. This will include an `EMPLOYEE` table with a few rows of data. Here's the DDL to create the needed objects on your DB2 LUW instance.

```
CREATE DATABASE DBHR AUTOMATIC STORAGE YES;

CREATE STOGROUP SGHR ON 'C:'
OVERHEAD 6.725
DEVICE READ RATE 100.0
DATA TAG NONE;

CREATE BUFFERPOOL BPHR
IMMEDIATE
ALL DBPARTITIONNUMS SIZE 1000
AUTOMATIC PAGESIZE 4096;

CREATE REGULAR TABLESPACE TSHR
IN DATABASE PARTITION GROUP IBMDEFAULTGROUP
```

```
PAGESIZE 4096
MANAGED BY AUTOMATIC STORAGE USING STOGROUP SGHR
AUTORESIZE YES
BUFFERPOOL BPHR
OVERHEAD INHERIT
TRANSFERRATE INHERIT
DROPPED TABLE RECOVERY
ON DATA TAG INHERIT;

CREATE TABLE HRSCHEMA.EMPLOYEE(
EMP_ID INT NOT NULL,
EMP_LAST_NAME VARCHAR(30) NOT NULL,
EMP_FIRST_NAME VARCHAR(20) NOT NULL,
EMP_SERVICE_YEARS INT NOT NULL WITH DEFAULT 0,
EMP_PROMOTION_DATE DATE,
PRIMARY KEY(EMP_ID))
IN TSHR;

INSERT INTO HRSCHEMA.EMPLOYEE
(EMP_ID,
 EMP_LAST_NAME,
 EMP_FIRST_NAME,
 EMP_SERVICE_YEARS,
 EMP_PROMOTION_DATE)
VALUES (3217,
'JOHNSON',
'EDWARD',
4,
'01/01/2017');

INSERT INTO HRSCHEMA.EMPLOYEE
VALUES (7459,
'STEWART',
'BETTY',
7,
'07/31/2016');
```

Now let's select the data from this table to ensure everything was added as we expected.

```
SELECT * FROM HRSCHEMA.EMPLOYEE;

EMP_ID      EMP_LAST_NAME   EMP_FIRST_NAME   EMP_SERVICE_YEARS       EMP_PROMOTION_DATE
------      -------------   --------------   -----------------       ------------------
  3217      JOHNSON         EDWARD                           4       2017-01-01
  7459      STEWART         BETTY                            7       2016-07-31
```

Now we are ready to enable federation support for our DB2 instance and then we'll configure the DB2FED database to access the EMPLOYEE table on the DBHR database.
Set the FEDERATED instance configuration parameter to YES.

```
db2 => update dbm cfg using FEDERATED YES
```

Stop and restart the DB2 instance

```
db2 => db2stop
db2 => db2start
```

Ok, now let's create the federated database that users will connect to in order to access the various data stores including the EMPLOYEE table. We'll call the federation database DB2FED.

```
CREATE DATABASE DB2FED
AUTOMATIC STORAGE YES;
```

Ok, connect to the DB2FED database.

```
CONNECT TO DB2FED
```

Now we need to create a "wrapper" for the DB2 server. The default wrapper for DB2 servers is named DRDA. Issue this command:

```
db2 =>  CREATE WRAPPER DRDA
```

Now we need to create a server using the **DRDA** wrapper with credentials for our DBHR database.

```
create server srvDBHR type DB2/UDB version '11.1' wrapper drda authorization
"rwinga01" PASSWORD "*********" options( DBNAME 'DBHR')
```

Next we can create a user mapping for instance DB2 on server srvDBHR

```
db2 CREATE USER MAPPING FOR HRSCHEMA SERVER SRVDBHR OPTIONS (REMOTE_AUTHID 'rob-
ert', REMOTE_PASSWORD '*********')
```

Finally, we create a nickname for table HRSCHEMA.EMPLOYEE.

```
db2 CREATE NICKNAME EMP_DATA FOR srvDBHR.HRSCHEMA.EMPLOYEE
```

Ok we are finished setting up the federation to access the DBHR database from the DB2FED database. Now when connected to DB2FED you can use the nickname EMP_DATA to retrieve the EMPLOYEE table data.

```
SELECT * from EMP_DATA

EMP_ID  EMP_LAST_NAME    EMP_FIRST_NAME  EMP_SERVICE_YEARS        EMP_PROMOTION_DATE
------  -------------    --------------  -----------------        ------------------
  3217  JOHNSON          EDWARD                          4        2017-01-01
  7459  STEWART          BETTY                           7        2016-07-31
```

If you do have two machines you can configure this federation as described as long as the data source server is visible on the network from the federated server.

Federation with non-DB2 Data Source

If you are using DB2 Express-C or any version of DB2 with a Community license, you will only be able to use federation for other Db2 servers or Informix. You must have a paid license to create wrappers for non-DB2 data sources. Still, we will do an example with MS SQL Server so you can see some of the differences between DB2 and non-DB2 data sources.

Let us assume an EMPLOYEE database which resides on a Microsoft SQL Server in a database named DBHR. If you want to follow along with this example, you can download and install the free MS SQL Server 2017 Express edition from Microsoft.

Here we've opened MS SQL Server Management Tools and we are viewing the data in the table. As you can see we have 6 rows. Once we've added this SQL Server database as a federated source, we'll be able to browse these rows from the federated server DB2FED.

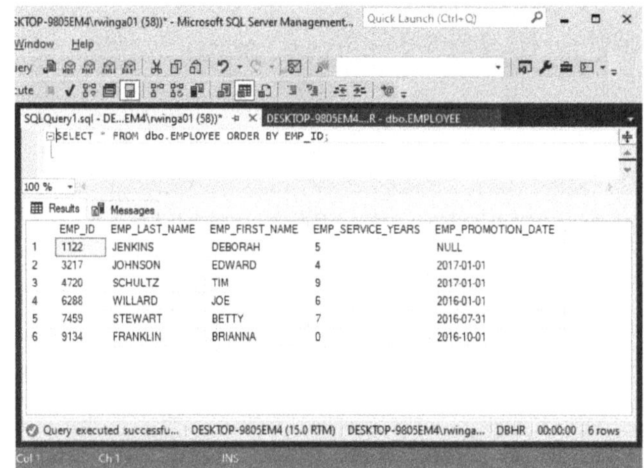

	EMP_ID	EMP_LAST_NAME	EMP_FIRST_NAME	EMP_SERVICE_YEARS	EMP_PROMOTION_DATE
1	1122	JENKINS	DEBORAH	5	NULL
2	3217	JOHNSON	EDWARD	4	2017-01-01
3	4720	SCHULTZ	TIM	9	2017-01-01
4	6288	WILLARD	JOE	6	2016-01-01
5	7459	STEWART	BETTY	7	2016-07-31
6	9134	FRANKLIN	BRIANNA	0	2016-10-01

First, you must make sure ODBC is set up for the MS SQL Server on your machine. Go to Control Panel - Administrative – ODBC Data Sources. Click on the `System DSN` tab. Create an entry for the SQL Server and database that the `EMPLOYEE` table resides in. As you can see, I already have the `DBHR` database configured with ODBC.

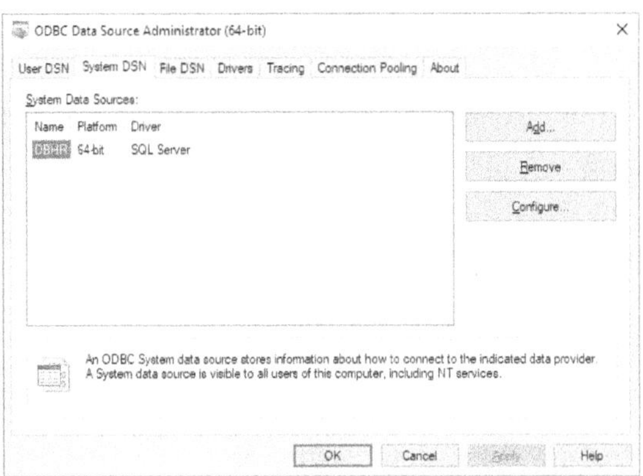

Next, create or update the ODBC environmental variables. You must have a file named db2dj.ini located in your DB2 path directory in folder CONFIG. Edit the file to ensure the correct location of the ODBC system files on your system. Here is the content of my db2dj.ini file:

```
ODBCINST=C:\Windows\odbc.ini
```

If you made any changes to the file, you'll need to stop and restart DB2:

```
db2 => db2stop
db2 => db2start
```

Now logon to the federated server/database `DB2FED` and register the wrapper for MS SQL Server. The default wrapper name for SQL Server is MSSQLODBC3:

```
CREATE WRAPPER MSSQLODBC3
LIBRARY 'db2mssql3.dll'
OPTIONS (ADD DB2_FENCED 'Y')
```

Then create the server definition for the SQL Server:

```
CREATE SERVER MSSQLDBHR
TYPE MSSQLSERVER
VERSION 14.0.2002.14
WRAPPER MSSQLODBC3
OPTIONS (NODE 'DBHR',
DBNAME 'DBHR');
```

Next, create the user mapping using the server name and your login credentials for the SQL server:

```
CREATE USER MAPPING FOR robert SERVER MSSQLDBHR
OPTIONS (REMOTE_AUTHID 'robert', REMOTE_PASSWORD '<yourpassword>');
```

Finally, create a `NICKNAME` for the SQL Server `EMPLOYEE` table. We'll use `SSRVR_EMPLOYEE` as the nickname. I am using my logonid as the schema but you could use `HRSCHEMA` or any other valid schema depending on your business requirement. If you don't specify a schema, then DB2 will use the userid of the person creating the nickname.

```
CREATE NICKNAME robert.SSRVR_EMPLOYEE
FOR MSSQLDBHR."dbo"."EMPLOYEE";
```

Let's run a query now, selecting all the data in the `EMPLOYEE` table:

```
SELECT * FROM robert.SSRVR_EMPLOYEE
ORDER BY EMP_ID;
```

```
EMP_ID     EMP_LAST_NAME   EMP_FIRST_NAME   EMP_SERVICE_YEARS       EMP_PROMOTION_DATE
------     -------------   --------------   -----------------       ------------------
  1122     JENKINS         DEBORAH                          5       NULL
  3217     JOHNSON         EDWARD                           4       2017-01-01
  4720     SCHULTZ         TIM                              9       2017-01-01
  6288     WILLARD         JOE                              6       2016-01-01
  7459     STEWART         BETTY                            7       2016-07-31
  9134     FRANKLIN        ROSEMARY                         0       NULL
```

It is also possible to do a "passthru" session in which you query the SQL Server directly. You would want to do this if your SQL used a function or feature that is not valid in DB2. To perform a passthru session, code:

```
SET PASSTHRU MSSQLDBHR;

SELECT * FROM dbo.EMPLOYEE
ORDER BY EMP_ID;

SET PASSTHRU RESET;
```

There are many other data sources that can be used with DB2 federation, including the following:

```
Informix
JDBC
BigSQL
Oracle
```

```
MS SQL Server
Hive
Impala
Spark
PostGreSQL
MySQL
MariaDB
SAP HANA
Greenplum
Netezza
Sybase
Teradata
MongoDB
CouchDB
Hadoop
```

For full configuration information by data source type, check out the IBM product support web site:

https://www.ibm.com/support/knowledgecenter/en/SSEPGG_11.1.0/com.ibm.data.fluidquery.doc/topics/iiyvfed_config_config_ds.html

Chapter Eight Questions

1. Consider the following stored procedure:

```
CREATE PROCEDURE GET_PATIENTS
(IN intHosp INTEGER)
DYNAMIC RESULT SETS 1
LANGUAGE SQL
P1: BEGIN

DECLARE cursor1 CURSOR
WITH RETURN FOR
SELECT PATIENT_ID,
LNAME,
FNAME
FROM PATIENT
WHERE PATIENT_HOSP = intHosp
ORDER BY PATIENT_ID ASC;

OPEN cursor1;
END P1
```

   Answer this question: How many parameters are used in this stored procedure?

   a. 0
   b. 1
   c. 2
   d. 3

2. When you want to create an external stored procedure, which of the following programming languages can NOT be used?

   a. COBOL
   b. C
   c. Fortran
   d. OLE

3. In order to invoke a stored procedure, which keyword would you use?

   a. RUN
   b. CALL
   c. OPEN
   d. TRIGGER

4. Which of the following is NOT a valid **return type** for a User Defined Function?

    a. Scalar
    b. Aggregate
    c. Column
    d. Row

5. Which of the following is NOT a valid type of user-defined function (UDF)?

    a. External sourced
    b. SQL sourced
    c. External table
    d. SQL table

6. Assume the following trigger DDL:

```
CREATE TRIGGER SAVE_EMPL
AFTER UPDATE ON EMPL
FOR EACH ROW
INSERT INTO EMPLOYEE_HISTORY
VALUES (EMPLOYEE_NUMBER,
EMPLOYEE_STATUS,
CURRENT TIMESTAMP)
```

   What will the result of this DDL be, provided the tables and field names are correctly defined?

    a. The DDL will create the trigger successfully and it will work as intended.
    b. The DDL will fail because you cannot use an INSERT with a trigger.
    c. The DDL will fail because the syntax of this statement is incorrect.
    d. The DDL will create the trigger successfully but it will fail when executed.

7. Which ONE of the following actions will NOT cause a trigger to fire?

    a. INSERT
    b. LOAD
    c. DELETE
    d. MERGE

8. If you use the "FOR EACH STATEMENT" granularity clause in a trigger, what type of timing can you use?

    a. BEFORE
    b. AFTER
    c. INSTEAD OF
    d. All of the above.

9. Assume you have a column FLD1 with a referential constraint that specifies that the FLD1 value must exist in another table. What type of field is FLD1?

    a. FOREIGN KEY
    b. PRIMARY KEY
    c. UNIQUE KEY
    d. INDEX KEY

10. Review the following and then answer the question.

```
CREATE TABLE "DBO"."HOSPITAL"  (
"HOSP_ID" INTEGER NOT NULL ,
"HOSP_NAME" CHAR(25) )
IN "USERSPACE1" ;

ALTER TABLE "DBO"."HOSPITAL"
ADD CONSTRAINT "PKeyHosp" PRIMARY KEY
("HOSP_ID");

CREATE TABLE "DBO"."PATIENT"  (
"PATIENT_ID" INTEGER NOT NULL ,
"PATIENT_NAME" CHAR(30) ,
"PATIENT_HOSP" INTEGER )
IN "USERSPACE1" ;

ALTER TABLE "DBO"."PATIENT"
ADD CONSTRAINT "PKeyPat" PRIMARY KEY
("PATIENT_ID");

ALTER TABLE "DBO"."PATIENT"
ADD CONSTRAINT "FKPatHosp" FOREIGN KEY
("PATIENT_HOSP")
REFERENCES "DBO"."HOSPITAL"
("HOSP_ID")
ON DELETE SET NULL;
```

If a row is deleted from the HOSPITAL table, what will happen to the PATIENT_HOSP rows that used the HOSP_ID value?

a. The PATIENT_HOSP field will be set to NULLS.
b. The PATIENT rows will be deleted from the PATIENT table.
c. Nothing will be changed on the PATIENT records.
d. The delete of the HOSPITAL record will fail.

11. Review the following DDL and then answer the question.

```
CREATE TABLE "DBO"."HOSPITAL"  (
"HOSP_ID" INTEGER NOT NULL ,
"HOSP_NAME" CHAR(25) )
IN "USERSPACE1" ;

ALTER TABLE "DBO"."HOSPITAL"
ADD CONSTRAINT "PKeyHosp" PRIMARY KEY
("HOSP_ID");

CREATE TABLE "DBO"."PATIENT"  (
"PATIENT_ID" INTEGER NOT NULL ,
"PATIENT_NAME" CHAR(30) ,
"PATIENT_HOSP" INTEGER )
IN "USERSPACE1" ;

ALTER TABLE "DBO"."PATIENT"
ADD CONSTRAINT "PKeyPat" PRIMARY KEY
("PATIENT_ID");

ALTER TABLE "DBO"."PATIENT"
ADD CONSTRAINT "FKPatHosp" FOREIGN KEY
("PATIENT_HOSP")
REFERENCES "DBO"."HOSPITAL"
("HOSP_ID")
ON DELETE CASCADE;
```

If a row is deleted from the HOSPITAL table, what will happen to the PATIENT_HOSP rows that used the HOSP_ID value?

a. The PATIENT_HOSP field will be set to NULLS.
b. The PATIENT rows will be deleted from the PATIENT table.
c. Nothing will be changed on the PATIENT records.
d. The delete of the HOSPITAL record will fail.

12. Review the following DDL, and then answer the question:

```
CREATE TABLE "DBO"."HOSPITAL"
("HOSP_ID" INTEGER NOT NULL,
"HOSP_NAME" CHAR(25) )  IN "USERSPACE1";

ALTER TABLE "DBO"."HOSPITAL"
ADD CONSTRAINT "PKeyHosp"
PRIMARY KEY ("HOSP_ID");

CREATE TABLE "DBO"."PATIENT"
("PATIENT_ID" INTEGER NOT NULL,
"PATIENT_NAME" CHAR(30) ,
"PATIENT_HOSP" INTEGER )   IN "USERSPACE1" ;

ALTER TABLE "DBO"."PATIENT"
ADD CONSTRAINT "PKeyPat"
PRIMARY KEY     ("PATIENT_ID");

ALTER TABLE "DBO"."PATIENT"
ADD CONSTRAINT "FKPatHosp"
FOREIGN KEY ("PATIENT_HOSP")
REFERENCES "DBO"."HOSPITAL" ("HOSP_ID")
ON DELETE RESTRICT;
```

If a row is deleted from the HOSPITAL table, what will happen to the PATIENT_HOSP rows that used the HOSP_ID value?

a. The PATIENT_HOSP field will be set to NULLS.
b. The PATIENT rows will be deleted from the PATIENT table.
c. Nothing will be changed on the PATIENT records.
d. The delete of the HOSPITAL record will fail.

13. What is the schema for a declared GLOBAL TEMPORARY table?

a. SESSION
b. DB2ADMIN
c. TEMP1
d. USERTEMP

14. If you create a temporary table and you wish to replace any existing temporary table that has the same name, what clause would you use?

    a. WITH REPLACE
    b. OVERLAY DATA ROWS
    c. REPLACE EXISTING
    d. None of the above.

15. What happens to the rows of a temporary table when the session that created it ends?

    a. The rows are deleted when the session ends.
    b. The rows are preserved in memory until the instance is restarted.
    c. The rows are held in the temp table space.
    d. None of the above.

16. Which of the following clauses DOES NOT allow you to pull data for a particular period from a version enabled table?

    a. FOR BUSINESS_TIME UP UNTIL
    b. FOR BUSINESS_TIME FROM ... TO ...
    c. FOR BUSINESS_TIME BETWEEN... AND...
    d. All of the above enable you to pull data for a particular period.

17. Assume you have an application that needs to aggregate and summarize data from several tables multiple times per day. One way to improve performance of that application would be to use a:

    a. Materialized query table
    b. View
    c. Temporary table
    d. Range clustered table

18. Assume you want to track employees in your company over time. Review the following DDL:

```
CREATE TABLE HRSCHEMA.EMPLOYZZ(
EMP_ID INT NOT NULL,
EMP_LAST_NAME VARCHAR(30) NOT NULL,
EMP_FIRST_NAME VARCHAR(20) NOT NULL,
EMP_SERVICE_YEARS INT
NOT NULL WITH DEFAULT 0,
EMP_PROMOTION_DATE DATE,
BUS_START   DATE  NOT NULL,
BUS_END     DATE  NOT NULL,

PERIOD BUSINESS_TIME(BUS_START, BUS_END),

PRIMARY KEY (EMP_ID, BUSINESS_TIME WITHOUT OVERLAPS));
```

What will happen when you execute this DDL?

a. It will fail because you cannot specify WITHOUT OVERLAPS in the primary key – the WITHOUT OVERLAPS clause belongs in the BUSINESS_TIME definition.
b. It will fail because you must specify SYSTEM_TIME instead of BUSINESS_TIME.
c. It will fail because the BUS_START has a syntax error.
d. It will execute successfully.

19. Given the previous question, assume there is a table named EMPLOYEE_HIST defined just like EMPLOYEE. What will happen when you execute the following DDL?

```
ALTER TABLE EMPLOYEE
ADD VERSIONING
USE HISTORY TABLE EMPLOYEE_HIST
```

a. The DDL will execute successfully and updates to EMPLOYEE will generate records in the EMPLOYEE_HIST table.
b. The DDL will succeed but you must still enable the history table.
c. The DDL will generate an error – only SYSTEM time enabled tables can use a history table.
d. The DDL will generate an error – only BUSINESS time enabled tables can use a history table.

20. For a system managed Materialized Query Table (MQT) named EMPMQT, how does the data get updated so that it becomes current?

    a. Issuing INSERT, UPDATE and DELETE commands against EMPMQT.
    b. Issuing the statement REFRESH TABLE EMPMQT.
    c. Issuing the statement MATERIALIZE TABLE EMPMQT.
    d. None of the above.

21. An XML index can be created on what column types?

    a. VARCHAR and XML.
    b. CLOB AND XML.
    c. XML only.
    d. Any of the above.

22. Which of the following can be used to validate an XML value according to a schema?

    a. Defining a column as type XML.
    b. Using XMLVALIDATE in the SQL.
    c. Both of the above.
    d. Neither of the above.

23. To determine whether an XML document has been validated, which function could you use?

    a. XMLXSROBJECTID.
    b. XMLDOCUMENT.
    c. XMLPARSE.
    d. None of the above.

24. Which of the following would enable parallel processing to improve performance of a query?

    a. Setting the CURRENT DEGREE special register to ANY
    b. Using the DEGREE(ANY) bind option
    c. Both a and b
    d. Neither a nor b

25. Which of the following is the least efficient type of predicate?

   a. Index bracketing
   b. Residual
   c. Data Sargable
   d. Index sargable

26. Which of the following would probably NOT improve query performance?

   a. Use indexable predicates in your queries.
   b. Execute the RUNSTATS utility and rebind application programs.
   c. Set the CURRENT DEGREE to the value 1
   d. All of the above could improve application performance.

27. If you find out that your application query is doing a table space scan, what changes could you make to improve the scan efficiency?

   a. Create one or more indexes on the query search columns.
   b. Load the data to a temporary table and query that table instead of the base table.
   c. If the table is partitioned, change it to a non-partitioned table.
   d. All of the above could improve the scan efficiency.

CHAPTER NINE: FINAL PROJECT

I want to finish out the text book with a project that applies some of the skills and knowledge we've learned, but in a different problem domain. Let's switch from the Human Resource domain to a simple frequent buyer domain (also known these days as loyalty programs). We'll provide a skeleton set of requirements, work through creating tables, indexes, stored procedures and so forth.

Project Requirements

At its most basic a frequent buyer system includes members, deposits and rewards. Let's use FB as an acronym for our frequent buyer program. Besides creating a new FB database and tablespace, we will also be creating a Members table, a Deposit table, and a Rewards table to track the various transactions in our Frequent Buyer system – let's call it FB for an acronym. In addition, we should be thinking about referential integrity. Let's plan on creating a Deposits Type table and a Rewards Type table that will control valid entries in the Deposits and Rewards table.

Here are the requirements for the aforementioned tables:

Member Table

Field	Type	Constraints
Member Number	Integer	NOT NULL, autogenerated
Last Name	Character up to 20	NOT NULL
First Name	Character up to 15	NOT NULL
Street	Character up to 30	NOT NULL
City	Character up to 20	NOT NULL
State	Character 2	NOT NULL
Zip Code	Big Integer	NOT NULL
Telephone	Big Integer	
Points Balance	Integer	NOT NULL

Deposits Table

Field	Type	Constraints
Member Number	Integer	NOT NULL, must exist on MEMBER table
Activity Date	Date	NOT NULL
Deposit Type	Character 3	NOT NULL, must exist in Deposit Type table
Deposit Amount	Integer	NOT NULL
Deposit Posted Date	Date	NOT NULL

Rewards Table

Field	Type	Constraints
Member Number	Integer	NOT NULL, must exist on MEMBER table
Reward Date	Date	NOT NULL
Reward Type	Char 3	NOT NULL, must exist in REWARD TYPE table
Reward Amount	Integer	NOT NULL

Deposit Type Table

Field	Type	Constraints
Deposit Type Code	Char 3	NOT NULL
Deposit Type Description	Char up to 20	NOT NULL
Maintenance Date	Date	NOT NULL

Reward Type Table

Field	Type	Constraints
Reward Type Code	Char 3	NOT NULL
Reward Type Description	Char up to 20	NOT NULL
Maintenance Date	Date	NOT NULL

We've made a few assumptions to keep things simple, such as that the credited points are always whole numbers (so we defined them as integers). Similarly the choice of DATE versus TIMESTAMP is somewhat arbitrary, and in a real production situation you could have reasons for choosing one over the other. TIMESTAMP of course is more precise but we'll stick with DATE.

Project DDL

Create the Tables

Now let's start to work on our DB2 system. Let's build the DDL to create the database, tablespace and a schema. We'll take the default storage group and bufferpool. Also since the member number is to be generated, we will create a sequence. Here's our DDL:

```
CREATE DATABASE DBFB
AUTOMATIC STORAGE YES;

CREATE REGULAR TABLESPACE TSFB
IN DATABASE PARTITION GROUP IBMDEFAULTGROUP
```

```
PAGESIZE 4096
MANAGED BY AUTOMATIC STORAGE
USING STOGROUP IBMSTOGROUP
AUTORESIZE YES
BUFFERPOOL IBMDEFAULTBP ;

CREATE SCHEMA FBSCHEMA
AUTHORIZATION robert;
```
← **This should be your DB2 id, whatever it is.**

Now let's create the sequence object. It should be auto-numbered beginning with 100100. Here is our sequence DDL:

```
CREATE SEQUENCE FBSCHEMA.MBRSEQ
START WITH 100100
INCREMENT BY 1
NO CYCLE;
```

Finally, let's create the DDL for our tables.

```
CREATE TABLE FBSCHEMA.MEMBER(
MBR_NBR        INT NOT NULL,
MBR_LAST_NAME  VARCHAR(20) NOT NULL,
MBR_FIRST_NAME VARCHAR(15) NOT NULL,
MBR_ADDRESS    VARCHAR(30) NOT NULL,
MBR_CITY       VARCHAR(20) NOT NULL,
MBR_STATE      CHAR(02)    NOT NULL,
MBR_ZIP        BIGINT      NOT NULL,
MBR_PHONE      BIGINT      NOT NULL,
MBR_BALANCE    INTEGER     NOT NULL,
   PRIMARY KEY(MBR_NBR)) IN TSFB;

 CREATE TABLE FBSCHEMA.DEPOSITS(
DEP_MBR_NBR    INT         NOT NULL,
DEP_ACT_DATE   DATE        NOT NULL,
DEP_TYPE       CHAR(03)    NOT NULL,
DEP_AMOUNT     INTEGER     NOT NULL,
DEP_POST_DATE  DATE        NOT NULL) IN TSFB;


CREATE TABLE FBSCHEMA.REWARDS(
RWD_MBR_NBR    INT         NOT NULL,
RWD_ACT_DATE   DATE        NOT NULL,
RWD_TYPE       CHAR(03)    NOT NULL,
RWD_AMOUNT     INTEGER     NOT NULL) IN TSFB;

CREATE TABLE FBSCHEMA.DEPOSIT_TYPE(
DEP_TYPE       CHAR(03)    NOT NULL,
DEP_DESC       VARCHAR(20) NOT NULL,
DEP_MAINT_DATE DATE        NOT NULL,
   PRIMARY KEY(DEP_TYPE) IN TSFB;
```

```
CREATE TABLE FBSCHEMA.REWARD_TYPE(
REW_TYPE        CHAR(03)    NOT NULL,
REW_DESC        VARCHAR(20) NOT NULL,
REW_MAINT_DATE  DATE        NOT NULL,
   PRIMARY KEY(REW_TYPE))IN TSFB;
```

Ok, now that we've created the tables and indexes, let's do the referential constraints. We'll need several. First, we need to make sure that no record can be added to the DEPOSITS or REWARDS tables if the type value on the new record does not have an entry in the DEPOSIT_TYPE or REWARD_TYPE table respectively. So here is the DDL for that:

```
ALTER TABLE FBSCHEMA.DEPOSITS
FOREIGN KEY FK_DEP_TYPE (DEP_TYPE)
REFERENCES FBSCHEMA.DEPOSIT_TYPE (DEP_TYPE)
ON DELETE RESTRICT;

ALTER TABLE FBSCHEMA.REWARDS
FOREIGN KEY FK_REW_TYPE (RWD_TYPE)
REFERENCES FBSCHEMA.REWARD_TYPE (REW_TYPE)
ON DELETE RESTRICT;
```

Next, we need to ensure that any member number entered on the DEPOSITS and REWARDS table actually exists in the MEMBER table. Here's the DDL for that.

```
ALTER TABLE FBSCHEMA.DEPOSITS
FOREIGN KEY FK_DEP_MBR (DEP_MBR_NBR)
REFERENCES FBSCHEMA.MEMBER (MBR_NBR)
ON DELETE RESTRICT;

ALTER TABLE FBSCHEMA.REWARDS
FOREIGN KEY FK_REW_MBR (RWD_MBR_NBR)
REFERENCES FBSCHEMA.MEMBER (MBR_NBR)
ON DELETE RESTRICT;
```

Ok that takes care of referential constraints. You could also add some check constraints if you wanted to restrict the value or format of certain columns. I'll leave that to your imagination and business requirements, and we'll just use what we have.

Initial Testing

We'll do some testing to ensure our tables, indexes and so forth have been set up and function correctly. Due to our referential constraints, let's first populate the DEPOSIT_TYPE and REWARD_TYPE tables. Let us say we have the entries below that we need to add, and we'll always use the current date as the maintenance date when we add or change a value.

Deposit Type Entries

DEP_TYPE	DEP_DESC
PUR	MEMBER PURCHASE
BON	BONUS POINTS
MGR	MANAGER DISCRETION

Reward Type Entries

REW_TYPE	REW_DESC
DEB	DEBITED DOLLAR AMOUNT REWARD
PRM	FREE PROMOTIONAL REWARD

Let's go ahead and add these entries and then verify them. First the DEPOSIT_TYPE entries:

```
INSERT INTO FBSCHEMA.DEPOSIT_TYPE
VALUES('PUR',
'MEMBER PURCHASE',
CURRENT DATE);

Updated 1 rows.

INSERT INTO FBSCHEMA.DEPOSIT_TYPE
VALUES('BON',
'BONUS POINTS',
CURRENT DATE);

Updated 1 rows.

INSERT INTO FBSCHEMA.DEPOSIT_TYPE
VALUES('MGR',
'MANAGER DISCRETION',
CURRENT DATE);

Updated 1 rows.


SELECT *
FROM FBSCHEMA.DEPOSIT_TYPE;

DEP_TYPE DEP_DESC           DEP_MAINT_DATE
-------- ------------------ --------------
 PUR      MEMBER PURCHASE    2017-11-01
 BON      BONUS POINTS       2017-11-01
 MGR      MANAGER DISCRETION 2017-11-01
```

And now the REWARD_TYPE entries:

```
INSERT INTO FBSCHEMA.REWARD_TYPE
VALUES('DEB',
'DOLLAR AMT REWARD',
CURRENT DATE);

Updated 1 rows.


INSERT INTO FBSCHEMA.REWARD_TYPE
VALUES('PRM',
'FREE PROMOTIONAL REW',
CURRENT DATE);

Updated 1 rows.


SELECT *
FROM FBSCHEMA.REWARD_TYPE;

REW_TYPE REW_DESC             REW_MAINT_DATE
-------- -------------------- --------------
DEB      DOLLAR AMT REWARD    2017-11-01
PRM      FREE PROMOTIONAL REW 2017-11-01
```

Ok, we have our control tables populated. Now let's add a member to the table, and then add the first deposit and withdrawal. Here's the DDL to add the first member:

```
INSERT INTO FBSCHEMA.MEMBER
(MBR_NBR,
 MBR_LAST_NAME,
 MBR_FIRST_NAME,
 MBR_ADDRESS,
 MBR_CITY,
 MBR_STATE,
 MBR_ZIP,
 MBR_PHONE,
 MBR_BALANCE)
 VALUES
 (NEXT VALUE FOR FBSCHEMA.MBRSEQ,
  'JEFFERSON',
  'RICHARD',
  '2497 MYRTLE LANE',
  'HOUSTON',
  'TX',
   77099,
  '2815683572',
  0);

  Updated 1 rows.
```

```
SELECT * FROM FBSCHEMA.MEMBER;

MBR_NBR   MBR_LAST_NAME  MBR_FIRST_NAME MBR_ADDRESS        MBR_CITY MBR_STATE MBR_ZIP  MBR_PHONE  MBR_BALANCE
-------- -------------  -------------- ------------------ --------- -------   -------  ---------- -----------
  100100 JEFFERSON      RICHARD        2497 MYRTLE LANE   HOUSTON   TX         77099   281568357            0
```

Now let's test our referential constraints on the deposits and awards table. We'll try adding a deposit for which the member number does not exist and for which the deposit type does not exist. We should get SQL errors indicating violation of the constraints.

```
INSERT INTO FBSCHEMA.DEPOSITS
(DEP_MBR_NBR,
 DEP_ACT_DATE,
 DEP_TYPE,
 DEP_AMOUNT,
 DEP_POST_DATE)
 VALUES
 (100222,
  '08/01/2017',
  'ZZZ',
  15,
  CURRENT DATE);
```

```
The insert or update value of the FOREIGN KEY "FBSCHEMA.DEPOSITS.FK_DEP_MBR" is
not equal to any value of the parent key of the parent table.. SQLCODE=-530,
SQLSTATE=23503, DRIVER=4.18.60
```

And as you can see there is a violation of the `FK_DEP_MBR` constraint. That's what we were expecting. Now let's use a good member number (100100) and try an invalid deposit type.

```
INSERT INTO FBSCHEMA.DEPOSITS
(DEP_MBR_NBR,
 DEP_ACT_DATE,
 DEP_TYPE,
 DEP_AMOUNT,
 DEP_POST_DATE)
 VALUES
 (100100,
  '08/01/2017',
  'ZZZ',
  15,
  CURRENT DATE);
```

```
The insert or update value of the FOREIGN KEY "FBSCHEMA.DEPOSITS.FK_DEP_TYPE"
is not equal to any value of the parent key of the parent table.. SQLCODE=-530,
SQLSTATE=23503, DRIVER=4.18.60
```

Ok good, again this is what we were expecting. Out insert failed with invalid foreign key in the deposit type field. Let's clean that up and do a good insert.

```
INSERT INTO FBSCHEMA.DEPOSITS
(DEP_MBR_NBR,
 DEP_ACT_DATE,
 DEP_TYPE,
 DEP_AMOUNT,
 DEP_POST_DATE)
 VALUES
 (100100,
  '08/01/2017',
  'PUR',
  15,
  CURRENT DATE);

Updated 1 rows.
```

And we can verify our result by querying the DEPOSIT table.

```
SELECT * FROM FBSCHEMA.DEPOSITS;

DEP_MBR_NBR    DEP_ACT_DATE DEP_TYPE DEP_AMOUNT DEP_POST_DATE
-----------    ------------ -------- ---------- -------------
     100100    2017-08-01   PUR              15 2017-11-01
```

Now let's check the `REWARDS` table for referential constraints. Try adding a reward record for a nonexistent member.

```
INSERT INTO FBSCHEMA.REWARDS
(RWD_MBR_NBR,
 RWD_ACT_DATE,
 RWD_TYPE,
 RWD_AMOUNT)
 VALUES
 (100199,
  '09/01/2017',
  'XXX',
  10);
```

```
The insert or update value of the FOREIGN KEY "FBSCHEMA.REWARDS.FK_REW_MBR" is
not equal to any value of the parent key of the parent table.. SQLCODE=-530, SQL-
STATE=23503, DRIVER=4.18.60
```

Let's fix the member and try again with a bad reward type.

```
INSERT INTO FBSCHEMA.REWARDS
(RWD_MBR_NBR,
 RWD_ACT_DATE,
 RWD_TYPE,
 RWD_AMOUNT)
 VALUES
```

```
 (100100,
  '09/01/2017',
  'XXX',
  10);

The insert or update value of the FOREIGN KEY "FBSCHEMA.REWARDS.FK_REW_TYPE" is
not equal to any value of the parent key of the parent table.. SQLCODE=-530,
SQLSTATE=23503, DRIVER=4.18.60
```

Finally, let's go ahead and do a good insert, and we see it works fine.

```
INSERT INTO FBSCHEMA.REWARDS
(RWD_MBR_NBR,
 RWD_ACT_DATE,
 RWD_TYPE,
 RWD_AMOUNT)
 VALUES
 (100100,
  '09/01/2017',
  'DEB',
  10);


SELECT * FROM FBSCHEMA.REWARDS;

RWD_MBR_NBR RWD_ACT_DATE RWD_TYPE RWD_AMOUNT
----------- ------------ -------- ----------
     100100 2017-09-01   DEB              10
```

Alright, we've reached an important milestone. We've created and tested our referential constraints. That should ensure data integrity and prevent unnecessary effort solving data discrepancies (both in testing and in production).

We should do additional testing, of course. Such as with 5-digit zip codes versus 9-digit. Also we should test incoming data to see what happens if someone sends a phone number with non-numeric values, such as dash separators (-). I'll leave these tasks to you to complete as an exercise, and we'll move on to loading and accessing data.

Stored Procedures and Programs for Data Access

Here we are going to create four stored procedures, all of which pertain to the MEMBER table. One procedure will be to retrieve data from the table. The second will be to add data to the table. The third will be to update data on the table. The fourth will be to delete data from the table. In all four cases we will use native SQL procedures, and I recommend that you do so whenever possible. External procedures are sometimes necessary to meet certain requirements, but they involve more complexity and overhead. Use native SQL procedures when you can.

Stored Procedure to Select

Let's begin by defining the retrieval procedure which we will name `GETMEM`. Here is the DDL.

```
 CREATE PROCEDURE FBSCHEMA.GETMEM
(IN  M_NBR        INTEGER,
 OUT M_LAST_NAME  VARCHAR(20),
 OUT M_FIRST_NAME VARCHAR(15),
 OUT M_ADDRESS    VARCHAR(30),
 OUT M_CITY       VARCHAR(20),
 OUT M_STATE      CHAR(02),
 OUT M_ZIP        INTEGER,
 OUT M_PHONE      BIGINT,
 OUT M_BALANCE    BIGINT)

 LANGUAGE SQL
 READS SQL DATA

 BEGIN
     SELECT MBR_LAST_NAME,
            MBR_FIRST_NAME,
            MBR_ADDRESS,
            MBR_CITY,
            MBR_STATE,
            MBR_ZIP,
            MBR_PHONE,
            MBR_BALANCE
       INTO M_LAST_NAME,
            M_FIRST_NAME,
            M_ADDRESS,
            M_CITY,
            M_STATE,
            M_ZIP,
            M_PHONE,
            M_BALANCE
     FROM FBSCHEMA.MEMBER
     WHERE M_NBR  = MBR_NBR ;

  END
```

You'll also need to grant security on your stored procedure as follows (grant it to your id, to whichever group you belong to, or in this case I am granting to PUBLIC).

```
GRANT EXECUTE ON PROCEDURE FBSCHEMA.GETMEM TO PUBLIC;
```

Now let's test our stored procedure. We can either test it in Data Studio or write a program to test it. Let's use Data Studio to expedite our testing.

Click **Application Objects** under the DBFB structure, and then click on **Stored Procedures**. When the listing comes up, right click on `GETMBR`, then select RUN from the popup menu. You will see this panel - enter 100100 as the member number. Click **Run.**

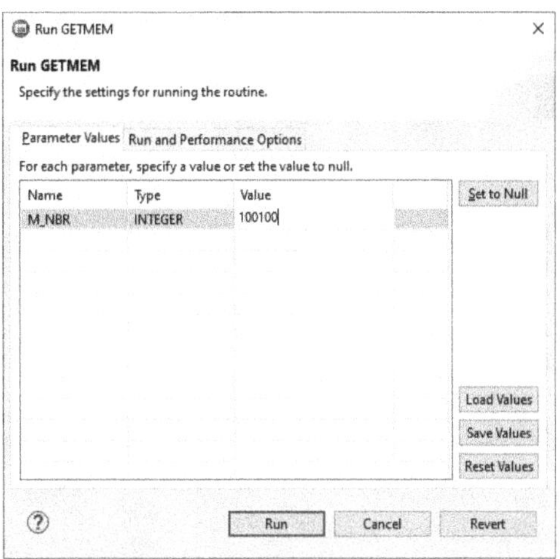

Here is the output in the SQL Results window.

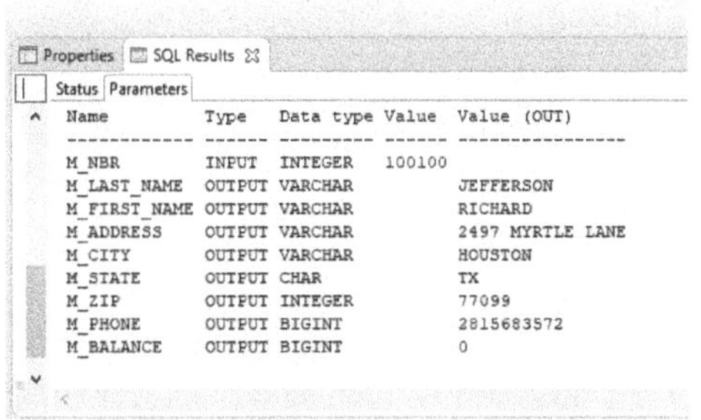

Here's a somewhat easier to read version of the parameters tab.

```
Name         Type   Data type Value  Value (OUT)
------------ ------ --------- ------ ----------------
M_NBR        INPUT  INTEGER   100100
M_LAST_NAME  OUTPUT VARCHAR          JEFFERSON
M_FIRST_NAME OUTPUT VARCHAR          RICHARD
M_ADDRESS    OUTPUT VARCHAR          2497 MYRTLE LANE
M_CITY       OUTPUT VARCHAR          HOUSTON
M_STATE      OUTPUT CHAR             TX
M_ZIP        OUTPUT INTEGER          77099
M_PHONE      OUTPUT BIGINT           2815683572
M_BALANCE    OUTPUT BIGINT           0
```

While I personally prefer doing initial stored procedure testing with Data Studio, you can also write a program to test the procedure. Here's a sample one:

```
package FreqBuyer;
import java.sql.*;

public class JAVFB1 {

        public static void main(String[] args) throws ClassNotFoundException
        {
            Class.forName("com.ibm.db2.jcc.DB2Driver");
            System.out.println("**** Loaded the JDBC driver");
            String url
               =jdbc:db2://localhost:50000/DBFB:retrieveMessagesFromServerOnGet-
        Message=true;";
            String user="rwinga01";
            String password="*********";

            Connection con=null;
            CallableStatement cs=null;

               try {
                    con=DriverManager.getConnection(url, user, password);
                    System.out.println("**** Created the connection");
                    int memNbr = 100100;
                    String lName;
                    String fName;
                    String address;
                    String city;
                    String state;
                    int zip;
                    long phone;
                    long balance;

                    System.out.println
                      ("Executing JAVFB1 to call stored procedure GETMEM");

                    cs = con.prepareCall("{call
                         FBSCHEMA.GETMEM(?,?,?,?,?,?,?,?,?)}");
                    cs.registerOutParameter(2, Types.VARCHAR);
                    cs.registerOutParameter(3, Types.VARCHAR);
                    cs.registerOutParameter(4, Types.VARCHAR);
                    cs.registerOutParameter(5, Types.VARCHAR);
                    cs.registerOutParameter(6, Types.CHAR);
                    cs.registerOutParameter(7, Types.INTEGER);
                    cs.registerOutParameter(8, Types.BIGINT);
                    cs.registerOutParameter(9, Types.BIGINT);

                    cs.setInt(1, memNbr);
                    cs.execute();

                    lName   = cs.getString(2);
```

```
                    fName   = cs.getString(3);
                    address = cs.getString(4);
                    city    = cs.getString(5);
                    state   = cs.getString(6);
                    zip     = cs.getInt(7);
                    phone   = cs.getLong(8);
                    balance = cs.getLong(9);

                    System.out.println("Member   ID : " + memNbr);
                    System.out.println("Last Name   : " + lName);
                    System.out.println("First Name  : " + fName);
                    System.out.println("Address     : " + address);
                    System.out.println("City        : " + city);
                    System.out.println("State       : " + state);
                    System.out.println("Zip         : " + zip);
                    System.out.println("Phone       : " + phone);
                    System.out.println("Balance     : " + balance);
                    System.out.println("Completed execution of JAVFB1");

                }

                catch (SQLException e) {
                      System.out.println(e.getMessage());
                      System.out.println(e.getErrorCode());
                      e.printStackTrace();

                }
        }
}
```

And here is the output:

```
**** Loaded the JDBC driver
**** Created the connection
Executing JAVFB1 to call stored procedure GETMEM
Member   ID : 100100
Last Name   : JEFFERSON
First Name  : RICHARD
Address     : 2497 MYRTLE LANE
City        : HOUSTON
State       : TX
Zip         : 77099
Phone       : 2815683572
Balance     : 0
Completed execution of JAVFB1
```

Here is the .NET solution in c#.

```
using System;
using System.Collections.Generic;
```

```
using System.Data;
using System.Linq;
using System.Text;
using System.Threading.Tasks;
using IBM.Data.DB2;
using IBM.Data.DB2Types;

/* Program to connect to DB2 for calling stored procedure  */
namespace NETFB1
{
    class NETFB1
    {
        static void Main(string[] args)
        {
            DB2Connection conn = null;
            DB2Command cmd = null;

            try
            {
                conn = new DB2Connection("Database=DBFB");
                conn.Open();
                Console.WriteLine("Successful connection!");
                cmd = conn.CreateCommand();
                String procName = "FBSCHEMA.GETMEM";
                cmd.CommandType = System.Data.CommandType.StoredProcedure;
                cmd.CommandText = procName;

             // Define parameters and establish direction for output parameters

                cmd.Parameters.Add(new DB2Parameter("@empid", 100100));
                cmd.Parameters.Add(new DB2Parameter("@lname",DB2Type.VarChar));
                cmd.Parameters.Add(new DB2Parameter("@fname", ""));
                cmd.Parameters.Add(new DB2Parameter("@address", ""));
                cmd.Parameters.Add(new DB2Parameter("@city", ""));
                cmd.Parameters.Add(new DB2Parameter("@state", ""));
                cmd.Parameters.Add(new DB2Parameter("@zip", DB2Type.Integer));
                cmd.Parameters.Add(new DB2Parameter("@phone", DB2Type.BigInt));
                cmd.Parameters.Add(new DB2Parameter("@balance",
                   DB2Type.BigInt));
                cmd.Parameters[1].Direction = ParameterDirection.Output;
                cmd.Parameters[2].Direction = ParameterDirection.Output;
                cmd.Parameters[3].Direction = ParameterDirection.Output;
                cmd.Parameters[4].Direction = ParameterDirection.Output;
                cmd.Parameters[5].Direction = ParameterDirection.Output;
                cmd.Parameters[6].Direction = ParameterDirection.Output;
                cmd.Parameters[7].Direction = ParameterDirection.Output;
                cmd.Parameters[8].Direction = ParameterDirection.Output;


                /* Execute the stored procedure */

                Console.WriteLine("Call stored procedure named " + procName);
                cmd.ExecuteNonQuery();
```

```
                /*  Now capture and display the results */

                String mbrId   = cmd.Parameters[0].Value.ToString();
                String lName   = cmd.Parameters[1].Value.ToString();
                String fName   = cmd.Parameters[2].Value.ToString();
                String address = cmd.Parameters[3].Value.ToString();
                String city    = cmd.Parameters[4].Value.ToString();
                String state   = cmd.Parameters[5].Value.ToString();
                String zip     = cmd.Parameters[6].Value.ToString();
                String phone   = cmd.Parameters[7].Value.ToString();
                String balance = cmd.Parameters[8].Value.ToString();


                Console.WriteLine("Member id    : " + mbrId);
                Console.WriteLine("Last Name    : " + lName);
                Console.WriteLine("First Name   : " + fName);
                Console.WriteLine("Address      : " + address);
                Console.WriteLine("City         : " + city);
                Console.WriteLine("State        : " + state);
                Console.WriteLine("Zip          : " + zip);
                Console.WriteLine("Phone        : " + phone);
                Console.WriteLine("Balance      : " + balance);

            }

            catch (Exception e)
            {
                Console.WriteLine(e.Message);
            }

            finally
            {
                conn.Close();
            }
        }
    }
}
```

And here is the output:

```
Successful connection!
Call stored procedure named FBSCHEMA.GETMEM
Member id    : 100100
Last Name    : JEFFERSON
First Name   : RICHARD
Address      : 2497 MYRTLE LANE
City         : HOUSTON
State        : TX
Zip          : 77099
Phone        : 2815683572
Balance      : 0
The program '[7768] NETFB1.exe' has exited with code 0 (0x0).
```

Finally, let's go back to Data Studio, change the searched member number to 123456 (which we know does not yet exist in the table), and then rerun. The results will show as follows:

```
Name         Type   Data type Value  Value (OUT)
------------ ------ --------- ------ -----------
M_NBR        INPUT  INTEGER   123456
M_LAST_NAME  OUTPUT VARCHAR          *NULL*
M_FIRST_NAME OUTPUT VARCHAR          *NULL*
M_ADDRESS    OUTPUT VARCHAR          *NULL*
M_CITY       OUTPUT VARCHAR          *NULL*
M_STATE      OUTPUT CHAR             *NULL*
M_ZIP        OUTPUT INTEGER          *NULL*
M_PHONE      OUTPUT BIGINT           *NULL*
M_BALANCE    OUTPUT BIGINT           *NULL*
```

The result indicates that the stored procedure ran successfully, but the parameter tab shows there was no data returned. Of course you could include the `GET DIAGNOSTICS` logic in your stored procedure and return a more detailed message. Check out the last version of the stored procedures we used for our HR application for details.

Ok, let's take a short break and then come back and add the calls for add, insert and delete.

Stored Procedure to INSERT

Alright, I'm back with a fresh cup of coffee – hope you are too. Let's create and test an INSERT stored procedure. You can use any earlier example in the HR system as a model. Give it a try on your own and then you can compare your DDL to mine.

.....

Ok, here is my DDL. Notice there is a single OUT parameter which will be used to return the value of the generated `MBR_NBR`. Your calling program will likely need that number – otherwise it won't know what account number was generated for the new member (recall that we are generating member numbers using a sequence object). Also note that we specify `MODIFIES SQL DATA` in the procedure definition.

```
CREATE PROCEDURE FBSCHEMA.ADD_MEM_INFO
(OUT M_NBR        INTEGER,
 IN M_LAST_NAME   VARCHAR(20),
 IN M_FIRST_NAME  VARCHAR(15),
 IN M_ADDRESS     VARCHAR(30),
 IN M_CITY        VARCHAR(20),
 IN M_STATE       CHAR(02),
 IN M_ZIP         INTEGER,
 IN M_PHONE       BIGINT,
 IN M_BALANCE     BIGINT)
```

```
LANGUAGE SQL
MODIFIES SQL DATA

BEGIN
   INSERT INTO FBSCHEMA.MEMBER
   (MBR_NBR,
    MBR_LAST_NAME,
    MBR_FIRST_NAME,
    MBR_ADDRESS,
    MBR_CITY,
    MBR_STATE,
    MBR_ZIP,
    MBR_PHONE,
    MBR_BALANCE)

    VALUES
    (NEXT VALUE FOR FBSCHEMA.MBRSEQ,
     M_LAST_NAME,
     M_FIRST_NAME,
     M_ADDRESS,
     M_CITY,
     M_STATE,
     M_ZIP,
     M_PHONE,
     M_BALANCE);

 SET M_NBR = PREVIOUS VALUE FOR FBSCHEMA.MBRSEQ;

 END
```

Now we must grant security on the stored procedure.

```
GRANT EXECUTE ON PROCEDURE FBSCHEMA.ADD_MEM_INFO TO PUBLIC;
```

Let's suppose we need to add a record with the following data:

Column	Value
Last Name	BROWN
First Name	OSCAR
Address	5162 HUNTINGTON RD
City	FRIENDSWOOD
State	TX
Zip	770822154
Phone	7139873472
Balance	0

Now let's create a Java program to call the INSERT stored procedure with these values.

```
package FreqBuyer;
import java.sql.*;

public class JAVFB2 {

        public static void main(String[] args) throws ClassNotFoundException
        {
                Class.forName("com.ibm.db2.jcc.DB2Driver");
                System.out.println("**** Loaded the JDBC driver");
            String url
                ="jdbc:db2://localhost:50000/DBFB:retrieveMessagesFromServerOn-
                GetMessage=true;";
            String user="rwinga01";
            String password="*********";

            Connection con=null;
            CallableStatement cs=null;

                try {
                        con=DriverManager.getConnection(url, user, password);
                        System.out.println("**** Created the connection");
                        int memNbr;
                        String lName;
                        String fName;
                        String address;
                        String city;
                        String state;
                        int zip;
                        long phone;
                        long balance;

                    System.out.println("Executing JAVFB2 to call stored procedure
                         FBSCHEMA.ADD_MEM_INFO");

                        cs = con.prepareCall("{call
                         FBSCHEMA.ADD_MEM_INFO(?,?,?,?,?,?,?,?,?)}");
                        cs.registerOutParameter(1, Types.INTEGER);

             //  Assign values to work variables

                    memNbr  = 0;
                    lName   = "BROWN";
                    fName   = "OSCAR";
                    address = "5162 HUNTINGTON RD";
                    city    = "FRIENDSWOOD";
                    state   = "TX";
                    zip     = 770822154;
                    phone   = 7139873472L;
                    balance = 0;

            //  Map work variables to parameters
```

```
            cs.setString(2, lName);
            cs.setString(3, fName);
            cs.setString(4, address);
            cs.setString(5, city);
            cs.setString(6, state);
            cs.setInt(7, zip);
            cs.setLong(8, phone);
            cs.setLong(9, balance);

        //  execute the SP

            cs.execute();

            // Capture the member number

            memNbr = cs.getInt(1);
           System.out.println("Member   ID : " + memNbr);
           System.out.println("Last Name   : " + lName);
           System.out.println("First Name  : " + fName);
           System.out.println("Address     : " + address);
           System.out.println("City        : " + city);
           System.out.println("State       : " + state);
           System.out.println("Zip         : " + zip);
           System.out.println("Phone       : " + phone);
           System.out.println("Balance     : " + balance);
           System.out.println("Completed execution of JAVFB2");
            }

             catch (SQLException e) {
                   System.out.println(e.getMessage());
                   System.out.println(e.getErrorCode());
                   e.printStackTrace();

            }

        }
    }
```

And here is the result of our run:

```
**** Loaded the JDBC driver
**** Created the connection
Executing JAVFB1 to call stored procedure FBSCHEMA.ADD_MEM_INFO
Member   ID : 100101
Last Name   : BROWN
First Name  : OSCAR
Address     : 5162 HUNTINGTON RD
City        : FRIENDSWOOD
State       : TX
Zip         : 770822154
Phone       : 7139873472
Balance     : 0
```

```
Completed execution of JAVFB1
```

Finally, let's do the .NET solution in c#, and here are the data values we'll use to create the new record.

Column	Value
Last Name	LEE
First Name	EVA
Address	1232 SCOTTSDALE DR
City	FOREST HILL
State	TX
Zip	761197352
Phone	8175683162
Balance	0

And here is the program:

```
using System;
using System.Collections.Generic;
using System.Data;
using System.Linq;
using System.Text;
using System.Threading.Tasks;
using IBM.Data.DB2;
using IBM.Data.DB2Types;

/* Program to connect to DB2 for calling stored procedure  */
namespace NETFB2
{
    class NETFB2
    {
        static void Main(string[] args)
        {
            DB2Connection conn = null;
            DB2Command cmd = null;

            try
            {   conn = new DB2Connection("Database=DBFB");
                conn.Open();
                Console.WriteLine("Successful connection!");
                cmd = conn.CreateCommand();
                String procName = "FBSCHEMA.ADD_MEM_INFO";
                cmd.CommandType = System.Data.CommandType.StoredProcedure;
                cmd.CommandText = procName;

             // Define parameters and establish direction for output parameters

                cmd.Parameters.Add(new DB2Parameter("@memId", DB2Type.Integer));
```

```
    cmd.Parameters.Add(new DB2Parameter("@lname", DB2Type.VarChar));
    cmd.Parameters.Add(new DB2Parameter("@fname", DB2Type.VarChar));
    cmd.Parameters.Add(new DB2Parameter("@address",
       DB2Type.VarChar));
    cmd.Parameters.Add(new DB2Parameter("@city", DB2Type.VarChar));
    cmd.Parameters.Add(new DB2Parameter("@state", DB2Type.Char));
    cmd.Parameters.Add(new DB2Parameter("@zip", DB2Type.Integer));
    cmd.Parameters.Add(new DB2Parameter("@phone", DB2Type.BigInt));
    cmd.Parameters.Add(new DB2Parameter("@balance",
       DB2Type.BigInt));
    cmd.Parameters[0].Direction = ParameterDirection.Output;

    /* Set parameter values  */
    cmd.Parameters[1].Value = "LEE";
    cmd.Parameters[2].Value = "EVA";
    cmd.Parameters[3].Value = "1232 SCOTTSDALE DR";
    cmd.Parameters[4].Value = "FOREST HILL";
    cmd.Parameters[5].Value = "TX";
    cmd.Parameters[6].Value = 761197352;
    cmd.Parameters[7].Value = 8175683162;
    cmd.Parameters[8].Value = 0;

    /* Execute the stored procedure */

    Console.WriteLine("Call stored procedure named " + procName);
    cmd.ExecuteNonQuery();

    /*  Now capture and display the results */

    String mbrId   = cmd.Parameters[0].Value.ToString();
    String lName   = cmd.Parameters[1].Value.ToString();
    String fName   = cmd.Parameters[2].Value.ToString();
    String address = cmd.Parameters[3].Value.ToString();
    String city    = cmd.Parameters[4].Value.ToString();
    String state   = cmd.Parameters[5].Value.ToString();
    String zip     = cmd.Parameters[6].Value.ToString();
    String phone   = cmd.Parameters[7].Value.ToString();
    String balance = cmd.Parameters[8].Value.ToString();

    /* Display all values including the new member number  */

    Console.WriteLine("SUCCESSFULLY ADDED RECORD " + mbrId);
    Console.WriteLine("Member id    : " + mbrId);
    Console.WriteLine("Last Name    : " + lName);
    Console.WriteLine("First Name   : " + fName);
    Console.WriteLine("Address      : " + address);
    Console.WriteLine("City         : " + city);
    Console.WriteLine("State        : " + state);
    Console.WriteLine("Zip          : " + zip);
    Console.WriteLine("Phone        : " + phone);
    Console.WriteLine("Balance      : " + balance);

}
```

```
                catch (Exception e)
                {
                    Console.WriteLine(e.Message);
                }

                finally
                {
                    conn.Close();
                }
            }
        }
    }
```

And here is the output:

```
Successful connection!
Call stored procedure named FBSCHEMA.ADD_MEM_INFO
SUCCESSFULLY ADDED RECORD 100102
Member id    : 100102
Last Name    : LEE
First Name   : EVA
Address      : 1232 SCOTTSDALE DR
City         : FOREST HILL
State        : TX
Zip          : 761197352
Phone        : 8175683162
Balance      : 0
The program '[8848] NETFB2.exe' has exited with code 0 (0x0).
```

The UPDATE and DELETE stored procedures follow the examples given earlier in the text. It is best if you work on these yourself first, and then compare to the code examples. Take some now to do that.

Stored Procedure to UPDATE

Here is the DDL to create the stored procedure to update the `MEMBER` table.

```
CREATE PROCEDURE FBSCHEMA.UPD_MEM_INFO
(IN M_NBR        INTEGER,
 IN M_LAST_NAME  VARCHAR(20),
 IN M_FIRST_NAME VARCHAR(15),
 IN M_ADDRESS    VARCHAR(30),
 IN M_CITY       VARCHAR(20),
 IN M_STATE      CHAR(02),
 IN M_ZIP        INTEGER,
 IN M_PHONE      BIGINT,
 IN M_BALANCE    BIGINT)
```

```
LANGUAGE SQL
MODIFIES SQL DATA

BEGIN
   UPDATE FBSCHEMA.MEMBER
   SET MBR_NBR        = M_NBR,
       MBR_LAST_NAME  = M_LAST_NAME,
       MBR_FIRST_NAME = M_FIRST_NAME,
       MBR_ADDRESS    = M_ADDRESS,
       MBR_CITY       = M_CITY,
       MBR_STATE      = M_STATE,
       MBR_ZIP        = M_ZIP,
       MBR_PHONE      = M_PHONE,
       MBR_BALANCE    = M_BALANCE
    WHERE MBR_NBR = M_NBR;

END
```

Don't forget to grant security on the procedure:

```
GRANT EXECUTE ON PROCEDURE FBSCHEMA.UPD_MEM_INFO TO PUBLIC;
```

Now here's a program to run the stored procedure. We'll just change the first name on the second record (100101) from Oscar to Osborne. You can change more fields if you like. You'll notice this program is very similar to the add program. Basically we are only changing the member number an input parameter rather than an output parameter.

Here is the program listing:

```
package FreqBuyer;

import java.sql.*;

public class JAVFB3 {

        public static void main(String[] args) throws ClassNotFoundException
        {
                Class.forName("com.ibm.db2.jcc.DB2Driver");
                System.out.println("**** Loaded the JDBC driver");
            String url
                ="jdbc:db2://localhost:50000/DBFB:retrieveMessagesFromServerOn-
        GetMessage=true;";
            String user="rwinga01";
            String password="*********";

            Connection con=null;
            CallableStatement cs=null;

                try {
                        con=DriverManager.getConnection(url, user, password);
```

```
    System.out.println("**** Created the connection");

    System.out.println("Executing JAVFB3 to call stored proce-
    dure FBSCHEMA.UPD_MEM_INFO");

    cs = con.prepareCall("{call FBSCHEMA.UPD_MEM_INFO(?,?,?,?,
    ?,?,?,?,?)}");

//  Assign data values to work variables

    int memNbr  = 100100;
    String lName   = "BROWN";
    String fName   = "OSBORNE";
    String address = "5162 HUNTINGTON RD";
    String city    = "FRIENDSWOOD";
    String state   = "TX";
    int zip     = 770822154;
    long phone   = 7139873472L;
    long balance = 0;

//  Map work variables to parameters

    cs.setInt(1, memNbr);
    cs.setString(2, lName);
    cs.setString(3, fName);
    cs.setString(4, address);
    cs.setString(5, city);
    cs.setString(6, state);
    cs.setInt(7, zip);
    cs.setLong(8, phone);
    cs.setLong(9, balance);

    //  execute the SP

    cs.execute();

    // Capture the member number

    System.out.println("** Successfully updated member "
    + memNbr);
     System.out.println("Member   ID : " + memNbr);
    System.out.println("Last Name   : " + lName);
    System.out.println("First Name  : " + fName);
    System.out.println("Address     : " + address);
    System.out.println("City        : " + city);
    System.out.println("State       : " + state);
    System.out.println("Zip         : " + zip);
    System.out.println("Phone       : " + phone);
    System.out.println("Balance     : " + balance);
    System.out.println("Completed execution of JAVFB3");

}
```

```
                catch (SQLException e) {
                      System.out.println(e.getMessage());
                      System.out.println(e.getErrorCode());
                      e.printStackTrace();

                }

          }

    }
```

Here is the output:

```
**** Loaded the JDBC driver
**** Created the connection
Executing JAVFB3 to call stored procedure FBSCHEMA.UPD_MEM_INFO
** Successfully updated member 100100
Member   ID : 100100
Last Name   : BROWN
First Name  : OSBORNE
Address     : 5162 HUNTINGTON RD
City        : FRIENDSWOOD
State       : TX
Zip         : 770822154
Phone       : 7139873472
Balance     : 0
Completed execution of JAVFB3
```

Let's do the .NET version of the update program, and then we'll take another break. Again, we need only make minor changes to the add program to turn it into an update program. Let's say we'll just change EVA LEE's given name to ELISA.

Here is our listing.

```
using System;
using System.Collections.Generic;
using System.Data;
using System.Linq;
using System.Text;
using System.Threading.Tasks;
using IBM.Data.DB2;
using IBM.Data.DB2Types;

/* Program to connect to DB2 for calling stored procedure  */

namespace NETFB3
{
    class NETFB3
    {
        static void Main(string[] args)
```

```
{
    DB2Connection conn = null;
    DB2Command cmd = null;

    try
    {   conn = new DB2Connection("Database=DBFB");
        conn.Open();
        Console.WriteLine("Successful connection!");
        cmd = conn.CreateCommand();
        String procName = "FBSCHEMA.UPD_MEM_INFO";
        cmd.CommandType = System.Data.CommandType.StoredProcedure;
        cmd.CommandText = procName;

        // Define parameters and establish direction for output parameters

        cmd.Parameters.Add(new DB2Parameter("@memId", DB2Type.Integer));
        cmd.Parameters.Add(new DB2Parameter("@lname", DB2Type.VarChar));
        cmd.Parameters.Add(new DB2Parameter("@fname", DB2Type.VarChar));
        cmd.Parameters.Add(new DB2Parameter("@address", DB2Type.VarChar));
        cmd.Parameters.Add(new DB2Parameter("@city", DB2Type.VarChar));
        cmd.Parameters.Add(new DB2Parameter("@state", DB2Type.Char));
        cmd.Parameters.Add(new DB2Parameter("@zip", DB2Type.Integer));
        cmd.Parameters.Add(new DB2Parameter("@phone", DB2Type.BigInt));
        cmd.Parameters.Add(new DB2Parameter("@balance", DB2Type.BigInt));

        /* Set parameter values  */
        cmd.Parameters[0].Value = 100102;
        cmd.Parameters[1].Value = "LEE";
        cmd.Parameters[2].Value = "ELISA";
        cmd.Parameters[3].Value = "1232 SCOTTSDALE DR";
        cmd.Parameters[4].Value = "FOREST HILL";
        cmd.Parameters[5].Value = "TX";
        cmd.Parameters[6].Value = 761197352;
        cmd.Parameters[7].Value = 8175683162;
        cmd.Parameters[8].Value = 0;

        /* Execute the stored procedure */

        Console.WriteLine("Call stored procedure named " + procName);
        cmd.ExecuteNonQuery();

        /*  Now capture and display the results */

        String mbrId = cmd.Parameters[0].Value.ToString();
        String lName = cmd.Parameters[1].Value.ToString();
        String fName = cmd.Parameters[2].Value.ToString();
        String address = cmd.Parameters[3].Value.ToString();
        String city = cmd.Parameters[4].Value.ToString();
        String state = cmd.Parameters[5].Value.ToString();
        String zip = cmd.Parameters[6].Value.ToString();
        String phone = cmd.Parameters[7].Value.ToString();
        String balance = cmd.Parameters[8].Value.ToString();
```

```
                /* Display all values including the new member number  */

                Console.WriteLine("SUCCESSFULLY ADDED RECORD " + mbrId);
                Console.WriteLine("Member id    : " + mbrId);
                Console.WriteLine("Last Name    : " + lName);
                Console.WriteLine("First Name   : " + fName);
                Console.WriteLine("Address      : " + address);
                Console.WriteLine("City         : " + city);
                Console.WriteLine("State        : " + state);
                Console.WriteLine("Zip          : " + zip);
                Console.WriteLine("Phone        : " + phone);
                Console.WriteLine("Balance      : " + balance);

            }

            catch (Exception e)
            {
                Console.WriteLine(e.Message);
            }

            finally
            {
                conn.Close();
            }
        }
    }
}
```

And here are the results.

```
Successful connection!
Call stored procedure named FBSCHEMA.UPD_MEM_INFO
SUCCESSFULLY ADDED RECORD 100102
Member id    : 100102
Last Name    : LEE
First Name   : ELISA
Address      : 1232 SCOTTSDALE DR
City         : FOREST HILL
State        : TX
Zip          : 761197352
Phone        : 8175683162
Balance      : 0
```

We can of course check to make sure all our changes really took place. Let's do a query now to validate:

```
SELECT * FROM FBSCHEMA.MEMBER;

MBR_NBR MBR_LAST_NAME MBR_FIRST_NAME MBR_ADDRESS        MBR_CITY    MBR_STATE MBR_ZIP   MBR_PHONE  MBR_BALANCE
------- ------------- -------------- ------------------ ----------- --------- --------- ---------- -----------
 100100 JEFFERSON     RICHARD        2497 MYRTLE LANE   HOUSTON     TX            77099 2815683572           0
 100101 BROWN         OSBORNE        5162 HUNTINGTON RD FRIENDSWOOD TX        770822154 7139873472           0
 100102 LEE           ELISA          1232 SCOTTSDALE DR FOREST HILL TX        761197352 8175683162           0
```

Ok let's take a break, and then we'll finish up with the delete SP.

.....

I'm back now. Let's create the delete stored procedure. This one is short and sweet:

```
CREATE PROCEDURE FBSCHEMA.DLT_MEM_INFO
(IN M_NBR INT)

 LANGUAGE SQL MODIFIES SQL DATA

 BEGIN
    DELETE FROM FBSCHEMA.MEMBER
    WHERE MBR_NBR = M_NBR;

 END
```

Again we will grant access to use the procedure:

```
GRANT EXECUTE ON PROCEDURE FBSCHEMA.DLT_MEM_INFO TO PUBLIC;
```

Let's just test this one using Data Studio. Normally this is what you'll do anyway until you or your teammates have written program code to use the stored procedure.

Let's delete member 100101. To do this we just need to open data studio, navigate to the stored procedures part of the `DBFB` object tree, right click on **`FBSCHEMA.DLT_MEM_INFO`**, and click on **Run.** Enter 100101 as the member number and click **Run.**

And we can verify that the record 100101 was deleted by listing the content of the `MEMBER` table:

```
SELECT MBR_NBR,
MBR_LAST_NAME,
MBR_FIRST_NAME
FROM FBSCHEMA.MEMBER;

MBR_NBR MBR_LAST_NAME MBR_FIRST_NAME
------- ------------- --------------
 100100 BROWN         OSBORNE
 100102 LEE           ELISA
```

As you can see, member 100101 is gone. Now we have created 4 data access stored procedures that we can use for data maintenance. That satisfies the requirements we established at the beginning of the project. Although this has been a mini-project, I hope you get the ideas about how to create and test DB2 components.

Special Project Wrap-up

There are many other things you could do with this project. Lacking a genuine business project, I encourage you to use your imagination. For example you might need a UDF to extract all member activity (deposits and rewards) for an online display on a web page. How about setting up archive tables for the `DEPOSIT_TYPE` and `REWARD_TYPE` tables? The archive records will come in handy when someone asked when a certain entry in these tables changed and what the previous entries were. Similarly you could set up a history table for the `MEMBER` table and make the latter a temporal table with system time. This way whenever a `MEMBER` record changes you will always have a copy of all previous versions.

Set up `DEPOSITS` and `REWARDS` as temporal tables to enable business time travel queries to show activity such as **`BALANCE`** as of a certain period of time. The possibilities are endless but the necessities will depend on your actual project requirements. Hopefully you have enough training from this text book to work through a full project, selecting the right components and techniques, and making your application execute efficiently and successfully.

We've come to the end, so let me close with a sincere "best of luck". It's been a pleasure walking you through these Db2 concepts. I truly hope you do exceptionally well as a Db2 for LUW developer, and that you have every success!

Best regards,

Robert Wingate

APPENDIXES

Answers for Chapter Questions and Exercises

Chapter Two Questions and Answers

1. Which of the following is NOT a valid data type for use as an identity column?

    a. INTEGER
    b. REAL
    c. DECIMAL
    d. SMALLINT

The correct answer is B. A REAL type cannot be used as an identity field because it is considered an approximation of a number rather than an exact value. Only numeric types that have an exact value can be used as an identity field. INTEGER, DECIMAL, and SMALLINT are all incorrect here because they CAN be used as identity fields.

2. You need to store numeric integer values of up to 5,000,000,000. What data type is appropriate for this?

    a. INTEGER
    b. BIGINT
    c. LARGEINT
    d. DOUBLE

The correct answer is B. BIGINT is an integer that can hold up to 9,223,372,036,854,775,807. INTEGER is not correct because an INTEGER can only hold up to 2,147,483,647. LARGEINT is an invalid type. DOUBLE could be used but since we are dealing with integer data, the double precision is not needed.

3. Which of the following is NOT a LOB (Large Object) data type?

   a. CLOB
   b. BLOB
   c. DBCLOB
   d. DBBLOB

The correct answer is D. There is no DBBLOB datatype in DB2. The other data types are valid. CLOB is a character large object with maximum length 2,147,483,647 bytes. A BLOB stores binary data and has a maximum size of 2,147,483,647. A DBCLOB stores double character data and has a maximum length of 1,073,741,824.

4. If you want to add an XML column VAR1 to table TBL1, which of the following would accomplish that?

   a. ALTER TABLE TBL1 ADD VAR1 XML
   b. ALTER TABLE TBL1 ADD COLUMN VAR1 XML
   c. ALTER TABLE TBL1 ADD COLUMN VAR1 (XML)
   d. ALTER TABLE TBL1 ADD XML COLUMN VAR1

The correct answer is B. The correct syntax is:

```
ALTER TABLE TBL1
ADD COLUMN VAR1 XML;
```

The other choices would result in a syntax error.

5. If you want rows that have similar key values to be stored physically close to each other, what keyword should you specify when you create an index?

   a. UNIQUE
   b. ASC
   c. INCLUDE
   d. CLUSTER

The correct answer is D - CLUSTER. Specifying a CLUSTER type index means that DB2 will attempt to physically store rows with similar keys close together. This is used for performance reasons when sequential type processing is needed according to the index. UNIQUE is incorrect because this keyword simply guarantees that there can be no more than one row with the same index key. ASC is incorrect because it has to do with the sort order for the index, and does not affect the physical storage of rows. INCLUDE specifies that a non-key field or fields will be stored with the index.

6. To ensure all records inserted into a view of a table are consistent with the view definition, you would need to include which of the following keywords when defining the view?

   a. UNIQUE
   b. WITH CHECK OPTION
   c. VALUES
   d. ALIAS

The correct answer is B. The WITH CHECK OPTION ensures that a record inserted via a view is consistent with the view definition. UNIQUE is a type of index and is unrelated to views. VALUES is part of an INSERT DML statement, but it does not ensure that the record is consistent with a view definition. An ALIAS is another name for a table that allows the table to be referenced without regard to the owning schema for that table.

7. If you want to determine various characteristics of a set of tables such as type (table or view), owner and status, which system catalog view would you query?

   a. SYSCAT.TABLES
   b. SYSIBM.SYSTABLES
   c. SYSCAT.OBJECTS
   d. SYSIBM.OBJECTS

The correct answer is A. Query the SYSCAT.TABLES view to get characteristics of the table such as table type (table or view), owner and status.

Chapter Two Exercises

1. Write a DDL statement to create a base table named EMP_DEPENDENT in schema HRSCHEMA and in tablespace TSHR. The columns should be named as follows and have the specified attributes. There is no primary key.

Field Name	Type	Attributes
EMP_ID	INTEGER	NOT NULL, PRIMARY KEY
EMP_DEP_LAST_NAME	VARCHAR(30)	NOT NULL
EMP_DEP_FIRST_NAME	VARCHAR(20)	NOT NULL
EMP_RELATIONSHIP	VARCHAR(15)	NOT NULL

```
CREATE TABLE HRSCHEMA.EMP_DEPENDENT
(EMP_ID INT NOT NULL,
EMP_DEP_LAST_NAME     VARCHAR(30) NOT NULL,
EMP_DEP_FIRST_NAME    VARCHAR(20) NOT NULL,
EMP_DEP_RELATIONSHIP VARCHAR(15) NOT NULL)
   IN TSHR;
```

You can add data to the table and retrieve it as follows:

```
INSERT INTO HRSCHEMA.EMP_DEPENDENT
VALUES(3217,'JOHNSON','ELENA','WIFE');


SELECT * FROM HRSCHEMA.EMP_DEPENDENT;

EMP_ID  EMP_DEP_LAST_NAME     EMP_DEP_FIRST_NAME    EMP_DEP_RELATIONSHIP
------  -----------------     ------------------    --------------------
  3217  JOHNSON               ELENA                 WIFE
```

2. Create a statement to create a referential constraint on table EMP_DEPENDENTS such that only employee ids which exist on the HRSCHEMA.EMPLOYEE table can have an entry in EMP_DEPENDENTS. If there is an attempt to delete an EMPLOYEE record that has EMP_DEPENDENTS records associated with it, then do not allow the delete to take place.

```
ALTER TABLE HRSCHEMA.EMP_DEPENDENT
   FOREIGN KEY FK_EMP_ID_DEP (EMP_ID)
      REFERENCES HRSCHEMA.EMPLOYEE(EMP_ID) ;
```

OR

```
ALTER TABLE HRSCHEMA.EMP_DEPENDENT
   FOREIGN KEY FK_EMP_ID_DEP (EMP_ID)
      REFERENCES HRSCHEMA.EMPLOYEE(EMP_ID)
         ON DELETE RESTRICT;
```

Both of the above have the same effect which is to prevent a parent record (EMPLOYEE) from being deleted if it has a related child record (in EMP_DEPENDENT).

Chapter Three Questions and Answers

1. Suppose you issue this INSERT statement:

```
UPDATE HRSCHEMA.EMPLOYEE
SET EMP_PROFILE
=(
‘<?xml version=”1.0”?>
<EMP_PROFILE>
<EMP_ID>4175</EMP_ID>
<EMP_ADDRESS>
<STREET>6161 MARGARET LANE</STREET>
<CITY>ERINDALE</CITY>
<STATE>AR</STATE>
<ZIP_CODE>72653</ZIP_CODE>
</EMP_ADDRESS>
<BIRTH_DATE>07/14/1991</BIRTH_DATE>
<EMP_PROFILE>
‘ )
WHERE EMP_ID = 4175;
```

What will the result be?

   a. An error – the specified XML version is incorrect
   b. An error – the document is not well formed
   c. The record will be inserted with a warning that it has not been validated
   d. The record will be inserted successfully

The correct answer is B. The XML document is not well formed – the last tag should be </EMP_PROFILE> rather than <EMP_PROFILE>.

2. Which of the following is NOT a special register in DB2 11?

   a. CURRENT DEGREE
   b. CURRENT RULES
   c. CURRENT ISOLATION
   d. All of the above are valid DB2 special registers.

The correct answer is B. There is no CURRENT RULES special register in DB2 11.1 LUW. CURRENT DEGREE indicates whether or not parallelism will be used for executing dynamic queries. CURRENT ISOLATION refers to the isolation level used with a dynamic SQL statement.

3. Which built-in routine returns the value of the first non-null expression?

   a. COALESCE
   b. ABS
   c. CEILING
   d. MIN

The correct answer is A. The COALESCE function returns the value of the first non-null expression. The ABS function returns the absolute value of a number. The CEILING function returns the smallest integer value that is greater than or equal to an expression. The MIN function returns the smallest value in a set of values in a group. For example:

```
SELECT COALESCE(SALARY,0)
```

Here, if the SALARY is not null then the COALESCE function returns the value of SALARY. Otherwise it returns the value zero and zero is included in the aggregation.

4. If you want the aggregate total for a set of values, which function would you use?

   a. COUNT
   b. SUM
   c. ABS
   d. CEIL

The correct answer is B - SUM. SUM returns the total of a set of numeric values. COUNT returns the number of records that meet the search criteria. ABS returns the absolute value of a numeric column. CEILING returns the smallest integer that is greater than or equal to an expression or value, e.g. CEILING(3.5) = 4.0.

5. If you want to return the first 3 characters of a 10-character column, which function would you use?

   a. SUBSTR
   b. LTRIM
   c. DIGITS
   d. ABS

The correct answer is A. SUBSTR returns a subset of the source string. The syntax is SUBSTR(X,Y, Z) where X is the source column, Y is the starting position, and Z is the length of the sub-string. Assuming a 10-character column FLD1, you would get the first three positions by coding SUBSTR(FLD1, 1,3).

The other answers are incorrect. LTRIM removes all leading blanks from a value, e.g., LTRIM(LASTNAME) would return the LASTNAME value minus any leading spaces. DIGITS returns a character representation of a numeric value. ABS returns the absolute value of a number and is unrelated to partial string values.

6. Assuming all referenced onjects are valid, what will the result of this statement be?

```
DELETE FROM EMPLOYEE;
```

   a. The statement will fail because there is no WHERE clause.

   b. The statement will fail because you must specify DELETE * .

   c. The statement will succeed and all rows in the table will be deleted.

   d. The statement will run but no rows in the table will be deleted.

The correct answer is C. Assuming all referenced objects are valid, running the DELETE FROM EMPLOYEE statement will succeed and all rows in the table will be deleted.

Chapter Three Exercises

1. Write a query to display the last and first names of all employees in the EMPLOYEE table. Display the names in alphabetic order by EMP_LAST_NAME.

```
SELECT EMP_LAST_NAME, EMP_FIRST_NAME
FROM HRSCHEMA.EMPLOYEE
ORDER BY EMP_LAST_NAME;
```

2. Write a query to change the first name of Edward Johnson (employee 3217) to Eddie.

```
UPDATE HRSCHEMA.EMPLOYEE
SET EMP_FIRST_NAME = 'EDDIE'
WHERE EMP_ID = 3217;
```

3. Write a query to produce the number of employees in the EMPLOYEE table.

```
SELECT COUNT(*)
FROM HRSCHEMA.EMPLOYEE;
```

Chapter Four Questions and Answers

1. Assume a table where certain columns contain sensitive data and you don't want all users to see these columns. Some other columns in the table must be made accessible to all users. What type of object could you create to solve this problem?

    a. INDEX
    b. SEQUENCE
    c. VIEW
    d. TRIGGER

    The correct answer is C. A view is a virtual table based upon a SELECT query that can include a subset of the columns in a table. So you can create multiple views against the same base table, and control access to the views based upon userid or group.

    The other answers do not address the problem of limiting access to specific columns. An INDEX is an object that stores the physical location of records and is used to improve performance and enforce uniqueness. A SEQUENCE allows for the automatic generation of sequential values, and has nothing to do with limiting access to columns in a table. A TRIGGER is an object that performs some predefined action when it is activated. A trigger is only activated by an INSERT, UPDATE or DELETE of a record in a particular table.

2. To grant a privilege to all users of the database, grant the privilege to whom?

    a. ALL
    b. PUBLIC
    c. ANY
    d. DOMAIN

    The correct answer is B. PUBLIC is a special "pseudo" group that means all users of the database. The other answers ALL, ANY, and DOMAIN are incorrect because they are not valid recipients of a grant statement.

3. Tara wants to grant CONTROL of table TBL1 to Bill, and also allow Bill to grant the same privilege to other users. What clause should Tara use on the GRANT statement?

    a. WITH CONTROL OPTION
    b. WITH GRANT OPTION
    c. WITH USE OPTION
    d. WITH REVOKE OPTION

The correct answer is B. Using the WITH GRANT OPTION permits the recipient of the grant to also grant this privilege to other users. The other choices WITH CONTROL OPTION, WITH USE OPTION, and WITH REVOKE OPTION are incorrect because they are not valid clauses on a GRANT statement.

Chapter Four Exercises

1. Write a DCL statement to grant SELECT access on table HRSCHEMA.EMPLOYEE to users HR001 and HR002.

```
GRANT SELECT ON HRSCHEMA.EMPLOYEE
TO HR001, HR002;
```

2. Write a DCL statement to revoke SELECT access on table HRSCHEMA.EMPLOYEE from user HR002.

```
REVOKE SELECT ON HRSCHEMA.EMPLOYEE
FROM HR002;
```

3. Write a DCL statement to grant SELECT, INSERT, UPDATE and DELETE access on table HRSCHEMA.EMP_PAY to user HRMGR01.

```
GRANT SELECT, INSERT, UPDATE, DELETE
ON HRSCHEMA.EMP_PAY
TO HRMGR01;
```

3. Tara wants to grant CONTROL of table TBL1 to Bill, and also allow Bill to grant the same privilege to other users. What clause should Tara use on the GRANT statement?

   a. WITH CONTROL OPTION
   b. WITH GRANT OPTION
   c. WITH USE OPTION
   d. WITH REVOKE OPTION

   The correct answer is B. Using the WITH GRANT OPTION permits the recipient of the grant to also grant this privilege to other users. The other choices WITH CONTROL OPTION, WITH USE OPTION, and WITH REVOKE OPTION are incorrect because they are not valid clauses on a GRANT statement.

Chapter Five Exercises

1. Write a Java or .NET program that creates a result set of all employees who have 5 years or more of service from the `HRSCHEMA.EMPLOYEE` table. Include logic to display any rows that are returned. Display the employee number, last name and first name of these employees.

<u>Java Solution</u>:

```
package employee;
import java.sql.*;

public class JAVEMP4 {

        public static void main(String[] args) throws ClassNotFoundException
        {
            Class.forName("com.ibm.db2.jcc.DB2Driver");
            System.out.println("**** Loaded the JDBC driver");
            String url="jdbc:db2://localhost:50000/DBHR";
            String user="rwinga01";
            String password="*********";
            Connection con=null;

                try {
                        con=DriverManager.getConnection(url, user, password);
                        System.out.println("**** Created the connection");

                        String query = "SELECT EMP_ID, "
                                             + "EMP_LAST_NAME,"
                                             + "EMP_FIRST_NAME "
                                             + "FROM HRSCHEMA.EMPLOYEE "
                                             + "WHERE EMP_SERVICE_YEARS >= 5";

                        Statement stmt;
                        stmt = con.createStatement();
                        ResultSet rs=stmt.executeQuery(query);
                        while(rs.next())
                        {
                            int intEmpId        = rs.getInt(1);
                                String strLastName  = rs.getString(2);
                            String strFirstName = rs.getString(3);
                            System.out.println("EMP_ID is " + intEmpId
                                        + " Last name is " + strLastName
                                        + " First name is " + strFirstName);
                        }

                } catch (SQLException e) {
                        System.out.println(e.getMessage());
                        System.out.println(e.getErrorCode());
                        e.printStackTrace();

                }
```

3. Tara wants to grant CONTROL of table TBL1 to Bill, and also allow Bill to grant the same privilege to other users. What clause should Tara use on the GRANT statement?

   a. WITH CONTROL OPTION
   b. WITH GRANT OPTION
   c. WITH USE OPTION
   d. WITH REVOKE OPTION

   The correct answer is B. Using the WITH GRANT OPTION permits the recipient of the grant to also grant this privilege to other users. The other choices WITH CONTROL OPTION, WITH USE OPTION, and WITH REVOKE OPTION are incorrect because they are not valid clauses on a GRANT statement.

Chapter Five Exercises

1. Write a Java or .NET program that creates a result set of all employees who have 5 years or more of service from the HRSCHEMA.EMPLOYEE table. Include logic to display any rows that are returned. Display the employee number, last name and first name of these employees.

Java Solution:

```
package employee;
import java.sql.*;

public class JAVEMP4 {

        public static void main(String[] args) throws ClassNotFoundException
        {
            Class.forName("com.ibm.db2.jcc.DB2Driver");
            System.out.println("**** Loaded the JDBC driver");
            String url="jdbc:db2://localhost:50000/DBHR";
            String user="rwinga01";
            String password="*********";
            Connection con=null;

                try {
                        con=DriverManager.getConnection(url, user, password);
                        System.out.println("**** Created the connection");

                        String query = "SELECT EMP_ID, "
                                     + "EMP_LAST_NAME,"
                                     + "EMP_FIRST_NAME "
                                     + "FROM HRSCHEMA.EMPLOYEE "
                                     + "WHERE EMP_SERVICE_YEARS >= 5";

                        Statement stmt;
                        stmt = con.createStatement();
                        ResultSet rs=stmt.executeQuery(query);
                        while(rs.next())
                        {
                            int intEmpId        = rs.getInt(1);
                                String strLastName  = rs.getString(2);
                            String strFirstName = rs.getString(3);
                            System.out.println("EMP_ID is " + intEmpId
                                        + " Last name is " + strLastName
                                        + " First name is " + strFirstName);
                        }

                } catch (SQLException e) {
                        System.out.println(e.getMessage());
                        System.out.println(e.getErrorCode());
                        e.printStackTrace();

                }
```

3. Tara wants to grant CONTROL of table TBL1 to Bill, and also allow Bill to grant the same privilege to other users. What clause should Tara use on the GRANT statement?

   a. WITH CONTROL OPTION
   b. WITH GRANT OPTION
   c. WITH USE OPTION
   d. WITH REVOKE OPTION

   The correct answer is B. Using the WITH GRANT OPTION permits the recipient of the grant to also grant this privilege to other users. The other choices WITH CONTROL OPTION, WITH USE OPTION, and WITH REVOKE OPTION are incorrect because they are not valid clauses on a GRANT statement.

Chapter Four Exercises

1. Write a DCL statement to grant SELECT access on table HRSCHEMA.EMPLOYEE to users HR001 and HR002.

```
GRANT SELECT ON HRSCHEMA.EMPLOYEE
TO HR001, HR002;
```

2. Write a DCL statement to revoke SELECT access on table HRSCHEMA.EMPLOYEE from user HR002.

```
REVOKE SELECT ON HRSCHEMA.EMPLOYEE
FROM HR002;
```

3. Write a DCL statement to grant SELECT, INSERT, UPDATE and DELETE access on table HRSCHEMA.EMP_PAY to user HRMGR01.

```
GRANT SELECT, INSERT, UPDATE, DELETE
ON HRSCHEMA.EMP_PAY
TO HRMGR01;
```

Chapter Five Questions and Answers

1. If you want to still reference data using a cursor after you issue a COMMIT, which clause would you use when you declare the cursor?

   a. WITH HOLD
   b. WITH RETAIN
   c. WITH STAY
   d. WITH REOPEN

   The correct answer is A. The WITH HOLD clause causes a commit to leave the cursor open and avoid releasing locks necessary to maintain the cursor's positioning. The other answers are invalid clauses.

2. Which of the following is not a valid Java object type?

   a. ResultSet
   b. Connection
   c. Statement
   d. Dataset

   The correct answer is D. A Dataset is a .NET object. The other specified objects: ResultSet, Connection and Statement are all valid Java objects.

3. Which of the following is not a valid .NET object type?

   a. DB2Stmt
   b. DB2Connection
   c. DB2Command
   d. All of the above are valid .NET objects.

   The correct answer is A. There is no DB2Stmt type object in .NET. There is a DB2Statement type. The other objects – DB2Connection and DB2Command – are also valid .NET objects.

Chapter Five Exercises

1. Write a Java or .NET program that creates a result set of all employees who have 5 years or more of service from the HRSCHEMA.EMPLOYEE table. Include logic to display any rows that are returned. Display the employee number, last name and first name of these employees.

Java Solution:

```
package employee;
import java.sql.*;

public class JAVEMP4 {

        public static void main(String[] args) throws ClassNotFoundException
        {
            Class.forName("com.ibm.db2.jcc.DB2Driver");
            System.out.println("**** Loaded the JDBC driver");
            String url="jdbc:db2://localhost:50000/DBHR";
            String user="rwinga01";
            String password="*********";
            Connection con=null;

                try {
                        con=DriverManager.getConnection(url, user, password);
                        System.out.println("**** Created the connection");

                        String query = "SELECT EMP_ID, "
                                          + "EMP_LAST_NAME,"
                                          + "EMP_FIRST_NAME "
                                          + "FROM HRSCHEMA.EMPLOYEE "
                                          + "WHERE EMP_SERVICE_YEARS >= 5";

                        Statement stmt;
                        stmt = con.createStatement();
                        ResultSet rs=stmt.executeQuery(query);
                        while(rs.next())
                        {
                            int intEmpId        = rs.getInt(1);
                                String strLastName  = rs.getString(2);
                            String strFirstName = rs.getString(3);
                            System.out.println("EMP_ID is " + intEmpId
                                        + " Last name is " + strLastName
                                        + " First name is " + strFirstName);
                        }

                } catch (SQLException e) {
                        System.out.println(e.getMessage());
                        System.out.println(e.getErrorCode());
                        e.printStackTrace();

                }
```

```
        }

}
```

Output is:

```
**** Loaded the JDBC driver
**** Created the connection
EMP_ID is 7459 Last name is STEWART First name is BETTY
EMP_ID is 3333 Last name is FORD First name is JAMES
EMP_ID is 4720 Last name is SCHULTZ First name is TIM
EMP_ID is 6288 Last name is WILLARD First name is JOE
EMP_ID is 1122 Last name is JENKINS First name is DEBORAH
EMP_ID is 1111 Last name is VEREEN First name is CHARLES
EMP_ID is 1112 Last name is YATES First name is JANENE
EMP_ID is 1113 Last name is DUGGAN First name is RITA
EMP_ID is 1114 Last name is MILLER First name is PHYLLIS
```

.NET Solution:

```
using System;
using System.Collections.Generic;
using System.Linq;
using System.Text;
using System.Threading.Tasks;
using IBM.Data.DB2;

namespace EMPLOYEE
{
    class NETEMP4
    {
        static void Main(string[] args)
        {
            DB2Connection conn = null;
            DB2Command cmd = null;
            DB2DataReader rdr = null;
            Boolean rows = false;
            int cols = 0;

            try
            {
                conn = new DB2Connection("Database=DBHR");
                conn.Open();
                Console.WriteLine("Successful connection!");
                cmd = conn.CreateCommand();
                cmd.CommandText
                    = "SELECT EMP_ID,"
                        + "EMP_LAST_NAME, "
                        + "EMP_FIRST_NAME "
                        + "FROM HRSCHEMA.EMPLOYEE "
                        + "WHERE EMP_SERVICE_YEARS >=5 ";
```

```
                    e.printStackTrace();

                }

        }

}
```

The output is:

```
**** Loaded the JDBC driver
**** Created the connection
One or more values in the INSERT statement, UPDATE statement, or foreign key up-
date caused by a DELETE statement are not valid because the primary key, unique
constraint or unique index identified by "1" constrains table "HRSCHEMA.EMPLOYEE"
from having duplicate values for the index key.. SQLCODE=-803, SQLSTATE=23505,
DRIVER=3.72.24
-803
com.ibm.db2.jcc.am.SqlIntegrityConstraintViolationException: One or more val-
ues in the INSERT statement, UPDATE statement, or foreign key update caused by
a DELETE statement are not valid because the primary key, unique constraint or
unique index identified by "1" constrains table "HRSCHEMA.EMPLOYEE" from hav-
ing duplicate values for the index key.. SQLCODE=-803, SQLSTATE=23505, DRIV-
ER=3.72.24
	at com.ibm.db2.jcc.am.hd.a(hd.java:809)
	at com.ibm.db2.jcc.am.hd.a(hd.java:66)
	at com.ibm.db2.jcc.am.hd.a(hd.java:140)
	at com.ibm.db2.jcc.am.ip.c(ip.java:2788)
	at com.ibm.db2.jcc.am.ip.d(ip.java:2776)
	at com.ibm.db2.jcc.am.ip.b(ip.java:2154)
	at com.ibm.db2.jcc.t4.bb.j(bb.java:233)
	at com.ibm.db2.jcc.t4.bb.c(bb.java:48)
	at com.ibm.db2.jcc.t4.p.b(p.java:38)
	at com.ibm.db2.jcc.t4.vb.h(vb.java:124)
	at com.ibm.db2.jcc.am.ip.ib(ip.java:2149)
	at com.ibm.db2.jcc.am.ip.a(ip.java:3322)
	at com.ibm.db2.jcc.am.ip.c(ip.java:762)
	at com.ibm.db2.jcc.am.ip.executeUpdate(ip.java:741)
	at employee.JAVEMP1.main(JAVEMP1.java:36)
```

.NET Solution:

```
using System;
using System.Collections.Generic;
using System.Linq;
using System.Text;
```

```
using System.Threading.Tasks;
using IBM.Data.DB2;

/* Program to connect to DB2 and insert a row    */
namespace NETEMP1
{
    class NETEMP1
    {
        static void Main(string[] args)
        {
            DB2Connection conn = null;
            DB2Command cmd = null;

            try
            {
                conn = new DB2Connection("Database=DBHR");
                conn.Open();
                Console.WriteLine("Successful connection!");
                cmd = conn.CreateCommand();
                cmd.CommandText
                    = "INSERT INTO HRSCHEMA.EMPLOYEE "
                    + "(EMP_ID, "
                    + "EMP_LAST_NAME, "
                    + "EMP_FIRST_NAME, "
                    + "EMP_SERVICE_YEARS, "
                    + "EMP_PROMOTION_DATE) "
                    + "VALUES (1111, "
                    + "'VEREEN', "
                    + "'CHARLES', "
                    + "12, "
                    + "'01/01/2017') ";

                int rowsAffected = cmd.ExecuteNonQuery();
                Console.WriteLine("\n Inserted Rows: " + rowsAffected + " \n");
            }

            catch (Exception e)
            {
                Console.WriteLine(e.Message);
            }

            finally
            {
                conn.Close();
            }

            DB2Connection connection = new DB2Connection("Database=HRDB");
        }
    }
}
```

The output is:

```
using System.Threading.Tasks;
using System.Data;
using IBM.Data.DB2;
using IBM.Data.DB2Types;


namespace NETEMP5
{
    class NETEMP5
    {
        static void Main(string[] args)
        {
            Console.WriteLine("Program NETEMP2 begins successfully");
            DB2Connection conn = new DB2Connection("Database=DBHR");

            DataSet EmployeesDataSet = new DataSet();
            DB2DataAdapter da;
            DB2CommandBuilder cmdBuilder = null;

            try
            {
                conn.Open();
                Console.WriteLine("Successful connection!");

                da = new DB2DataAdapter("SELECT EMP_ID, EMP_LAST_NAME "
                            + "FROM HRSCHEMA.EMPLOYEE "
                            + "WHERE EMP_ID IN (1113, 1114)", conn);

                cmdBuilder = new DB2CommandBuilder(da);

                da.Fill(EmployeesDataSet, "Employees");

                foreach (DataTable DT in EmployeesDataSet.Tables)
                {
                    foreach (DataRow DR in DT.Rows)
                    {
                        string empId = DR[0].ToString();
                        string lName = DR[1].ToString();
                        Console.WriteLine("Employee " + empId
                            + " Last Name " + lName + " is being deleted");
                        DR.Delete();
                    }

                    // Update the table using the delete actions in the dataset

                    da.UpdateCommand = cmdBuilder.GetUpdateCommand();
                    da.Update(EmployeesDataSet, "Employees");
                }

            }

            catch (Exception e)
            {
```

```
using System.Threading.Tasks;
using IBM.Data.DB2;

/* Program to connect to DB2 and insert a row    */
namespace NETEMP1
{
    class NETEMP1
    {
        static void Main(string[] args)
        {
            DB2Connection conn = null;
            DB2Command cmd = null;

            try
            {
                conn = new DB2Connection("Database=DBHR");
                conn.Open();
                Console.WriteLine("Successful connection!");
                cmd = conn.CreateCommand();
                cmd.CommandText
                    = "INSERT INTO HRSCHEMA.EMPLOYEE "
                    + "(EMP_ID, "
                    + "EMP_LAST_NAME, "
                    + "EMP_FIRST_NAME, "
                    + "EMP_SERVICE_YEARS, "
                    + "EMP_PROMOTION_DATE) "
                    + "VALUES (1111, "
                    + "'VEREEN', "
                    + "'CHARLES', "
                    + "12, "
                    + "'01/01/2017') ";

                int rowsAffected = cmd.ExecuteNonQuery();
                Console.WriteLine("\n Inserted Rows: " + rowsAffected + " \n");
            }

            catch (Exception e)
            {
                Console.WriteLine(e.Message);
            }

            finally
            {
                conn.Close();
            }

            DB2Connection connection = new DB2Connection("Database=HRDB");
        }
    }
}
```

The output is:

```
Successful connection!
Exception thrown: 'IBM.Data.DB2.DB2Exception' in IBM.Data.DB2.dll
ERROR [23505] [IBM][DB2/NT64] SQL0803N  One or more values in the INSERT state-
ment, UPDATE statement, or foreign key update caused by a DELETE statement are
not valid because the primary key, unique constraint or unique index identified
by "1" constrains table "HRSCHEMA.EMPLOYEE" from having duplicate values for the
index key.  SQLSTATE=23505
```

3. Write a Java or .NET program to delete employee number 1114.

Java Solution:

There are a few ways to do this. We could simply create a statement object and run the SQL. In this case, let's create a result set so that we can record the actions and the employee numbers that we are deleting.

```
package employee;
import java.sql.*;

public class JAVEMP5 {

        public static void main(String[] args) throws ClassNotFoundException
        {
            Class.forName("com.ibm.db2.jcc.DB2Driver");
            System.out.println("**** Loaded the JDBC driver");

            String url
               ="jdbc:db2://localhost:50000/DBHR:retrieveMessagesFromServerOnGet
        Message=true;";
            String user="rwinga01";
            String password="*********";

            Connection con=null;

               try {
                       con=DriverManager.getConnection(url, user, password);
                       System.out.println("**** Created the connection");

                       String query = "SELECT EMP_ID, EMP_LAST_NAME "
                                    + "FROM HRSCHEMA.EMPLOYEE "
                                    + "WHERE EMP_ID = 1114" ;

                       Integer intEmpNo;
                       String strLastName;
                       Statement stmt;
```

```
                stmt
                   = con.createStatement(ResultSet.TYPE_SCROLL_SENSITIVE,
                     ResultSet.CONCUR_UPDATABLE);
                ResultSet rs=stmt.executeQuery(query);
                while(rs.next())
                {
                    intEmpNo     = rs.getInt(1);
                    strLastName  = rs.getString(2);
                    rs.deleteRow();
                    System.out.println("Employee " + intEmpNo
                              + " Last name is: " + strLastName
                              + " has been deleted");
                }
                con.commit();

         }

         catch (SQLException e) {
                System.out.println(e.getMessage());
                System.out.println(e.getErrorCode());
                e.printStackTrace();

         }

         finally {

                System.out.println("** JAVEMP5 finished");
         }

    }

}
```

Output is:

```
    **** Loaded the JDBC driver
    **** Created the connection
    Employee 1113 Last name is: DUGGAN has been deleted
    Employee 1114 Last name is: MILLER has been deleted
          ** JAVEMP5 finished
```

NET Solution:

```
using System;
using System.Collections.Generic;
using System.Linq;
using System.Text;
```

```
using System.Threading.Tasks;
using System.Data;
using IBM.Data.DB2;
using IBM.Data.DB2Types;


namespace NETEMP5
{
    class NETEMP5
    {
        static void Main(string[] args)
        {
            Console.WriteLine("Program NETEMP2 begins successfully");
            DB2Connection conn = new DB2Connection("Database=DBHR");

            DataSet EmployeesDataSet = new DataSet();
            DB2DataAdapter da;
            DB2CommandBuilder cmdBuilder = null;

            try
            {
                conn.Open();
                Console.WriteLine("Successful connection!");

                da = new DB2DataAdapter("SELECT EMP_ID, EMP_LAST_NAME "
                             + "FROM HRSCHEMA.EMPLOYEE "
                             + "WHERE EMP_ID IN (1113, 1114)", conn);

                cmdBuilder = new DB2CommandBuilder(da);

                da.Fill(EmployeesDataSet, "Employees");

                foreach (DataTable DT in EmployeesDataSet.Tables)
                {
                    foreach (DataRow DR in DT.Rows)
                    {
                        string empId = DR[0].ToString();
                        string lName = DR[1].ToString();
                        Console.WriteLine("Employee " + empId
                            + " Last Name " + lName + " is being deleted");
                        DR.Delete();
                    }

                    // Update the table using the delete actions in the dataset

                    da.UpdateCommand = cmdBuilder.GetUpdateCommand();
                    da.Update(EmployeesDataSet, "Employees");
                }

            }

            catch (Exception e)
            {
```

```
                Console.WriteLine(e.Message);
            }

            finally
            {
                conn.Close();
                Console.WriteLine("Program NETEMP2 concluded successfully");

            }

        }
    }

}
```

The output is:

```
Successful connection!
Employee 1113 Last Name DUGGAN is being deleted
Employee 1114 Last Name MILLER is being deleted
Program NETEMP2 concluded successfully
```

4. Write a Java or .NET program to merge the following employee information into the `EMPLOYEE` table.

Employee Number	: 1114
Employee Last Name	: MILLER
Employee First Name	: PHILLIS
Employee Years of Service	: 11
Employee Promotion Date	: 01/01/2017

Employee Number	: 1115
Employee Last Name	: JENSON
Employee First Name	: PAUL
Employee Years of Service	: 8
Employee Promotion Date	: 01/01/2016

Java Solution:

The MERGE can operate with multiple VALUES clauses, so we can do two records with one statement. The SQL is:

```
MERGE INTO HRSCHEMA.EMPLOYEE AS TARGET
USING (VALUES(1114,
       "MILLER",
       "PHYLLIS",
       11,
       "01/01/2017),
      (1115,
      "JENSON",
       "PAUL",
       8,
       "01/01/2016))


       AS SOURCE(EMP_ID,
       EMP_LAST_NAME,
       EMP_FIRST_NAME,
       EMP_SERVICE_YEARS,
       EMP_PROMOTION_DATE)
       ON TARGET.EMP_ID = SOURCE.EMP_ID

       WHEN MATCHED THEN UPDATE
          SET TARGET.EMP_LAST_NAME
                  = SOURCE.EMP_LAST_NAME,
             TARGET.EMP_FIRST_NAME
                  = SOURCE.EMP_FIRST_NAME
             TARGET.EMP_SERVICE_YEARS
                  = SOURCE.EMP_SERVICE_YEARS
               TARGET.EMP_PROMOTION_DATE
                  = SOURCE.EMP_PROMOTION_DATE

       WHEN NOT MATCHED THEN INSERT
         (EMP_ID,
          EMP_LAST_NAME,
          EMP_FIRST_NAME,
          EMP_SERVICE_YEARS,
          EMP_PROMOTION_DATE)
          VALUES
          (SOURCE.EMP_ID,
          SOURCE.EMP_LAST_NAME,
          SOURCE.EMP_FIRST_NAME,
          SOURCE.EMP_SERVICE_YEARS,
          SOURCE.EMP_PROMOTION_DATE);
```

A possible Java program solution is:

```
package employee;
```

```
import java.sql.*;

public class JAVEMP6 {

        public static void main(String[] args) throws ClassNotFoundException
        {
                Class.forName("com.ibm.db2.jcc.DB2Driver");
            String url
                ="jdbc:db2://localhost:50000/DBHR:retrieveMessagesFromServerOn-
        GetMessage=true;";
            String user="rwinga01";
            String password="*********";

            Connection con=null;

                try {
                        con=DriverManager.getConnection(url, user, password);
                        System.out.println("**** Created the connection");

                        String query = "MERGE INTO HRSCHEMA.EMPLOYEE AS TARGET"
                              + " USING (VALUES(1114, "
                              +    "'MILLER', "
                        +   "'PHYLLIS', "
                              +    "11, "
                              +    "'01/01/2017'),"
                                 +   "(1115, "
                                 +   "'JENSON' ,"
                              +    "'PAUL' ,"
                              +    "8, "
                              +    "'01/01/2016'))"
                              +    " AS SOURCE(EMP_ID, "
                              +    "     EMP_LAST_NAME, "
                              +    "     EMP_FIRST_NAME, "
                              +    "     EMP_SERVICE_YEARS, "
                              +    "     EMP_PROMOTION_DATE) "
                              +    "     ON TARGET.EMP_ID = SOURCE.EMP_ID "

                              +    "     WHEN MATCHED THEN UPDATE "
                              +    "        SET TARGET.EMP_LAST_NAME "
                              +    "          = SOURCE.EMP_LAST_NAME, "
                                 +      "             TARGET.EMP_FIRST_NAME "
                              +    "          = SOURCE.EMP_FIRST_NAME, "
                              +   "            TARGET.EMP_SERVICE_YEARS "
                              +    "          = SOURCE.EMP_SERVICE_YEARS, "
                              +    "            TARGET.EMP_PROMOTION_DATE "
                              +    "          = SOURCE.EMP_PROMOTION_DATE "

                              +    "     WHEN NOT MATCHED THEN INSERT "
                              +    "       (EMP_ID, "
                              +    "        EMP_LAST_NAME, "
                              +    "        EMP_FIRST_NAME, "
                              +    "        EMP_SERVICE_YEARS, "
                              +    "        EMP_PROMOTION_DATE) "
```

```
                              +   "       VALUES "
                              +   "       (SOURCE.EMP_ID, "
                              +   "       SOURCE.EMP_LAST_NAME, "
                              +   "       SOURCE.EMP_FIRST_NAME, "
                              +   "       SOURCE.EMP_SERVICE_YEARS, "
                              +   "       SOURCE.EMP_PROMOTION_DATE)";


                  Statement stmt;
                  stmt = con.createStatement();
                  stmt.executeUpdate(query);
                  System.out.println("Successful MERGE of employees 1114 and
1115");


            } catch (SQLException e) {
                  System.out.println(e.getMessage());
                  System.out.println(e.getErrorCode());
                  e.printStackTrace();

            }

      }

      }
```

The output is:

```
**** Loaded the JDBC driver
**** Created the connection
Successful MERGE of employees 1114 and 1115
```

<u>NET Solution</u>:

```
using System;
using System.Collections.Generic;
using System.Linq;
using System.Text;
using System.Threading.Tasks;
using IBM.Data.DB2;

namespace NETEMP6
{
    class NETEMP6
    {
        static void Main(string[] args)
        {
            DB2Connection conn = null;
```

```
DB2Command cmd = null;

try
{
    conn = new DB2Connection("Database=DBHR");
    conn.Open();
    Console.WriteLine("Successful connection!");
    cmd = conn.CreateCommand();
    cmd.CommandText
      = "MERGE INTO HRSCHEMA.EMPLOYEE AS TARGET"
      + " USING (VALUES(1114, "
      + "'MILLER', "
      + "'PHYLLIS', "
      + "11, "
      + "'01/01/2017'),"
      + "(1115, "
        + "'JENSON' ,"
      + "'PAUL' ,"
      + "8, "
      + "'01/01/2016'))"
      + " AS SOURCE(EMP_ID, "
      + "    EMP_LAST_NAME, "
      + "    EMP_FIRST_NAME, "
      + "    EMP_SERVICE_YEARS, "
      + "    EMP_PROMOTION_DATE) "
      + "    ON TARGET.EMP_ID = SOURCE.EMP_ID "

      + "    WHEN MATCHED THEN UPDATE "
      + "       SET TARGET.EMP_LAST_NAME "
      + "         = SOURCE.EMP_LAST_NAME, "
      + "           TARGET.EMP_FIRST_NAME "
      + "         = SOURCE.EMP_FIRST_NAME, "
      + "           TARGET.EMP_SERVICE_YEARS "
      + "         = SOURCE.EMP_SERVICE_YEARS, "
      + "           TARGET.EMP_PROMOTION_DATE "
      + "         = SOURCE.EMP_PROMOTION_DATE "

      + "    WHEN NOT MATCHED THEN INSERT "
      + "      (EMP_ID, "
      + "       EMP_LAST_NAME, "
      + "       EMP_FIRST_NAME, "
      + "       EMP_SERVICE_YEARS, "
      + "       EMP_PROMOTION_DATE) "
      + "       VALUES "
      + "       (SOURCE.EMP_ID, "
      + "       SOURCE.EMP_LAST_NAME, "
      + "       SOURCE.EMP_FIRST_NAME, "
      + "       SOURCE.EMP_SERVICE_YEARS, "
      + "       SOURCE.EMP_PROMOTION_DATE)";


    int rowsAffected = cmd.ExecuteNonQuery();
    Console.WriteLine("\n Merged Rows: " + rowsAffected + " \n");
```

```
            }

            catch (Exception e)
            {
                Console.WriteLine(e.Message);
            }

            finally
            {
                conn.Close();
            }

            DB2Connection connection = new DB2Connection("Database=HRDB");
        }
    }
}
```

The output is:

```
Successful connection!

Merged Rows: 2
```

Chapter Six Questions and Answers

1. To end a transaction without making the changes permanent, which DB2 statement should be issued?

    a. COMMIT
    b. BACKOUT
    c. ROLLBACK
    d. NO CHANGE

    The correct answer is C. Issuing a ROLLBACK statement will end a transaction without making the changes permanent.

2. If you want to maximize data concurrency without seeing uncommitted data, which isolation level should you use?

    a. RR
    b. UR
    c. RS
    d. CS

    The correct answer is D (Cursor Stability). CURSOR STABILITY (CS) only locks the row where the cursor is placed, thus maximizing concurrency compared to RR or RS. REPEATABLE READ (RR) ensures that a query issued multiple times within the same unit of work will produce the exact same results. It does this by locking ALL rows that could affect the result, and does not permit any changes to the table that could affect the result. With READ STABILITY(RS), all rows that are returned by the query are locked. UNCOMMITTED READ (UR) is incorrect because it permits reading of uncommitted data and the question specifically disallows that.

3. Assume you have a long running process and you want to commit results after processing every 500 records, but to be able to undo any work that has taken place after the commit point. One mechanism that would allow you to do this is to issue a:

    a. SAVEPOINT
    b. COMMITPOINT
    c. BACKOUT
    d. None of the above

The correct answer is A. Issuing a SAVEPOINT enables you to execute several SQL statements as a single executable block. You can then undo changes back out to that savepoint by issuing a ROLLBACK TO SAVEPOINT statement.

4. A procedure that commits transactions independent of the calling procedure is known as a/an:

   a. External procedure.
   b. SQL procedure.
   c. Autonomous procedure.
   d. Independent procedure.

The correct answer is C. Autonomous procedures run with their own units of work, separate from the calling program. Autonomous procedures were introduced in DB2 11.

External procedures are written in a host language. They can contain SQL statements. SQL procedures are written entirely in SQL statements. There is no "independent" type of procedure.

5. To end a transaction and make the changes visible to other processes, which statement should be issued?

   a. ROLLBACK
   b. COMMIT
   c. APPLY
   d. CALL

The correct answer is B. The COMMIT statement ends a transaction and makes the changes visible to other processes.

6. Order the isolation levels, from greatest to least impact on performance.

   a. RR, RS, CS, UR
   b. UR, RR, RS, CS
   c. CS, UR, RR, RS
   d. RS, CS, UR, RR

The correct answer is A - RR, RS, CS, UR. Repeatable Read has the greatest impact on performance because it incurs the most overhead and locks the most

rows. It ensures that a query issued multiple times within the same unit of work will produce the exact same results. It does this by locking all rows that could affect the result, and does not permit any adds/changes/deletes to the table that could affect the result. Next, READ STABILITY locks for the duration of the transaction those rows that are returned by a query, but it allows additional rows to be added to the table. CURSOR STABILITY only locks the row that the cursor is placed on (and any rows it has updated during the unit of work). UN-COMMITTED READ permits reading of uncommitted changes which may never be applied to the database and does not lock any rows at all unless the row(s) is updated during the unit of work.

7. Which isolation level is most appropriate when few or no updates are expected to a table?

   a. RR
   b. RS
   c. CS
   d. UR

The correct answer is D. UR - Uncommitted Read uses less overhead than the other isolation levels and is most appropriate for read-only access of tables. The other isolation levels acquire locks that are unnecessary in a read-only environment. Repeatable Read is required to obtain locks to ensure that a query issued multiple times within the same unit of work will produce the exact same results. With READ STABILITY any rows that are returned by the query are locked. CURSOR STABILITY locks the row that the cursor is placed on, which is not necessary in a primarily read-only environment.

Chapter Six Exercises

1. Write the Java or .NET concurrency code that ensures no records which have been retrieved during a unit of work can be changed by other processes until the unit of work completes. However, it is ok for other (new) records to be added to the table during the unit of work.

   Assuming a connection call "con", in Java:

   ```
   con.setTransactionIsolation(con.TRANSACTION_REPEATABLE_READ);
   ```

   In .NET, you define the concurrency level in the connection string:

   ```
   DB2Connection conn = new DB2Connection("Database=DBHR;IsolationLevel=ReadCommitted");
   ```

2. Write the Java or .NET code to end a unit of work and make the changes visible to other processes.

   Assuming a DB2Connection object named "con":

   In Java:

   ```
   con.commit();
   ```

   In .NET:

   You have to use the transaction object to issue a commit. Assuming a connection named conn:

   ```
   DB2Transaction tran = null;

   tran = conn.BeginTransaction();

   //  DO SOME UPDATES HERE

   tran.Commit();
   ```

3. Write the Java or .NET code to end a unit of work and discard any updates that have been made since the last commit point.

Assuming a DB2Connection object named "con":

In Java:

```
con.rollback();
```

In .NET:

You have to use the transaction object to issue a rollback. Assuming a connection named conn:

```
DB2Transaction tran = null;

tran = conn.BeginTransaction();

//  DO SOME UPDATES HERE

tran.Rollback();
```

Chapter Seven Questions and Answers

1. Suppose you have created a test version of a production table, and you want to to use the UNLOAD utility to extract the first 1,000 rows from the production table to load to the test version. Which keyword would you use in the UNLOAD statement?

    a. WHEN
    b. SELECT
    c. SAMPLE
    d. SUBSET

    The correct answer is C. You can specify SAMPLE n where n is the number of rows to unload. For example you can limit the unloaded rows to the first 5,000 by specifying:

    ```
    SAMPLE 1000
    ```

    WHEN is used to specify rows that meet a criteria such as: WHEN (EMP_SALARY < 90000).

    SELECT and SUBSET are invalid clauses and would cause an error.

2. Which of the following is an SQL error code indicating a DB2 package is not found within the DB2 plan?

    a. -803
    b. -805
    c. -904
    d. -922

    The correct answer is B. The -805 SQLCODE indicates that the specified package was not found in the DB2 plan (or the named package with the correct timestamp could not be found). Typically this requires a BIND PACKAGE action to resolve.

    The -803 SQLCODE means a violation of a unique index. A -904 SQLCODE means an unavailable resource such as a tablespace. A -922 is a security violation such as when an unauthorized user tries to access a DB2 plan (usually this requires a GRANT action to correct the problem).

3. Which DB2 utility updates the statistics used by the DB2 Optimizer to choose a data access path?

a. REORG
b. RUNSTATS
c. REBIND
d. OPTIMIZE

The correct answer is B. Executing RUNSTATS provides the DB2 Optimizer with the latest information on the tables and could improve the access path chosen for the DB2 plan. Of course after RUNSTATS is executed, the related application packages need to be rebound in order for the DB2 Optimizer to use this new infiormation.

4. Which of the following is NOT a way you could test a DB2 SQL statement?

a. Running the statement from the DB2 command line processor.
b. Running the statement from a Java program.
c. Running the statement from IBM Data Studio.
d. All of the above are valid ways to test an SQL statement.

The correct answer is D. Any of these three methods could be used to test a DB2 SQL statement.

Chapter Seven Exercises:

1. Write an UNLOAD control statement to unload a sample of 5000 records from the HRSCHEMA.EMP_PAY table for use in loading a test table.

```
UNLOAD DATA
FROM TABLE HRSCHEMA.EMP_PAY
SAMPLE 5000
```

2. Suppose that the HRSCHEMA.EMPLOYEE table exists on a system whose location name is DENVER. Write a query that uses three part naming to implicitly connect to DENVER and then retrieve all information for EMP_ID 3217.

```
SELECT *
FROM DENVER.HRSCHEMA.EMPLOYEE
WHERE EMP_ID = 3217;
```

Chapter Eight Questions and Answers

1. Consider the following stored procedure:

```
CREATE PROCEDURE GET_PATIENTS
(IN intHosp INTEGER)
DYNAMIC RESULT SETS 1
LANGUAGE SQL
P1: BEGIN

DECLARE cursor1 CURSOR
WITH RETURN FOR
SELECT PATIENT_ID,
LNAME,
FNAME
FROM PATIENT
WHERE PATIENT_HOSP = intHosp
ORDER BY PATIENT_ID ASC;

OPEN cursor1;
END P1
```

Answer this question: How many parameters are used in this stored procedure?

a. 0
b. 1
c. 2
d. 3

The correct answer is B (1). intHosp is accepted as an IN parameter that is used in the query. Parameters can be IN (they pass a value to the stored procedure), OUT (they return a value from a stored procedure), or INOUT (they pass a value to and return a value from a stored procedure). The answers 0, 2, and 3 are incorrect because exactly 1 parameter is used in this stored procedure.

2. When you want to create an external stored procedure, which of the following programming languages can NOT be used?

a. COBOL
b. C
c. Fortran
d. OLE

The correct answer is C. You cannot use Fortran to create an external stored

procedure. The valid programming languages for creating an external stored procedure are:

- C
- C++
- COBOL
- JAVA
- OLE (windows operating system only)

3. In order to invoke a stored procedure, which keyword would you use?

   a. RUN
   b. CALL
   c. OPEN
   d. TRIGGER

The correct answer is B. CALL is the correct statement to invoke a stored procedure. The syntax is CALL <procedure-name>. RUN is not a valid DB2 statement or command unless you are issuing the command to RUN(DSN). The OPEN statement opens a cursor in an embedded SQL program. A TRIGGER is an object that performs some predefined action when it is activated by an INSERT, UPDATE or DELETE of a record in a particular table.

4. Which of the following is NOT a valid **return type** for a User Defined Function?

   a. Scalar
   b. Aggregate
   c. Column
   d. Row

The correct answer is C. There is no "column" return type for a UDF. A user-defined function can be any of the following which returns the specific data type:

- **A scalar function, which returns a single value each time it is called.**
- **An aggregate function, which is passed a set of like values and returns a single value for the set.**
- **A row function, which returns one row.**
- **A table function, which returns a table.**

5. Which of the following is NOT a valid type of user-defined function (UDF)?

    a. External sourced
    b. SQL sourced
    c. External table
    d. SQL table

The correct answer is B. There is no SQL sourced user-defined function type. Valid types of user defined functions include:

- **External scalar**
- **External table**
- **External sourced**
- **SQL scalar**
- **SQL table**

An external scalar function is written in a programming language and returns a scalar value. An external table function is written in a programming language and returns a table to the subselect from which it was started. A sourced function is implemented by starting another function that exists at the server. An SQL scalar function is written exclusively in SQL statements and returns a scalar value. An SQL table function is written exclusively as an SQL RETURN statement and returns a set of rows.

6. Assume the following trigger DDL:

```
CREATE TRIGGER SAVE_EMPL
AFTER UPDATE ON EMPL
FOR EACH ROW
INSERT INTO EMPLOYEE_HISTORY
VALUES (EMPLOYEE_NUMBER,
EMPLOYEE_STATUS,
CURRENT TIMESTAMP)
```

What will the result of this DDL be, provided the tables and field names are correctly defined?

    a. The DDL will create the trigger successfully and it will work as intended.
    b. The DDL will fail because you cannot use an INSERT with a trigger.
    c. The DDL will fail because the syntax of this statement is incorrect.
    d. The DDL will create the trigger successfully but it will fail when executed.

The correct answer is A. The DDL is syntactically correct and will sucessfully create a trigger that will work as intended.

7. Which ONE of the following actions will NOT cause a trigger to fire?

   a. INSERT
   b. LOAD
   c. DELETE
   d. MERGE

   The correct answer is B. The LOAD action does not (by default) causes triggers to fire. The INSERT, DELETE and MERGE do cause INSERT, UPDATE and DELETE triggers to fire.

8. If you use the "FOR EACH STATEMENT" granularity clause in a trigger, what type of timing can you use?

   a. BEFORE
   b. AFTER
   c. INSTEAD OF
   d. All of the above.

   The correct answer is B. You cannot use a FOR EACH STATEMENT with BEFORE or INSTEAD OF timing. FOR EACH STATEMENT means your trigger logic is to be applied only once after the triggering statement finishes processing the affected rows.

9. Assume you have a column FLD1 with a referential constraint that specifies that the FLD1 value must exist in another table. What type of field is FLD1?

   a. FOREIGN KEY
   b. PRIMARY KEY
   c. UNIQUE KEY
   d. INDEX KEY

   The correct answer is A. A foreign key relationship is a constraint that says a column must only contain values that exist in another (parent) table. A PRIMARY KEY ensures that all values in a table are unique and non-null. A UNIQUE key ensures that no two rows have the same value in the column defined as being

UNIQUE. INDEX KEY is incorrect because no information in the question suggested that an index was involved.

10. Review the following and then answer the question.

```
CREATE TABLE "DBO"."HOSPITAL"  (
"HOSP_ID" INTEGER NOT NULL ,
"HOSP_NAME" CHAR(25) )
IN "USERSPACE1" ;

ALTER TABLE "DBO"."HOSPITAL"
ADD CONSTRAINT "PKeyHosp" PRIMARY KEY
("HOSP_ID");

CREATE TABLE "DBO"."PATIENT"  (
"PATIENT_ID" INTEGER NOT NULL ,
"PATIENT_NAME" CHAR(30) ,
"PATIENT_HOSP" INTEGER )
IN "USERSPACE1" ;

ALTER TABLE "DBO"."PATIENT"
ADD CONSTRAINT "PKeyPat" PRIMARY KEY
("PATIENT_ID");

ALTER TABLE "DBO"."PATIENT"
ADD CONSTRAINT "FKPatHosp" FOREIGN KEY
("PATIENT_HOSP")
REFERENCES "DBO"."HOSPITAL"
("HOSP_ID")
ON DELETE SET NULL;
```

If a row is deleted from the HOSPITAL table, what will happen to the PATIENT_HOSP rows that used the HOSP_ID value?

a. The PATIENT_HOSP field will be set to NULLS.
b. The PATIENT rows will be deleted from the PATIENT table.
c. Nothing will be changed on the PATIENT records.
d. The delete of the HOSPITAL record will fail.

The correct answer is A - the PATIENT_HOSP field will be set to NULL because of the ON DELETE SET NULL clause in the foreign key definition. You can specify the action to take in the foreign key definition by including the ON DELETE clause. Besides SET NULL, there are other possibilities for the ON DELETE clause. If ON DELETE CASCADE is specified, then references in

the child table to the parent record being deleted will cause the child records to also be deleted. If no action is specified, or if RESTRICT is specified in the ON DELETE clause, then the parent record cannot be deleted unless all child records which reference that record are first deleted.

11. Review the following DDL and then answer the question.

```
CREATE TABLE "DBO"."HOSPITAL"  (
"HOSP_ID" INTEGER NOT NULL ,
"HOSP_NAME" CHAR(25) )
IN "USERSPACE1" ;

ALTER TABLE "DBO"."HOSPITAL"
ADD CONSTRAINT "PKeyHosp" PRIMARY KEY
("HOSP_ID");

CREATE TABLE "DBO"."PATIENT"  (
"PATIENT_ID" INTEGER NOT NULL ,
"PATIENT_NAME" CHAR(30) ,
"PATIENT_HOSP" INTEGER )
IN "USERSPACE1" ;

ALTER TABLE "DBO"."PATIENT"
ADD CONSTRAINT "PKeyPat" PRIMARY KEY
("PATIENT_ID");

ALTER TABLE "DBO"."PATIENT"
ADD CONSTRAINT "FKPatHosp" FOREIGN KEY
("PATIENT_HOSP")
REFERENCES "DBO"."HOSPITAL"
("HOSP_ID")
ON DELETE CASCADE;
```

If a row is deleted from the HOSPITAL table, what will happen to the PATIENT_HOSP rows that used the HOSP_ID value?

a. The PATIENT_HOSP field will be set to NULLS.
b. The PATIENT rows will be deleted from the PATIENT table.
c. Nothing will be changed on the PATIENT records.
d. The delete of the HOSPITAL record will fail.

The correct answer is B - when ON DELETE CASCADE is specified, then references in the child table to the parent record being deleted will cause the child records to also be deleted. You can specify the action to take in the foreign key

definition by including the ON DELETE clause. If no action is specified, or if RESTRICT is specified in the ON DELETE clause, then the parent record cannot be deleted unless all child records which reference that record are first deleted. If the clause ON DELETE SET NULL was present in the foreign key definition, the PATIENT_HOSP field will be set to NULL.

12. Review the following DDL, and then answer the question:

```
CREATE TABLE "DBO"."HOSPITAL"
("HOSP_ID" INTEGER NOT NULL,
"HOSP_NAME" CHAR(25) )  IN "USERSPACE1";

ALTER TABLE "DBO"."HOSPITAL"
ADD CONSTRAINT "PKeyHosp"
PRIMARY KEY ("HOSP_ID");

CREATE TABLE "DBO"."PATIENT"
("PATIENT_ID" INTEGER NOT NULL,
"PATIENT_NAME" CHAR(30) ,
"PATIENT_HOSP" INTEGER )   IN "USERSPACE1" ;

ALTER TABLE "DBO"."PATIENT"
ADD CONSTRAINT "PKeyPat"
PRIMARY KEY     ("PATIENT_ID");

ALTER TABLE "DBO"."PATIENT"
ADD CONSTRAINT "FKPatHosp"
FOREIGN KEY ("PATIENT_HOSP")
REFERENCES "DBO"."HOSPITAL" ("HOSP_ID")
ON DELETE RESTRICT;
```

If a row is deleted from the HOSPITAL table, what will happen to the PATIENT_HOSP rows that used the HOSP_ID value?

a. The PATIENT_HOSP field will be set to NULLS.
b. The PATIENT rows will be deleted from the PATIENT table.
c. Nothing will be changed on the PATIENT records.
d. The delete of the HOSPITAL record will fail.

The correct answer is D - the delete of the HOSPITAL record will fail because ON DELETE RESTRICT has been specified. You can specify the action to take in the foreign key definition by including the ON DELETE clause. If no action is specified, or if RESTRICT is specified in the ON DELETE clause, then the parent record cannot be deleted unless all child records which reference that

record are first deleted. If ON DELETE CASCADE is specified, then references in the child table to the parent record being deleted will cause the child records to also be deleted. If ON DELETE SET NULL is specified, then the foreign key field will be set to NULL for corresponding rows that reference the parent record to be deleted.

13. What is the schema for a declared GLOBAL TEMPORARY table?

    a. SESSION
    b. DB2ADMIN
    c. TEMP1
    d. USERTEMP

The correct answer is A. The schema for a GLOBAL TEMPORARY table is always SESSION. The schema for a GLOBAL TEMPORARY table cannot be DB2ADMIN, TEMP1, or USERTEMP.

14. If you create a temporary table and you wish to replace any existing temporary table that has the same name, what clause would you use?

    a. WITH REPLACE
    b. OVERLAY DATA ROWS
    c. REPLACE EXISTING
    d. None of the above.

The correct answer is A. If you create a temporary table and you wish to replace any existing temporary table that has the same name, use the WITH REPLACE clause. OVERLAY DATA ROWS and REPLACE EXISTING are fictitious clauses that would result in an error.

15. What happens to the rows of a temporary table when the session that created it ends?

    a. The rows are deleted when the session ends.
    b. The rows are preserved in memory until the instance is restarted.
    c. The rows are held in the temp table space.
    d. None of the above.

The correct answer is A. When the session that created a temporary table ends, any rows in the table are deleted, along with the table and table definition.

16. Which of the following clauses DOES NOT allow you to pull data for a particular period from a version enabled table?

    a. FOR BUSINESS_TIME UP UNTIL
    b. FOR BUSINESS_TIME FROM ... TO ...
    c. FOR BUSINESS_TIME BETWEEN... AND...
    d. All of the above enable you to pull data for a particular period.

    The correct answer is A. There is no UP UNTIL clause in DB2 temporal data management. The other two clauses may be used to specify the time period on a query against a version enabled table.

17. Assume you have an application that needs to aggregate and summarize data from several tables multiple times per day. One way to improve performance of that application would be to use a:

    a. Materialized query table
    b. View
    c. Temporary table
    d. Range clustered table

    The correct answer is A, a materialized query table (MQT) is a table whose definition is based upon the result of a query, similar to a view. The difference is that the query on which a view is based must generate the resultset each time the view is referenced. In contrast, an MQT stores the query results as table data, and you can work with the data that is in the MQT instead of incurring the overhead of running a query to generate the data each time you run it. An MQT can thereby significantly improve performance for applications that need summarized, aggregated data.

18. Assume you want to track employees in your company over time. Review the following DDL:

```
CREATE TABLE HRSCHEMA.EMPLOYZZ(
EMP_ID INT NOT NULL,
EMP_LAST_NAME VARCHAR(30) NOT NULL,
EMP_FIRST_NAME VARCHAR(20) NOT NULL,
EMP_SERVICE_YEARS INT
NOT NULL WITH DEFAULT 0,
EMP_PROMOTION_DATE DATE,
BUS_START   DATE  NOT NULL,
```

```
BUS_END      DATE  NOT NULL,
PERIOD BUSINESS_TIME(BUS_START, BUS_END),
PRIMARY KEY (EMP_ID, BUSINESS_TIME WITHOUT OVERLAPS));
```

What will happen when you execute this DDL?

a. It will fail because you cannot specify WITHOUT OVERLAPS in the primary key – the WITHOUT OVERLAPS clause belongs in the BUSINESS_TIME definition.
b. It will fail because you must specify SYSTEM_TIME instead of BUSINESS_TIME.
c. It will fail because the BUS_START has a syntax error.
d. It will execute successfully.

The correct answer is D. It will execute successfully.

19. Given the previous question, assume there is a table named EMPLOYEE_HIST defined just like EMPLOYEE. What will happen when you execute the following DDL?

```
ALTER TABLE EMPLOYEE
ADD VERSIONING
USE HISTORY TABLE EMPLOYEE_HIST
```

a. The DDL will execute successfully and updates to EMPLOYEE will generate records in the EMPLOYEE_HIST table.
b. The DDL will succeed but you must still enable the history table.
c. The DDL will generate an error – only SYSTEM time enabled tables can use a history table.
d. The DDL will generate an error – only BUSINESS time enabled tables can use a history table.

The correct answer is C. Only SYSTEM time enabled tables can use a history table.

20. For a system managed Materialized Query Table (MQT) named EMPMQT, how does the data get updated so that it becomes current?

a. Issuing INSERT, UPDATE and DELETE commands against EMPMQT.
b. Issuing the statement REFRESH TABLE EMPMQT.
c. Issuing the statement MATERIALIZE TABLE EMPMQT.
d. None of the above.

The correct answer is B. The REFRESH TABLE statement refreshes the data in a materialized query table. This is the only way to refresh a system-managed MQT. For example:

```
REFRESH TABLE EMPMQT;
```

For a user defined MQT you can use INSERT, UPDATE, DELETE, TRUNCATE, MERGE or LOAD to make the data current. There is no MATERIALIZE TABLE statement.

21. An XML index can be created on what column types?

    a. VARCHAR and XML.
    b. CLOB AND XML.
    c. XML only.
    d. Any of the above.

The correct answer is C. An XML index can be created only on an XML type column.

22. Which of the following can be used to validate an XML value according to a schema?

    a. Defining a column as type XML.
    b. Using XMLVALIDATE in the SQL.
    c. Both of the above.
    d. Neither of the above.

The correct answer is B. To validate the XML according to a schema, you must use the XMLVALIDATE function in your SQL (or else arrange for a trigger to intercept and force the XMLVALIDATE when the INSERT or UPDATE action occurs). Defining a column with the XML type modifier ensures only that XML documents stored in an XML column are well formed, not that they are validated.

23. To determine whether an XML document has been validated, which function could you use?

    a. XMLXSROBJECTID.
    b. XMLDOCUMENT.
    c. XMLPARSE.
    d. None of the above.

The correct answer is A. The XMLXSROBJECTID function returns the XSR object identifier of the XML schema that was used to validate the specified XML document. If the value returned is zero, then the document was not validated.

The XMLDOCUMENT function returns an XML value with a single document node and its children nodes (if any). The XMLPARSE function parses an argument as an XML document and returns an XML value.

24. Which of the following would enable parallel processing to improve performance of a query?

    a. Setting the CURRENT DEGREE special register to ANY
    b. Using the DEGREE(ANY) bind option
    c. Both a and b
    d. Neither a nor b

The correct answer is C. The DEGREE option determines whether to run a query using parallel processing to maximize performance. Setting the CURRENT DEGREE special register to ANY enables parallel processing. Specifying DEGREE(ANY) when binding a package also enables parallel processing.

25. Which of the following is the least efficient type of predicate?

    a. Index bracketing
    b. Residual
    c. Data Sargable
    d. Index sargable

The correct answer is B. Residual predicates are the least optimal in terms of performance because they require additional processing beyond access to a base table. Examples of these include use of ANY, ALL, SOME or IN. Reading LOB data is also residual if the data is stored in a LOB table space (which is typical).

26. Which of the following would probably NOT improve query performance?

    a. Use indexable predicates in your queries.
    b. Execute the RUNSTATS utility and rebind application programs.
    c. Set the CURRENT DEGREE to the value 1
    d. All of the above could improve application performance.

The correct answer is C. CURRENT DEGREE specifies the number of processors that will used to perform parallel processing. If the value is set to 1, then parallel processing will not be used and will not improve performance. The other choices could all potentially improve performance in an application. All indexable predicates refer to table indexes (which could improve performance). Executing RUNSTATS and rebinding application programs uses the latest information on the tables and could improve the access path chosen by the DB2 optimizer.

27. If you find out that your application query is doing a table space scan, what changes could you make to improve the scan efficiency?

    a. Create one or more indexes on the query search columns.
    b. Load the data to a temporary table and query that table instead of the base table.
    c. If the table is partitioned, change it to a non-partitioned table.
    d. All of the above could improve the scan efficiency.

The correct answer is A. You could create one or more indexes on the search columns so that an index scan would occur instead of the tablespace scan. The other choices would not improve scan efficiency. Temporary tables do not allow for indexes and so a full table scan would still occur. It would not make sense to change partitioned tables to non-partitioned if you were trying to improve scan efficiency. Partitioned tables have some advantage over non-partitioned if the table is partitioned by one of the fields being searched on. In that case the search could be automatically limited to only certain partitions instead of the entire table. Also, partitioned tables enable other performance improving techniques such as parallel processing.

Index

Other Titles by Robert Wingate

COBOL Basic Training Using VSAM, IMS and DB2

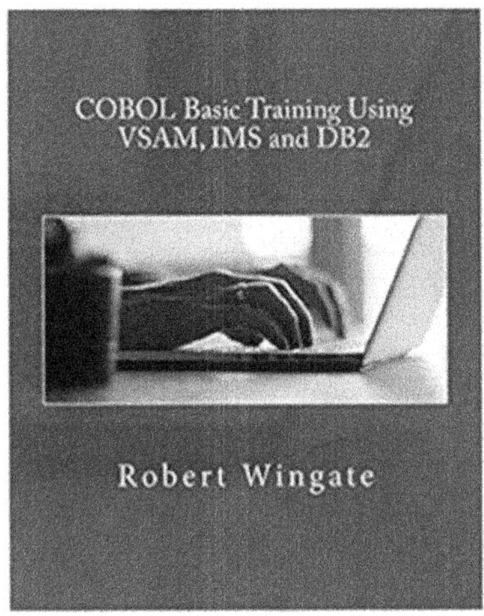

ISBN-13: 978-1720820710

This book will teach you the basic information and skills you need to develop applications with COBOL on IBM mainframes running z/OS. The instruction, examples and sample programs in this book are a fast track to becoming productive as quickly using COBOL. The content is easy to read and digest, well organized and focused on honing real job skills.

CICS Basic Training for Application Developers Using DB2 and VSAM

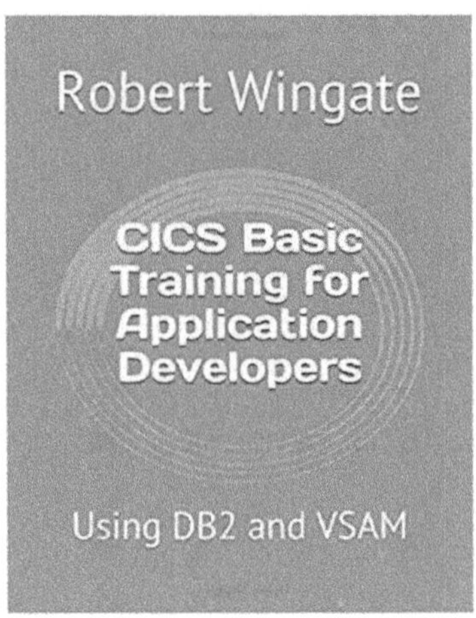

ISBN-13: 978-1794325067

This book will teach you the basic information and skills you need to develop applications with CICS on IBM mainframe computers running z/OS. The instruction, examples and sample programs in this book are a fast track to becoming productive as quickly as possible using CICS with the COBOL programming language. The content is easy to read and digest, well organized and focused on honing real job skills.

Quick Start Training for IBM z/OS Application Developers, Volume 1

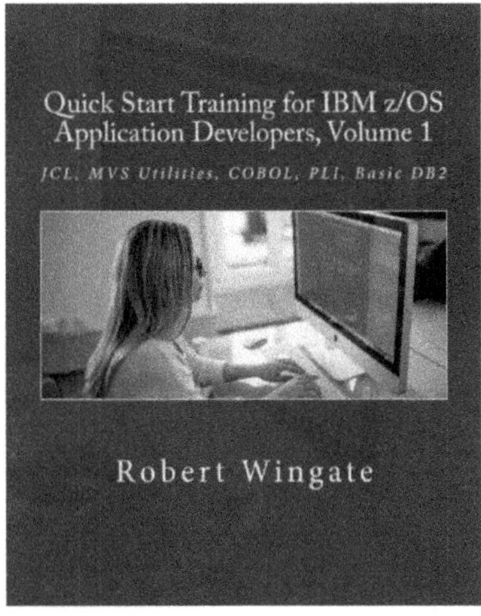

ISBN-13: 978-1986039840

This book will teach you the basic information and skills you need to develop applications on IBM mainframes running z/OS. The instruction, examples and sample programs in this book are a fast track to becoming productive as quickly as possible in JCL, MVS Utilities, COBOL, PLI and DB2. The content is easy to read and digest, well organized and focused on honing real job skills. IBM z/OS Quick Start Training for Application Developers is a key step in the direction of mastering IBM application development so you'll be ready to join a technical team.

Quick Start Training for IBM z/OS Application Developers, Volume 2

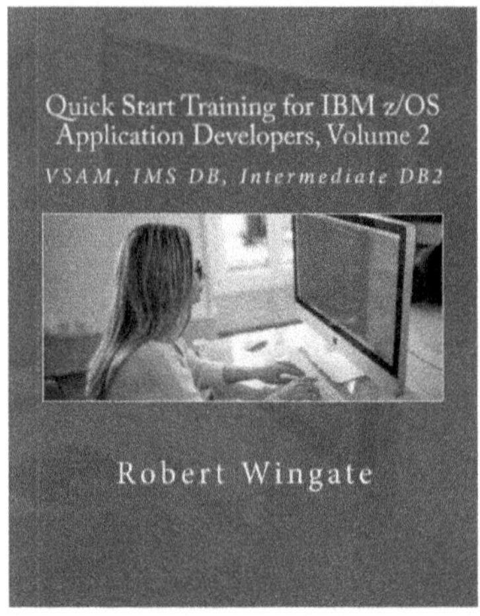

ISBN-13: 978-1717284594

This book will teach you the basic information and skills you need to develop applications on IBM mainframes running z/OS. The instruction, examples and sample programs in this book are a fast track to becoming productive as quickly as possible in VSAM, IMS and DB2. The content is easy to read and digest, well organized and focused on honing real job skills. IBM z/OS Quick Start Training for Application Developers is a key step in the direction of mastering IBM application development so you'll be ready to join a technical team.

Teradata Basic Training for Application Developers

Teradata
Basic Training
for
Application
Developers

Robert Wingate

ISBN-13: 978-1082748882

This book will help you learn the basic information and skills you need to develop applications with Teradata. The instruction, examples and questions/answers in this book are a fast track to becoming productive as quickly as possible. The content is easy to read and digest, well organized and focused on honing real job skills. Programming examples are coded in both Java and C# .NET. Teradata Basic Training for Application Developers is a key step in the direction of mastering Teradata application development so you'll be ready to join a technical team.

IMS Basic Training for Application Developers

ISBN-13: 978-1793440433

This book will teach you the basic information and skills you need to develop applications with IMS on IBM mainframe computers running z/OS. The instruction, examples and sample programs in this book are a fast track to becoming productive as quickly as possible using IMS with COBOL and PLI. The content is easy to read and digest, well organized and focused on honing real job skills.

DB2 Exam C2090-313 Preparation Guide

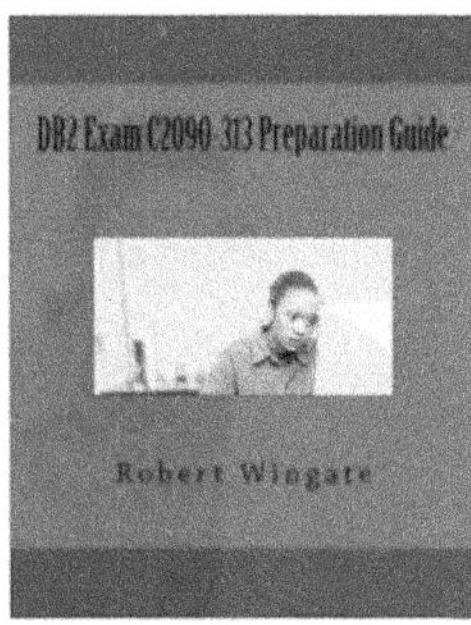

ISBN 13: 978-1548463052

This book will help you pass IBM Exam C2090-313 and become an IBM Certified Application Developer - DB2 11 for z/OS. The instruction, examples and questions/answers in the book offer you a significant advantage by helping you to gauge your readiness for the exam, to better understand the objectives being tested, and to get a broad exposure to the DB2 11 knowledge you'll be tested on.

DB2 Exam C2090-320 Preparation Guide

ISBN 13: 978-1544852096

This book will help you pass IBM Exam C2090-320 and become an IBM Certified Database Associate - DB2 11 Fundamentals for z/OS. The instruction, examples and questions/answers in the book offer you a significant advantage by helping you to gauge your readiness for the exam, to better understand the objectives being tested, and to get a broad exposure to the DB2 11 knowledge you'll be tested on. The book is also a fine introduction to DB2 for z/OS!

DB2 Exam C2090-313 Practice Questions

ISBN 13: 978-1534992467

This book will help you pass IBM Exam C2090-313 and become an IBM Certified Application Developer - DB2 11 for z/OS. The 180 questions and answers in the book (three full practice exams) offer you a significant advantage by helping you to gauge your readiness for the exam, to better understand the objectives being tested, and to get a broad exposure to the DB2 11 knowledge you'll be tested on.

DB2 Exam C2090-615 Practice Questions

ISBN 13: 978-1535028349

This book will help you pass IBM Exam C2090-615 and become an IBM Certified Database Associate (DB2 10.5 for Linux, Unix and Windows). The questions and answers in the book offer you a significant advantage by helping you to gauge your readiness for the exam, to better understand the objectives being tested, and to get a broad exposure to the knowledge you'll be tested on.

About the Author

Robert Wingate is a computer services professional with over 30 years of IBM mainframe programming experience. He holds several IBM certifications, including IBM Certified Application Developer - DB2 11 for z/OS, and IBM Certified Database Administrator for LUW. He lives in Fort Worth, Texas.

www.ingramcontent.com/pod-product-compliance
Lightning Source LLC
Chambersburg PA
CBHW081501050326
40690CB00015B/2877

* 9 7 8 1 7 3 4 5 8 4 7 1 4 *